D1068234

Springer Series in Electrophysics
Volume 21

Edited by Walter Engl

Springer Series in Electrophysics

Editors: D. H. Auston G. Ecker W. Engl L. B. Felsen

Volume 1 **Structural Pattern Recognition** By T. Pavlidis

Volume 2 **Noise in Physical Systems** Editor: D. Wolf

Volume 3 **The Boundary-Layer Method in Diffraction Problems**
By V. M. Babič, N. Y. Kirpičnikova

Volume 4 **Cavitation and Inhomogeneities** in Underwater Acoustics
Editor: W. Lauterborn

Volume 5 **Very Large Scale Integration (VLSI)**
Fundamentals and Applications
Editor: D. F. Barbe 2nd Edition

Volume 6 **Parametric Electronics** An Introduction
By K.-H. Löcherer, C. D. Brandt

Volume 7 **Insulating Films on Semiconductors**
Editors: M. Schulz, G. Pensl

Volume 8 **Theoretical Fundamentals of the Ocean Acoustics**
By L. Brekhovskikh, Y. P. Lysanov

Volume 9 **Principles of Plasma Electrodynamics**
By A. F. Alexandrov, L. S. Bogdankevich, A. A. Rukhadze

Volume 10 **Ion Implantation Techniques**
Editors: H. Ryssel, H. Glawischnig

Volume 11 **Ion Implantation: Equipment and Techniques**
Editors: H. Ryssel, H. Glawischnig

Volume 12 **Fundamentals of VLSI Technology**
Editor: Y. Tarui

Volume 13 **Physics of Highly Charged Ions**
By R. K. Janev, L. P. Presnyakov, V. P. Shevelko

Volume 14 **Plasma Physics for Physicists**
By I. A. Artsimovich, R. Z. Sagdeev

Volume 15 **Relativity and Engineering**
By J. Van Bladel

Volume 16 **Elements of Random Process Theory**
By S. M. Rytov, Y. A. Kravtsov, V. I. Tatarsky

Volume 17 **Concert Hall Acoustics**
By Y. Ando

Volume 18 **Planar Circuits** for Microwaves and Lightwaves
by T. Okoshi

Volume 19 **Physics of Shock Waves in Gases and Plasmas**
By M. A. Liberman, A. L. Velikovich

Volume 20 **Kinetic Theory of Particles and Photons**
By J. Oxenius

Volume 21 **Picosecond Electronics and Optoelectronics**
Editors: G. A. Mourou, D. M. Bloom, C.-H. Lee

Picosecond Electronics and Optoelectronics

Proceedings of the Topical Meeting
Lake Tahoe, Nevada, March 13–15, 1985

Editors:
G. A. Mourou, D. M. Bloom, and C.-H. Lee

With 202 Figures

Springer-Verlag
Berlin Heidelberg New York Tokyo

Dr. Gerard Albert Mourou
The University of Rochester, Department of Physics, Rochester, NY 14627, USA

Dr. David M. Bloom
Edward L. Ginzton Laboratory, Stanford University, Stanford, CA 94305, USA

Dr. Chi-H. Lee
Department of Electrical Engineering, University of Maryland,
College Park, MD 20742, USA

ISBN 3-540-15884-7 Springer-Verlag Berlin Heidelberg New York Tokyo
ISBN 0-387-15884-7 Springer-Verlag New York Heidelberg Berlin Tokyo

Library of Congress Cataloging in Publication Data. Main entry under title: Picosecond electronics and optoelectronics. (Springer series in electrophysics ; v. 21), Proceedings of the first Picosecond Electronics and Optoelectronics Conference. 1. Optoelectronics–Congresses. 2. Laser pulses, Ultrashort–Congresses. I. Mourou, Gerard. II. Bloom, D. M. (David M.), 1948-. III. Lee, Chi H. IV. Picosecond Electronics and Optoelectronics Conference (1st : 1985 : Lake Tahoe, Nev.) V. Series. TA1673.P53 1985 621.36 85-20804

Offset printing: Beltz Offsetdruck, 6944 Hemsbach/Bergstr.
Bookbinding: J. Schäffer OHG, 6718 Grünstadt
2153/3130-543210

Preface

Over the past decade, we have witnessed a number of spectacular advances in the fabrication of crystalline semiconductor devices due mainly to the progress of the different techniques of heteroepitaxy. The discovery of two-dimensional behavior of electrons led to the development of a new breed of ultrafast electronic and optical devices, such as modulation doped FETs, permeable base transistors, and double heterojunction transistors. Comparable progress has been made in the domain of cryoelectronics, ultrashort pulse generation, and ultrafast diagnostics. Dye lasers can generate 8 fs signals after compression, diode lasers can be modulated at speeds close to 20 GHz and electrical signals are characterized with subpicosecond accuracy via the electro-optic effect. Presently, we are experiencing an important interplay between the field of optics and electronics; the purpose of this meeting was to foster and enhance the interaction between the two disciplines.

It was logical to start the conference by presenting to the two different audiences, i.e., electronics and optics, the state-of-the-art in the two respective fields and to highlight the importance of optical techniques in the analysis of physical processes and device performances. One of the leading techniques in this area is the electro-optic sampling technique. This optical technique has been used to characterize transmission lines and GaAs devices. Carrier transport in semiconductors is of fundamental importance and some of its important aspects are stressed in these proceedings. Because of its role in communications, a major part of the conference was devoted to laser diodes and transducers operating in the picosecond range. We also thought it important to cover the field of cryoelectronics where impressive results in high-speed signal processing and sampling have been obtained. Since heteroepitaxy and its applications will have a key role in the next generation of laser diodes, devices and transducers, so the techniques of MBE, MOCVD and their impact on the conception of new devices were presented.

This first Picosecond Electronics and Optoelectronics Conference was held at Lake Tahoe, Nevada, March 13-15, 1985, and was attended by more than 240 scientists. It was sponsored by the Optical Society of Americal in cooperation with the Lasers and Electro-Optics Society and Electron Devices Society of the Institute of Electrical Engineers.

We would like to gratefully acknowledge the generous grants from AFOSR and ONR which provided the major source of funding. Finally, we would like to specially thank Mary Ellen Malzone from the Optical Society of America for her dynamic assistance and total dedication in setting up this meeting.

Rochester, July 1985 *G.A. Mourou · D.M. Bloom · C.H. Lee*

Contents

Part II High-Speed Phenomena in Bulk Semiconductors

Part III Quantum Structures and Applications

Part IV Picosecond Diode Lasers

Part V Optoelectronics and Photoconductive Switching

Part I

Ultrafast Optics and Electronics

Ultrafast Optical Electronics: From Femtoseconds to Terahertz

D.H. Auston, K.P. Cheung, J.A. Valdmanis, and P.R. Smith
AT&T Bell Laboratories, Murray Hill, NJ 07974, USA

The application of femtosecond optics to the measurement and characterization of the electronic properties of materials and devices is reviewed with emphasis on new approaches to the generation and detection of extremely fast electromagnetic transients. Specific topics discussed are: (i) the use of the inverse electro-optic effect for electrical pulse generation, (ii) the intrinsic speed of response of electro-optic materials, and (iii) a new approach to far-infrared spectroscopy which uses time-domain measurements of single-cycle pulses.

1. INTRODUCTION

The use of optical pulses for the measurement and control of high speed electronic devices, circuits, and materials, is a topic that has grown from its inception approximately ten years ago to a field that now encompasses a wide range of activities[1]. This approach which we refer to as optical electronics has some unique features which distinguish it from conventional electronics and opto-electronics. Most important is the use of optics to control specific functions in an electronic device or circuit. Conceptionally, optical electronics goes beyond conventional opto-electronics by using optical signals as sources of power and timing in addition to their traditional role as carriers of information. This more complete marriage of optics and electronics results in a substantial improvement in speed and increased flexibility.

An essential aspect of this approach is the use of high speed transducers to couple optical and electronic signals. Two specific material systems have been used for this purpose. These are photoconductors and electro-optic materials. In this paper, we address some of the key problems related to the use of these materials in the limit of very high speeds. In particular, we find that the increased spectral content of sub-picosecond electrical transients requires some new approaches to methods of generation, detection and transmission. The method we have explored is to treat these electromagnetic transients as radiating free waves, rather than to attempt to confine them to conventional transmission structures. This not only results in greater speed of response, but also adds a flexibility that enables the full spectral content of these pulses to be used for far-infrared spectroscopy.

2. BASIC CONCEPTS OF OPTICAL ELECTRONICS

Before describing the details of our current work, we first review some of the basic concepts of optical electronics as a method for achieving very high speeds in electronic systems. An example of this general approach is illustrated schematically in figure 1.

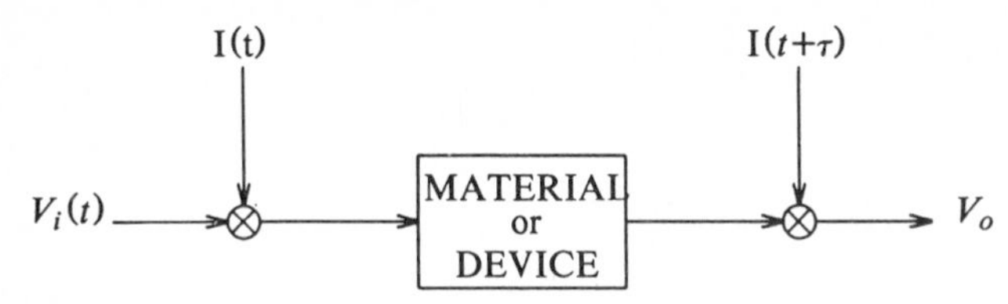

Fig. 1 Schematic diagram of an optical electronics instrumentation system for the measurement of the response of high speed electronic devices and materials.

In this case, two optical signals, I(t) and $I(t+\tau)$, having a precise relative variable delay, τ, between them are used to measure the response of a test device or material. This is accomplished by using the optical signal, I(t), to interact with an input electrical signal, $V_i(t)$ to produce an electrical transient which is then applied to the test sample. The specific nature of this interaction is not specified, although one of the most effective is a convolution of the two inputs. For example in the case where the optical signal is a short pulse, the electrical input V_i is a dc level, and the interaction is a perfect convolution, an electrical pulse is generated which has the same timing and duration as the optical pulse. In other, more general situations, $V_i(t)$, could be time varying , such as a microwave signal, and the optical signal could be a longer pulse or a sequence of pulses which is used to initiate or gate V_i. The response of the test system to this transient is measured by a second convolution with the delayed optical pulse, $I(t+\tau)$. This acts to sample the electrical response producing an output whose amplitude is proportional to the electrical signal at the instant of arrival of the optical pulse, i,e, τ. Since the speed of response is determined by the duration of the optical pulse, it is not necessary to time-resolve the detected signal. It is sufficient to integrate it using low frequency electronics. By varying τ, is is then possible to map out the precise waveform of the response. The ability to precisely control the relative delay, τ, by varying optical path lengths, enables a jitter-free measurement in which the speed of response is controlled by the optical pulses.

Photoconductors are commonly used as sources of electrical pulses. In this case, the incoming electrical signal, V_i is used as a bias on the photoconductor. This has proven to be an effective method of producing extremely fast electrical transients with negligible jitter. The detection process can be accomplished with photoconductors or with electro-optic materials. In the case where photoconductors are used for detection, the electrical response of the system under test is used to bias the photoconductor. The output signal, V_o, is a charge packet proportional to the amplitude of the sampled waveform. When electro-optic materials are used for detection, the information about the amplitude of the signal waveform is conveyed on the transmitted optical pulse as a small change in polarization ellipticity.

3. THE ELECTRO-OPTIC CHERENKOV EFFECT

We have recently shown that electro-optic materials can also be used as effective sources of very short electromagnetic transients[2]. This is accomplished with the use of the inverse electro-optic effect, which produces a Cherenkov cone of pulsed radiation having a duration of approximately one cycle and a frequency in the THz range. This is due to the radiation from a moving dipole moment arising from the second order nonlinear polarization:

$$P_i(\omega = \omega_1 - \omega_2) = \chi_{ijk} E_j(\omega_1) E_k^*(\omega_2) \, . \tag{1}$$

The nonlinear susceptibility, χ_{ijk}, is related to the electro-optic tensor, r_{kij}, by the expression:

$$\chi_{kij} = -\frac{\epsilon_0}{4} n_i^2 n_j^2 r_{ijk} \tag{2}$$

Due to the additional contribution to the low frequency dielectric response from the infrared lattice vibrations, the velocity of the source exceeds the radiation velocity. This produces a characteristic cone of radiation in the form of a shock wave. Unlike classical Cherenkov radiation, however, the radiation source is spatially extended, being proportional to the intensity envelope of the optical pulse. Consequently the details of the radiation field depend sensitively on both the duration and beam waist of the optical pulse.

For our experiments femtosecond optical pulses were obtained from a colliding pulse mode-locked ring dye laser[3], modified by the insertion of four prisms for control of the group velocity dispersion[4]. The pulse duration, measured by second harmonic autocorrelation, was 50 fs (full width at half maximum), and their center wavelength was 625 nm. A relatively low pulse energy of only 10^{-10} J at a repetition rate of 125 MHz was sufficient for the experiments described in this paper.

The detection of these pulses was accomplished with the use of a second femtosecond optical pulse which was synchronized and delayed with respect to the generating pulse. This configuration exploits a novel symmetry arising from the use of the electro-optic effect for both generation and detection. In the case of the generating pulse, it is the inverse electro-optic effect that produces the nonlinear polarization responsible for the radiation field, and in the detection process, it is the direct electro-optic effect that is used to measure the small birefringence produced by the electric field of the radiation pulse. This latter technique, known as electro-optic sampling[5], has previously been used to measure sub-picosecond electrical transients in traveling-wave electro-optic transmission lines.

An example of the waveform produced by the electro-optic Cherenkov effect in lithium tantalate is shown in figure 2. In this particular case, an extremely fast signal was produced by using very tight focusing and reducing the distance between the generating and detecting optical beams to approximately $15 \mu m$ to minimize the effects of far-infrared absorption.

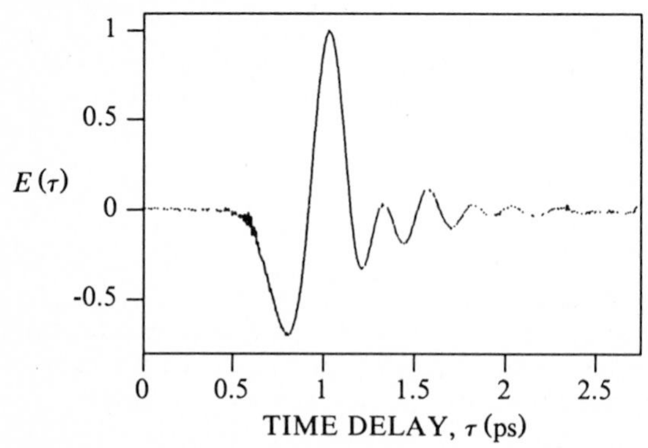

Fig. 2 Far-infrared pulse generated by optical rectification of femtosecond optical pulse in lithium tantalate.

A crucial aspect of the detection method is that it is coherent; i.e. it measures the electric field amplitude, not the intensity profile of the pulse. Also, the lack of time jitter between the generation and detection processes means that small time displacements have physical significance. This means that the waveform is well defined with respect to both its amplitude and "phase". Consequently, we can Fourier transform time-domain waveforms to obtain their complex frequency spectra. The spectral content of this pulse is extremely broad, extending from dc to 4.5 THz.

4. INTRINSIC RESPONSE OF ELECTRO-OPTIC MATERIALS

An important feature of the waveform in figure 2 is the high frequency ringing on the trailing edge. This is due to the resonant contribution to the electro-optic tensor from infrared lattice vibrations. This effect, which was first discussed by Faust and Henry[6], is expected to play an important role in determining the intrinsic speed of response of electro-optic materials. The rate of decay of this ringing is determined by the damping of the lattice vibrations. We separately measured the absorption and dispersion of lithium tantalate over the spectral range contained in these pulses. These results can be used to determine the resonant frequency and damping rate of the lowest transverse optic lattice mode. We find a center frequency of 6 THz and a damping rate of 1.3 THz. The dispersion of the electro-optic tensor can also be estimated from Faust and Henry's model. This result is plotted in figure 3.

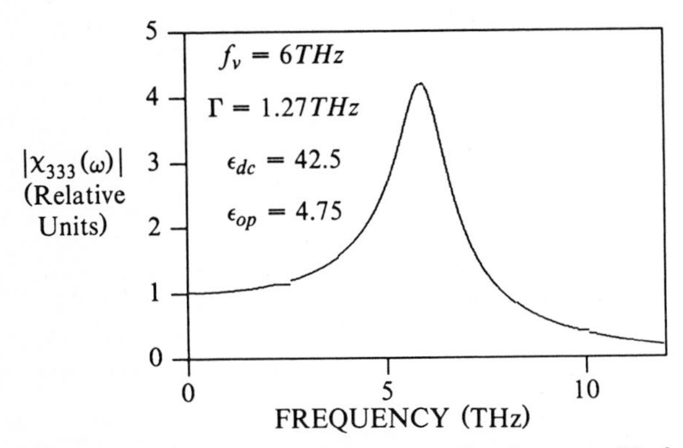

Fig. 3 Dispersion of the second order nonlinearity responsible for both optical rectification and the electro-optic effect in lithium tantalate(normalized to dc value). The resonance is caused by strong coupling to the TO lattice vibration.

We have used the data of Boyd and Pollack[7] to estimate the relative contributions from the ionic and electronic effects in lithium tantalate. As expected, χ_{333} shows a strong resonance at the TO frequency. The resonant enhancement relative to the dc value is a factor of 4.2. In our experiment, this factor will be squared due to the use of the electro-optic effect both for generation and detection. Above the resonance, the magnitude of χ_{333} falls off due to the much smaller electronic contribution to the electro-optic effect(in our simplified model we have neglected higher resonances).

5. COHERENT TIME-DOMAIN SPECTROSCOPY

The concept of coherent time-domain spectroscopy is not new. It is implicit in electronic measurements which use impulse excitation to determine the dispersive properties of a circuit or device. In this case, the response manifests itself as a distortion or broadening of the pulse as measured in real time on an oscilloscope. Since the oscilloscope is a "coherent detector", the frequency response can be determined by a simple numerical Fourier transform. This approach, however, is usually not possible in optics since most optical detectors are not coherent: they measure only the intensity envelope of the signal and do not trace out the true electric field amplitude. The essential information about the phase of the carrier is usually lost. If, however, as in our case, a coherent detection scheme can be employed, the full equivalency of time and frequency applies. This means that the complex frequency response of a material is uniquely determined by its temporal response to a short pulse. Each frequency component contained in the pulse can be treated as an independent signal. The short pulse is thus equivalent to a tunable monochromatic source covering the same range of frequencies.

We have applied this concept to the measurement of the far-infrared properties of materials using the broad spectrum pulses produced by the electro-optic Cherenkov effect. The basic approach is as follows: A material whose dielectric response is desired is placed in contact with the lithium tantalate crystal in which the far-infrared signals are generated and measured. The Cherenkov wavefront is then reflected from the interface between the two materials. The probing optical beam measures both the incident waveform and at a delayed time, the reflected waveform. At first thought, this would appear to provide sufficient information to determine the dielectric response of the material of interest. A difficulty arises however due to the fact that the probing optical pulse measures the incident and reflected waveforms at a finite distance from the interface. Although this distance can be made relatively small(typically 100 μm), some distortion and attenuation results from propagating through this additional distance in the lithium tantalate. Also, the absolute value of this distance is difficult to determine which introduces an uncertainty into the timing of the reflected waveform and consequently the phase of its spectrum is not defined. An expedient solution to this problem is to do a second measurement under identical conditions of a reference sample whose dielectric properties are known. This reference waveform accurately defines the timing of the reflection and enables the essential phase information to be determined. The frequency spectra of the reflected waveforms are then determined numerically. The reflectivity of the reference is calculated from the Fresnel formula for a single interface using its known complex dielectric constant. Once the reflectivity spectrum of the sample is known, its complex dielectric functions can be determined by inverting the Fresnel reflectivity expression. The full power of this procedure is apparent in its ability to obtain both the real and imaginary parts of the dielectric response over a broad range of frequencies from a single waveform(plus a reference). This can be compared to incoherent methods which measure the magnitude of the reflectivity and must resort to Kramers-Kronig analyses to determine the complex dielectric function.

We have applied the technique of coherent time-domain far-infrared spectroscopy to the determination of the momentum relaxation times of free electrons in semiconductors[8]. Some preliminary results are summarized in the following table:

Table 1. Values of the electron momentum relaxation time at room temperature in semiconductor samples measured by coherent time-domain far-infrared spectroscopy.

DAMPING TIMES		
Material	Experiment	Theory
n-Germanium	225 fs	265 fs
n-Silicon	180-220 fs	200fs
GaAs/GaAlAs MQW	190 fs	190 fs

In this table the values listed as "theoretical" momentum relaxation times are deduced from mobility measurements. The agreement is extremely good. This method gives an independent measure of electron mobility which does not require electrical contacts on the sample. This is an important advantage for materials such as the GaAs/GaAlAs multiquantum well sample to which it is difficult to make good quality ohmic contacts. A more detailed description of this approach to far-infrared spectroscopy is given in reference [8].

6. CONCLUSIONS

Our conclusions about our recent work on the generation, detection, and transmission of extremely fast electromagnetic transients for applications to high speed measurements of electronic materials and devices can be summarized as follows:

[1] The electro-optic Cherenkov effect is an extremely effective method of producing very fast electromagnetic transients.

[2] The spectral content of these pulses is sufficiently broad to permit their use for coherent far-infrared spectroscopy by measuring their change in shape due to transmission through or reflection from dispersive materials.

[3] The intrinsic speed of electro-optic materials is strongly influenced by coupling to lattice vibrations which introduce absorption and a resonant enhancement of the nonlinearity.

[4] The most expedient method of transmitting femtosecond duration electrical transients is to treat them as free propagating waves rather than attempt to guide them on conventional transmission lines.

Clearly, the main applications of sub-picosecond electronics is for the measurement of the electronic properties of materials, since at present there are no devices with temporal responses in this range. These measurements, however, are critically important for the development and understanding of the next generation of high speed devices. In this regard, an important extension of the approach outlined here is the measurement of the non-equilibrium and non-linear transient response of electronic materials.

REFERENCES

[1] For a review, the reader is referred to the articles in this volume and to the book, *Picosecond Opto-electronic Devices,* ed. C. H. Lee, Academic Press (1984).

[2] D. H. Auston, K. P. Cheung, J. A. Valdmanis, and D. A. Kleinman, Phys. Rev. Lett., *53*, 1555 (1984).

[3] R. L. Fork, B. I. Greene, and C. V. Shank, Appl. Phys. Lett. *38*, 671 (1981).

[4] J. A. Valdmanis, R. L. Fork, and J. P. Gordon, Opt. Lett., *10*, 11 (1985).

[5] J. A. Valdmanis, G. A. Mourou, and C. W. Gabel, IEEE J. Quant. Electr., *QE-19*, 664 (1983).

[6] W. L. Faust, C. H. Henry and R. H. Eick, Phys. Rev., *173*, 781 (1968).

[7] G. D. Boyd and M. A. Pollack, Phys. Rev. B, *7*, 5345 (1973).

[8] D. H. Auston and K. P. Cheung, to be published in J. Opt. Soc. Am., B, April (1985).

Prospects of High-Speed Semiconductor Devices

Naresh Chand and Hadis Morkoç
Coordinated Science Laboratory, 1101 W. Springfield Avenue, University of Illinois at Urbana-Champaign, Urbana, IL 61801, USA

Concepts, technologies, and performance of high speed devices are reviewed. Among the devices treated are the modulation doped GaAs/AlGaAs heterojunction transistors (MODFET), heterojunction bipolar transistors (HBT), resonant tunneling devices (RTD) and hot electron transistors (HET) with semiconductor or metal type base. A contrast is drawn between the intrinsic switching speed and the extrinsic speed which is dominated by, but often overlooked, parasitic resistances and capacitances. It is shown that because of parasitics, not just the carrier velocity but the device current needs optimization.

I. INTRODUCTION

Search for high speed has over the years focused on miniaturization and high performance semiconducting materials. Scaling up in speed reaches a saturation as the dimensions are reduced, because parasitics do not entirely scale down, particularly in integrated circuits. Scaling dimensions down also reduces current carrying capability, which increases resistance R in the time constant CR. Also, the dimensions can not be reduced indefinitely due to lithographic limitations. As a result GaAs which has significant advantages over Si such as its five times higher electron mobility, its high peak (threshold) velocity, and availability of semi-insulating substrate which substantially reduces parasitic capacitances and also allows integration of both lumped and distributed microwave components, has received a great deal of attention. For comparable performance, the minimum geometries in GaAs can be twice those in silicon. InGaAs is another III-V material which has even better transport properties, but due to difficulty in growing its lattice matched widegap InP by MBE, research in this material for high speed applications is in its infancy. With the possibility of growing $Al_{0.52}In_{0.48}As$, a substitute for InP, by MBE the situation may change in the future.

Besides commercially available GaAs MESFET's, newly emerged devices are the GaAlAs/GaAs modulation doped field effect transistor (MODFET also called SDHT, HEMT or TEGFET) [1]-[3] and heterojunction bipolar transistor (HBT) [4]-[6]. Resonant tunneling device (RTD) [7]-[10] and different variations of hot electron transistors (HET's) [11]-[17], which are based on old concepts, are receiving renewed attention. Intrinsically, the tunneling is much faster than drift (FET) and diffusion (HBT).

The propagation delay in Si devices (both bipolars and FET's) has perhaps reached a lower limit at around 30 ps while the GaAs devices operate at the lowest propagation delays, 9.4 ps at 77K and 11 ps at 300K for AlGaAs/GaAs MODFET's [2], and 30 ps at 300K for HBT's. The current gain cutoff frequencies f_T, are 70 GHz and 40 GHz for MODFET's [2] and HBT's [5], respectively.

The properties that make a device fast are the low junction and parasitic capacitance C and resistance R, large current density and high transconductance g_m. The g_m/C ratio measures how fast the device can drive itself, and the magnitude of the transconductance determines how well an external capacitor can be driven. The presence of parasitics reduces the effective transconductance and increases the time constant of the circuit. While an intrinsic delay of about 1 ps should be possible, progress below 10 ps in a given material is very difficult because of parasitics. Even at the present size of transistors, VLSI circuits are already becoming interconnection limited. The solution of this problem demands invention of devices performing higher order primitive functions which are optimized to minimize interconnects. One possibe solution [3] could be an array of the quantum-coupled devices, in which at least nearest neighboring quantum wells are coupled by tunneling provided the charge leakage is negligible. Another possible solution could be to process optical signal instead of electrical signal by combining optical and electronic devices on the same chip at least for special ultra high speed applications like optical communication, which will surely receive a lot of attentio in the near future. Operating the device at low temperatures can also improve the properties of the interconnects. Metal resistance decreases about linearl and electromigration exponentially with reducing temperature.

II. MODULATION DOPED FIELD TRANSISTORS (MODFET)

In a high speed switching field effect transistor, carrier transit time must be minimized and the current level I_d and transconductance g_m must be maximized. This demands small gate length, large values of carrier concentration n, carrier saturated velocity v_s, and large gate width to length ratio. Increase of carrier concentration beyond 5×10^{18} cm^{-3} while maintaining high velocity, can not be met in a MESFET as an increase of the former reduces the latter due to ionized impurity scattering. To meet the requirements of larger n and v_s modulation doped structures [1] are used. A typical MODFET structure along with the energy band diagram is shown in Fig. 1. The structure consists of an undoped GaAs, an undoped $Al_xGa_{1-x}As$ and a Si doped $Al_xGa_{1-x}As$ layers grown sequentially on a semi-insulating substrate by MBE. To form the gate, a Schottky barrier is formed on the doped AlGaAs. The band gap of $Al_xGa_{1-x}As$ is larger than that of GaAs and depends on the Al content x. Due to the lower energy level of the GaAs conduction band, electrons diffuse over from the dope

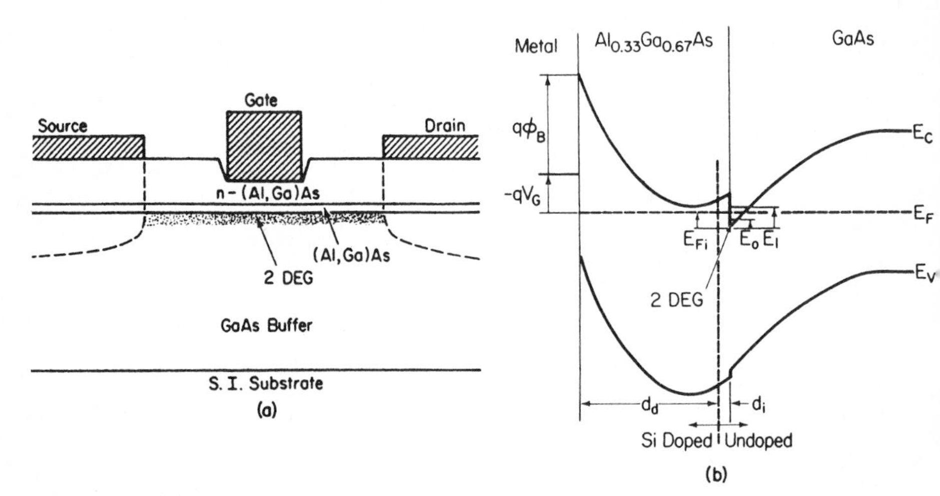

Figure 1. (a) Cross-section view of a MODFET, and (b) the corresponding energy band diagram under applied bias

AlGaAs into the doped GaAs. These electrons neither drift away from the interface in GaAs due to attractive coulomb force of the depleted donors nor do they go back into AlGaAs due to potential barrier at the heterojunction. With the result, they accumulate as a two-dimensional electron gas (2DEG) in a potential well at the interface and form the channel of the FET. Their concentration is modulated by the gate potential. The doping densities and dimensions are so selected that AlGaAs is fully depleted of free carriers and the conduction is mainly due to 2DEG, the density of which depends on the doping, and the Al content in AlGaAs and the thickness of the undoped AlGaAs spacer layer.

Since the electrons in the channel move in high purity GaAs and are separated from their donors by a thin AlGaAs undoped layer, they experience no ionized impurity scattering. Consequently, the carrier mobility in the channel is very high, especially at low temperatures. For example, the typical mobility at 77K is 50000 $cm^2V^{-1}s^{-1}$ as against 5000 $cm^2V^{-1}s^{-1}$ in doped GaAs. This, coupled with small separation (~ 300 Å) between the gate and the conducting channel leads to extremely high transconductance (e.g., 500 mS mm at 77K), large current carrying capability, small source resistance, and very low noise figure. In a MODFET amplifier at 77K a 0.25 dB noise figure has been measured at 10 GHz and 0.35 dB at 18 GHz. With large g_m, the current gain cut-off frequency, $f_T(g_m/2\pi\ C_{gs}$, where C_{gs} is the gate to source capacitance) is also large. The maximum frequency of oscillations (unity power gain), f_{max}, on the other hand is influenced considerably by the parasitics. Although f_{max}, is proportional to the low-frequency transconductance, it is inversely proportional to rf drain conductance, feedback capacitance and some time constants determined by the product of input capacitance and parasitic resistences e.g. gate resistance and source resistance. It is, therefore, extremely important to reduce these parasitic elements if the potential of this device is to be harnessed. The output rf conductance can be reduced by increasing the aspect ratio, the ratio of gate length divided by the gate to 2DEG distance. This ratio is maintained in MODFET's at about 10 or so, even in devices with 0.3 μm gate length, which is why this device is attractive for high speed applications. The problems associated with the feedback capacitance and source resistance will have to be addressed before full realization of potential extremely high frequency operation.

III. HETEROJUNCTION BIPOLAR TRANSISTORS (HBT)

In applications with large load capacitances, bipolars are preferred over FET's because of their large current carrying capability, high transconductance and excellent threshold voltage control. The principle motive in developing a heterojunction bipolar transistor [4] is to reduce the base resistance R_B which severely limits the high speed performance of bipolar transistors in both digital and microwave circuits. For example, f_{max} in an HBT is affected by R_B as follows:

$$f_{max} = 1/4\pi[(R_BC_C)\tau_{ec}]^{1/2} \quad (1)$$

where C_C is the collector junction capacitance, which also need to be minimized by reducing collector junction area. τ_{ec} is the total emitter to collector delay time and is given by [18]

$$\tau_{ec} = \tau_e + \tau_B + \tau_d + \tau_c \quad (2)$$

where t_e is emitter junction capacitance charging time, τ_B is the base transit time, τ_d is the collector depletion region transit time and τ_c is the collector capacitance charging time. τ_e is inversely proportional to the emitter current density, and so a large current density improves the frequency response.

The base resistance R_B can also have a dominant effect on the noise and distortion performance of the device, and it can be reduced by increasing the base width W_B and/or base doping N_B. Increasing W_B increases τ_B and reduces the base transport factor (hence the gain). Increasing N_B in a homojunction transistor increases unwanted carrier injection from base into the emitter and this decreases the emitter injection efficiency. In an npn heterojunction transistor, however, due to the presence of dissimilar energy barriers for injections of electrons and holes, the barrier for hole injections being larger the injection efficiency is nearly independent of the doping density in the base and the gain is limited by the base transport factor. With the result the base in an HBT can be heavily doped without affecting the gain significantly. The dopings in the emitter and collector can be adjusted to minimize junction capacitances and series resistances. This provides the designer a great degree of freedom.

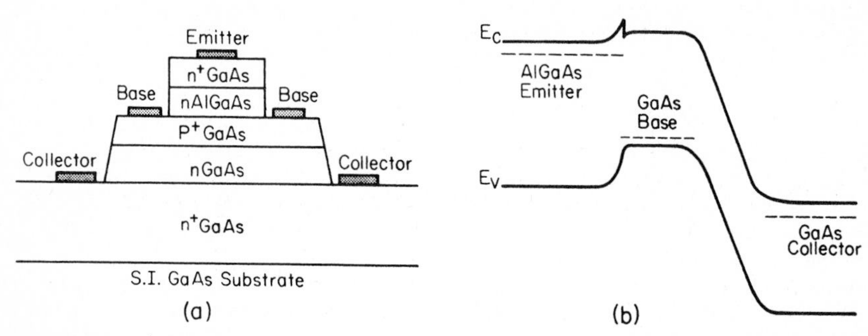

Figure 2. (a) Cross-sectional view of an HBT, and (b) the corresponding energy band diagram under applied bias

A schematic crosssection of an npn HBT is shown in Fig. 2 along with the energy band diagram. Electrons are injected from the widegap AlGaAs emitter ($n \sim 10^{17}$ cm^{-3}) into the narrow gap GaAs base ($p \sim 10^{18} - 10^{19}$ cm^{-3}). For (a) suppression of hole injection into the collector in saturating logic, (b) emitter collector interchangeability leading to an improvement in VLSI circuit design, packing density and interconnections, and (c) a better control of the collector-emitter offset voltage, the collector can also be a widegap AlGaAs resulting in a double heterojunction transistor (DHBT) [6]. In a DHBT τ_d may, however increase significantly as the carriers now move in AlGaAs with lower saturation velocity.

The conduction band spike at the emitter-base junction shown in Fig. 2 is due to abrupt AlGaAs/GaAs interface and it is smoothed out by employing a compositional grading [19]. The spike can be used, however, for near ballistic injection of electrons into the base to reduce the base transit time τ_B from the present 1 ps to ~ 0.2 ps for a base width of 0.1 μm. For the best reported f_T ($=1/2\pi\tau_{ec}$) [5] of 40 GHz (without ballistic injection) for 1.6 μm wide emitter, τ_{ec} is 4 ps. Therefore, with ballistic injection f_T of this device can be increased by 25%. To further reduce τ_{ec}, the device area and parasitics have to be reduced and current level needs to be increased. Ultimate values of f_T and f_{max} of more than 100 GHz are expected from an optimized AlGaAs/GaAs HBT. For f_T = 40 GHz, f_{max} obtained was only 25 GHz due to large R_B. A great deal of effort is, therefore, needed to reduce R_B. In fact, the efforts in reducing R_B gave birth to the concept of metal base transistor [13].

IV. HOT ELECTRON TRANSISTORS (HET)

These devices are based on an old concept first proposed in 1960 by Mead [11] with the primary objectives of reducing base resistance and carrier transit times, and increasing the current density for high frequency performance. Various metal, insulator and semiconductor structures were proposed [12] but due to difficulty in preparing these materials on one another, the success was limited.

With the ability to grow very thin layers by MBE, availability of GaAs and its lattice matched larger bandgap $Al_xGa_{1-x}As$ which when undoped has high resistivity and provides a barrier whose height can be tuned by varying x, and the use of modulation doping which can be used to form metal-like two-dimensional electron gas (as in a MODFET) in GaAs base, there are renewed interests in HET's [14]. As a result,quite a few AlGaAs/GaAs HET's structures have already emerged [15]-[17], [20]. The main difference in these structures is the method by which the electrons are injected into the base. All of them, however, involve injection of hot electron beams (electrons with energy more than a few kT above the conduction band or Fermi level) into the base either by means of tunneling or injection over the barrier. For base thicknesses smaller than the mean free path of hot electrons, a large fraction of electrons are collected by the collector where they thermalize to the lattice temperature. The electrons lost in the base constitute base current and need to be removed fast, which, provided R_B is low, can be done efficiently as these devices involve only majority carrier electrons,and the base current does not depend on the slow process of electron-hole recombination as in an HBT.

The conduction band energies of the structures being studied in various laboratories are shown in Figs. 3,4, and 5. All the structures have GaAs base whose thickness is less than 0.1 μm. In all of the structures, electrons in the base, being hot, travel at such a high velocity (~ 5 x 10^7 cms^{-1}) that the base transit time τ_B is negligible,and the speed is expected to be limited by the emitter capacitance charging time through the emitter resistance,whose value depends on the current injection mechanism and is larger for tunneling [12]. Although the base conductivity in these structures can be made very high,either by heavy doping or 2DEG it may be difficult to make good ohmic contacts to the base. In some cases nonalloyed ohmic contacts will have to be used,or contacts will have to be made to the undoped base layer. This will increase the base resistance,which may limit the high speed performance. In

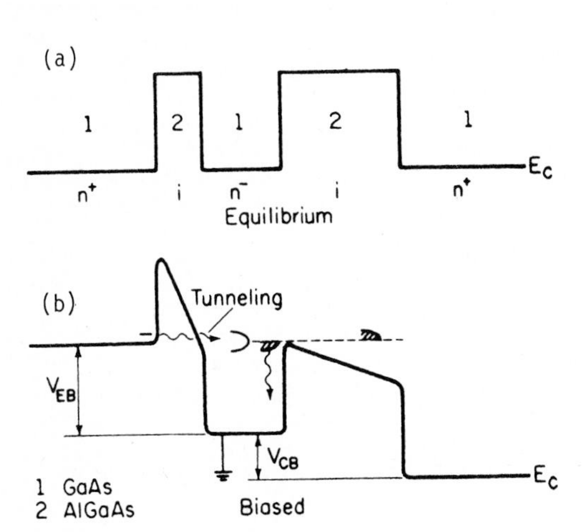

Figure 3. Conduction band energy profile of a tunneling HET (a) in equilibrium, and (b) under operating bias

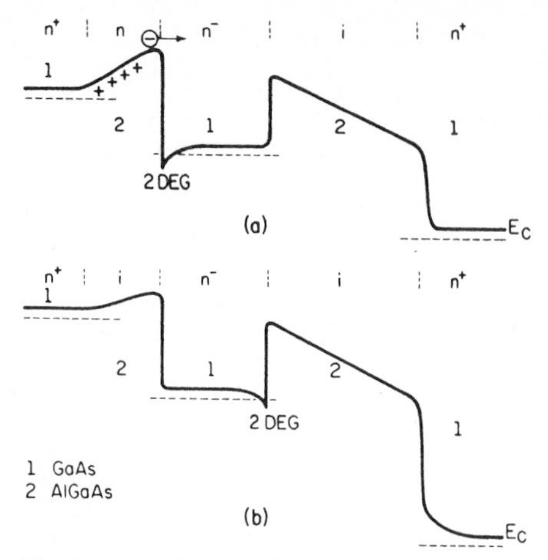

Figure 4. Conduction band energy profile of modified hot electron transistors with 2DEG in the base which is formed in (a) by modulation doping and in (b) by collector base voltage

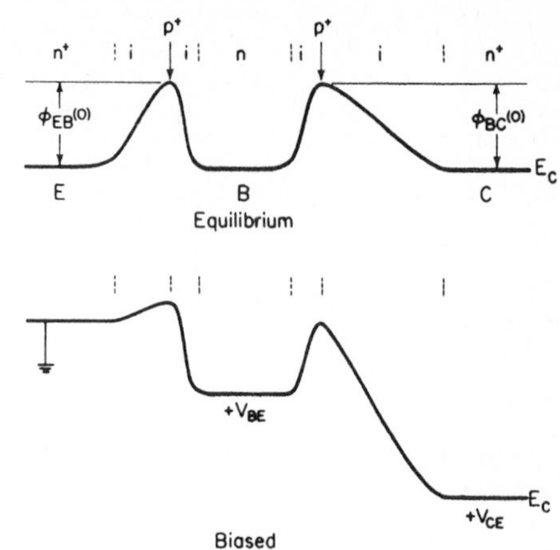

Figure 5. Conduction band energy profile of a planar doped transistor (a) in equilibrium, and (b) under operating bias

addition,it will not be possible to operate these devices at room temperature due to excessive leakage currents through the thin barriers of small heights (~0.25 eV as against ~1.3 eV in HBT's). The barrier heights in GaAs base devices cannot be increased more than 0.28 eV as this will transfer the electrons from high mobility Γ valley to low mobility indirect L valley. Although these devices after optimization have been predicted to operate in subpicosecond range [14], their predicted superior performance to that of HBT's is yet to be established.

Shown in Fig. 3 is a tunneling HET in which n-doped GaAs emitter, base and collector are separated by thin undoped AlGaAs potential barriers. As shown, the carrier injection relies on the tunneling through the AlGaAs barrier,which occurs when the base is biased positive with respect to the emitter. If the collector barrier height is smaller than the energy of hot electrons, the electrons are collected by the collector. The effective barrier widths for electron tunneling can be varied by the applied biases. The emitter barrier has to be thin enough to allow tunneling,and the collector barrier thick to minimize the leakage current. The electron transfer ratio, or the common base current gain, α will be unity if the losses due to spread of energy of the tunneling electrons, scattering in the base and reflection from the collector barrier are prevented. A first device based on this structure was reported recently [15] with a common emitter gain h_{FE}, of 1.3 at 40K. The device showed excessive leakage current at 120K. The low gain indicates significant loss of carriers in the base,which can be minimized by reducing the base width and/or collector barrier height. Reduced base width will, however, increase the base resistance.

Two modified structures of HET's which have the potential of very small base resistance are shown in Figs. 4(a) and (b). In both cases, the GaAs base is undoped or lightly doped,and the AlGaAs barrier between the emitter and base is of triangular shape,which is doped with donor impurities in Fig.4(a) and undoped in Fig. 4(b). A two-dimensional electron gas (2DEG) is formed in the base by modulation doping (as in a MODFET) in Fig. 4(a) and by base-collector voltage (as in a MOSFET) in Fig. 4(b). Carriers are injected over the barrier and not by tunneling. Since the base is undoped, the carrier mobility of the 2DEG in the base will be very high, about 3000 $cm^2V^{-1}s^{-1}$ at 300K and will increase greatly at 77K because of reduced impurity scattering. With such a high electron mobility, the base will behave like a metal, and the devices will work like a metal base transistor. These structures, however, suffer from a serious drawback. Since the base is intrinsicially undoped or lightly doped and is thin, making good base ohmic contact may not be easy and the resulting increase in base resistance may more than compensate the advantage obtained by forming the 2DEG in the base. However, if the base is doped heavily where the contacts are to be made, very small base resistance can be obtained.

Shown in Fig. 5 is a planar doped barrier transistor, which is another kind of hot electron transistor made entirely of GaAs. It employs two planar-doped homojunction triangular barriers [17]. A triangular barrier is formed by placing a thin fully-depleted p^+ layer in an intrinsic region bounded on each side by n^+ layers [16]. Upon forward biasing the emitter junction, hot electrons are injected over the barrier into the base and the rest of the device operation is similar to that of earlier structures. In a first experimental device current gain α of 0.75 was measured at 77K with base width of 870 Å. Microwave measurements at 77K revealed a maximum frequency of oscillation f_{max} of about 5 GHz and an extrapolated value of f_T of roughly 40 GHz for an emitter current of 10^4 A cm^{-2}. The low f_{max} was stated to be due to large base resistance ~ 190 Ω. These initial results are quite impressive. The value of f_T is comparable to the best value reported for HBT's. This device too, however, suffers from the disadvantages of excessive leakage current and needs to operate at low temperatures, and in addition demands strigent control on the placement of the p-doped plane in the intrinsic region.

Another device under investigation is the permeable base transistor [20] which has been predicted to lead an f_T above 200 GHz, an f_{max} near 1000 GHz and a power-delay product below 1 fJ. An f_{max} of ~100 GHz and a gain of 16 dB at 18 GHz have already been achieved. This device is a kind of metal base transistor and has a potential of extremely low base resistance, and for this reason its predicted f_{max} is so high. The device reported [20] employed a tungsten grating which is embedded inside a single crystal of GaAs. The embedded grating forms a Schottky barrier with GaAs and is the base of the transistor. The structure consists of an n^+ GaAs substrate, an n^- GaAs emitter, the tungsten base grating, and an n^- GaAs collector. The carrier densities in the emitter and collector layers are so adjusted that depletion region due to the embedded tungsten-GaAs Schottky barriers extends across the openings in the grating. The flow of electron from emitter to collector is only through the tungsten grating, and is controlled by varying the tungsten base potential.

V. RESONANT TUNNELING DEVICES (RTD)

Since transport in conventional devices is limited by drift (FET's) and diffusion (HBT's), faster phenomena e.g. tunneling become attractive. Resonant tunneling device is one where the intrinsic speed can be as low as 10^{-4} ps and where an oscillation frequency of a few hundred GHz is expected as against a 100 GHz barrier faced by Gunn, Impatt and Esaki diodes. At present this is a two terminal negative differential device (NDR) and the device has already been used successfully to generate oscillations at 18 GHz [9] with an output power of 5 μW at 200K. Detecting and mixing studies at 2.5 THz demonstrated that the charge transport was faster than 10^{-1} ps [8]. The device has also been used to generate oscillations at 300K [10].

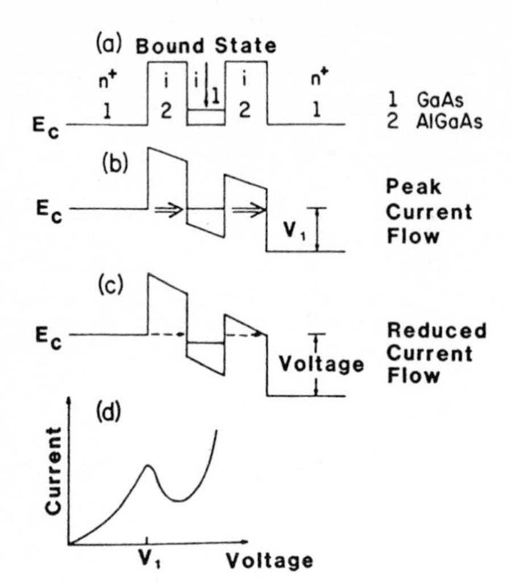

Figure 6. Schematic diagrams illustrating the working of a resonant tunneling device

This device relies on the quantum mechanical tunneling through a multi-barrier structure consisting of alternating layers of potential barriers (e.g. AlGaAs) and quantum wells (e.g. GaAs) and on the fact that the maximum tunneling current occurs when the injected carriers have certain resonant energies which are determined by the Fermi energy in the doped cap region and the electron energy levels in the quantum well(s) [7]-[10]. The energy band diagram of a double barrier AlGaAs/GaAs structure and how the resonance occurs in shown in Fig. 6. Electrons originate from the conduction band of the left doped GaAs which is at ground potential and tunnel through the AlGaAs barrier into the GaAs well, and finally tunnel through the second barrier into unoccupied states of the doped GaAs at positive potential. Resonance occurs, i.e. the tunneling current peaks when the energy of the electrons injected into the well becomes equal to the discrete quantum states in the well (Fig.6.(b)). The tunneling probability (or the current), however, drops rapidly. Fig. 6(c)) as the discrete energy level in the well drops below the conduction band edge of the left GaAs layer which happens on increasing the applied bias. This phenomenon, therefore, leads to negative differential resistance (NDR), in the I-V characteristics (Fig. 6(d)). The NDR is more visible at low temperatures and can be used for microwave generation and amplification.

VI. CONCLUSIONS

It is evident that one super electron is not simply capable of producing a high speed device. This is because of the presence of capacitances, both depletion and charge accumulation types, interconnects etc. which demand high current. This simply means we need as many electrons as possible to travel at high speeds, meaning a large current. Since the power dissipation is always a problem, we need to obtain high current levels at voltages as small as possible which implies, among other things, small parasitic resistances. Cryogenic temperature operation aids in improving electron velocities and reducing parasitic resistances. The ultimate, if possible, would be to harness a new concept that bypasses some of these parasitic resistance and capacitance problems. Combination of electron optics and electronics should be explored for this purpose.

References

1. P. M. Solomon and H. Morkoc,, IEEE Trans. Electron Devices, ED-31, pp. 1015, (1984).
2. T. J. Drummond, W. T. Masselink, H. Morkoç, Proc. IEEE, to be published.
3. C. A. Liechti, "GaAs IC Technology-Impact on the Semiconductor Industry", IEEE 1984 Int. Electron Dev. Mtg. Proc., pp. 13-18.
4. H. Kroemer, Proc. IEEE 70, p. 13, (1982).
5. P. M. Asbeck, A. K. Gupta, F. J. Ryan, D. L. Miller, R. J. Anderson, C. A. Liechti, and F. H. Eisen, "Microwave Performance of GaAs/(GaAl)As Heterojunction Bipolar Transistors", IEEE 1984, Int. Electron Dev. Mtg. Proc., pp. 864-865.
6. S. L. Su, R. Fischer, W. G. Lyons, O. Tejavadi, D. Arnold, J. Klem, and H. Morkoç, J. Appl. Phys. 54, 6725 (1983).
7. R. Tsu and L. Esaki, Appl. Phys. Lett. 22, 562 (1973).
8. T. C. L. G. Sollner, W. D. Goodhue, P. E. Tannenwald, C. D. Parker and D. D. Peck, Appl. Phys. Lett., 43, 588 (1983)
9. T. C. L. G. Sollner, P.E. Tannenwald, D. D. Peck, and W. D. Goodhue, Appl. Phys. Lett., 45, 1319 (1984).
10. T. J. Shewchuk, P. C. Chaplin, P. D. Coleman, W. Kopp, R. Fischer, and H. Morkoç, Appl. Phys. Lett. 46, 508 (1985).
11. C. A. Mead, Proc. IRE, 48, 359 (1980).
12. S. M. Sze, Physics of Semiconductor Devices (John Wiley, New York 1969).
13. S. M. Sze and H. K. Gummel, Solid State Electron, 9, 751 (1966).
14. M. Heiblum, Solid State Electron, 24, 343 (1961).
15. N. Yokoyama, K. Imamura, T. Ohshima, H. Nishi, S. Muto, K. Kondo and S. Hiyamizu, Jpn. J. Appl. Phys. 23, L311 (1984).
16. R. J. Malik, T. R. Aucoin, R. L. Ross, K. Board, C. E. C. Wood and L. F. Eastman, Electron. Lett. 16, 22 (1980).
17. M. A. Hollis, S. C. Palmateer, L. F. Eastman, N. V. Dandekar, and P. M. Smith, IEEE Electron Device Lett. EDL-4, 440 (1983).
18. H. F. Cooke, Proc. IEEE 59, 1163 (1971).
19. N. Chand and H. Morkoç, IEEE Trans. Electron Devices, ED-32. to be published in July 1985.
20. C. O. Bozler and G. D. Alley, IEEE Trans. Electron Devices, ED-27, 1128 (1980).

The Role of Ultrashort Optical Pulses in High-Speed Electronics

C.V. Shank
AT&T Bell Laboratories, Holmdel, NJ 07733, USA

As we look to the future of high-speed electronic and optoelectronic devices, we are contemplating devices that work in the picosecond time domain. Optical techniques have the potential of an ever increasing importance in analyzing both device performance and the physical processes necessary for understanding high-speed device operation. Optical methods have the advantage of ultrashort time resolution that exceeds all other measurement techniques. Optics may influence electronics in ways that go beyond measurement and characterization. A new class of devices appears possible that would integrate optics into the electronic function. Such devices would incorporate the high-speed information capacity of light waves into the device and circuit level.

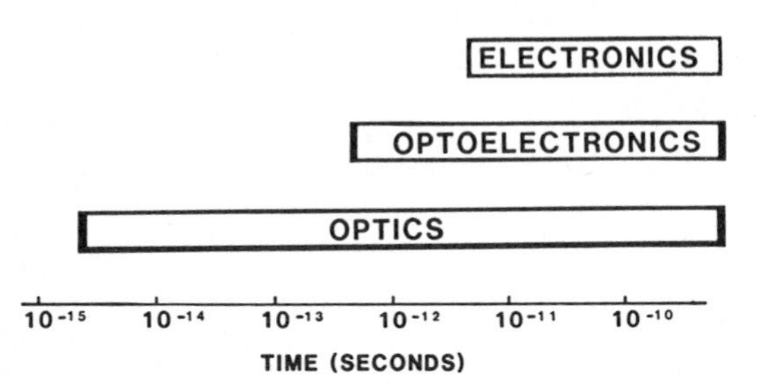

Figure 1 - Time Scale of Technologies

In Figure 1 we compare the time scales of existing high-speed technologies. Clearly, the speed of optics exceeds that of all other high-speed technologies. Although not electronic, techniques involving short optical pulses have pushed our ability to resolve events in time to a fraction of a picosecond and are valuable tools for fundamental investigations of high-speed electronic devices. Optoelectronics combines the fields of optics and electronics and shows considerable promise for the development of high-speed devices that can join some of the high-speed capabilities of optics with electronics. Current limits of optical pulse generation have reached 8 femtoseconds[1] and optoelectronic measurements have been performed in several laboratories with a resolution of less than a picosecond.

We can gain some insight into the speed of a semiconductor device by examining the performance of an idealized circuit containing only a block of semiconductor and a load capacitor, C. From elementary circuit analysis we can calculate the time, t, to switch the voltage from an initial voltage Vo, to a voltage V:

$$t = \int_{V_o}^{V} \frac{Cdv}{I(V)}$$

The current, I(V), is the transfer current-voltage relationship of the device. Clearly an important factor in determining the speed of the circuit is the capacitance, C, that must be charged by the device current. This capacitance is minimized by reducing the physical size of the device and other parasitic capacitances. Another very fundamental factor governing the switching speed is the current-voltage relationship which is governed by the properties of the semiconductor itself. The dynamics of the drift of electrons across the active region can provide a limit to device performance. As devices become smaller and faster, the physics of nonequilibrium carrier transport in the semiconductor material becomes more important.

The motion of carriers through a semiconductor under the influence of an electric field ultimately determines the material limit for high-speed device performance. Ultrashort optical pulses provide the means for experimentally investigating energy and momentum relaxation in semiconductors. The cooling of optically excited carriers in GaAs and GaAs multiquantum wells has been measured directly using picosecond spectroscopic techniques [2].

In a semiconductor at low electric fields the velocity of an electron is given by

$$v = \mu E$$

where $\mu = e/m^*$. In the region where the mobility is constant, current flow is governed by Ohm's law. As the electric field is increased, deviation from simple ohmic behavior is seen as the carrier velocity approaches saturation. For a semiconductor like GaAs, the nonequilibrium transport differs considerably from steady state transport [3]. Let us consider the transport of electrons injected into the central valley of the conduction band. When an electric field is rapidly applied, the carriers are heated and can be scattered to a satellite valley. The effective mass in the satellite valley is larger than that in the central valley. During the first few picoseconds, while the electrons are in the central valley, the average electron velocity can exceed or overshoot its quasi-equilibrium value as the carriers move from the central valley (small effective mass). Monte Carlo calculations have established that these transients are important in GaAs [4] during the first few picoseconds after the field is switched on. Electron carriers travel less than 1 micron in this time period. Direct observation of this effect is beyond the range of current electronic measurement methods but can be investigated with optical techniques.

An all-optical approach for measuring carrier velocities [5] with subpicosecond time resolution is shown in Figure 2. A GaAs layer approximately 2 microns thick was grown between two heavily doped $Ga_{0.3}Al_{0.7}As$ layers. A voltage applied across this sandwich structure creates an electric field E in the GaAs layer. The electric field modifies the optical absorption in a manner first described by FRANZ [6] and KELDYSH [7]. The field-induced absorption change appears as a displacement of the exponential band tail and exhibits an oscillatory behavior for photons with energies in excess of the band gap.

The basic concept of the measurement technique is to optically inject carriers near the band edge and use the Franz-Keldysh effect to monitor the evolution of the electric field as the holes and electrons drift to opposite ends of the region of applied field. As the carriers drift, a small space-charged field opposite in sign to the applied fields grows in time until the carriers reach their respective contacts. The screening field ΔE must be kept small compared to the applied field E. The small space-charge field induced by the carriers perturbs the optical absorption through the Franz-Keldysh effect. By measuring the induced change in optical absorption as a function of time, we can determine the carrier velocity.

A diagram of the experimental apparatus is shown in Figure 2. The experiment is performed by exciting the GaAs layer with a short pulse from a

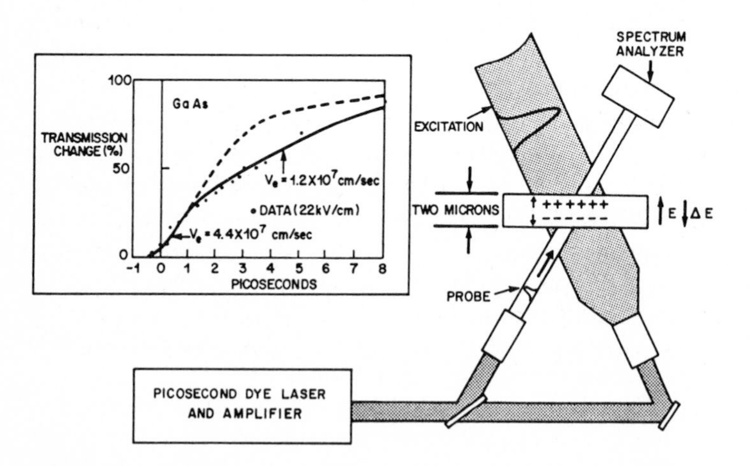

Figure 2 - Diagram of apparatus for making a time resolved measurement of velocity overshoot on GaAs. (Inset) Plot of measured transmission change against time. The dashed line is a single velocity fit of 4.4×10^7 cm/sec and the solid line is a computer generated two velocity fit.

passively mode-locked dye laser. A second optical pulse, which is spectrally broadened is used to probe the absorption spectrum at delayed times. The optical absorption spectrum is measured as a function of time delay after excitation by the pump pulse.

The carrier velocity is determined by plotting the change in the optical absorption as a function of time following excitation. Near t=0 the velocity can be determined approximately by the slope function

$$\frac{\partial \Delta\alpha(t)}{\partial \Delta\alpha(\infty)} \cong \frac{v_e + v_n}{d}$$

where $\Delta\alpha$ is the absorbence change, v_e and v_h are the electron and hole velocities, and d is the length. After t=0 a computer fit to the data is required. The dashed curve in Figure 2 is a continuation of a single-velocity computer fit to the data near t=0 for $v_e = 4.4 \times 110^7$ *cm/sec*. Clearly, this single-velocity curve diverges dramatically from the data for times greater than 2 picoseconds where the velocity has slowed down to 1.2×10^7 *cm/sec*.

Highspeed optics has the potential to play a more significant role in the future of electronics beyond measurement. A new class of devices appears possible which integrate optics into the function of what is now a purely electronic device. Optical pulses are an ideal way to communicate on chip and possibly the device level. We might envision for example that the output of a transistor be a light modulator rather than a source of current to drive an electrical transmission line. Since transmission lines are very lossy and dispersive for electronic pulses of a picosecond or less, a severe limitation is placed upon interconnecting circuit elements. We might consider connecting circuits with what amounts to a local area network on a chip. The network would act to interconnect regions of high speed operation with optics and provide a means of inputing and outputing data to the circuit. A number of components that are not currently available are needed to make this concept a reality. For example sources, light modulators and detectors need to be developed which are small in size and consume little power.

In conclusion, optical techniques hold great promise for high-speed electronics in the future. Optics should be particularly important for investigating material dynamics and could form the basis for a new class of electronic devices that integrate optics into the electronic device function.

REFERENCES

1. W.H.Knox, R.L.Fork, M.C.Downer, R.H.Stolen, C.V.Shank, J.Valdmanis, to be publ. in Applied Physics Letters.

2. C.V.Shank, R.L.Fork, R.F.Leheny, and J.Shah, Physical Review Letters, *42,* 112 (1979).

3. H.Kroemer, Solid State Electron., 2161 (1978).

4. T.J.Maloney and J.Frey, Journal of Applied Physics, *48,* 781 (1977).

5. C.V.Shank, R.L.Fork, B.I.Greene, F.K.Reinhart, and R.A.Logan, Applied Physics Letters, *38,* 104 (1981).

6. W.Franz, Z.Naturforsch., *13,* 484 (1958).

7. L.V.Keldysh, Journal of Soviet Physics, JETP, *7,* 788 (1958).

GaAs Integrated Circuit Technology for High Speed Analog and Digital Electronics

R. Castagne
IEF - Univ. d'ORSAY, F-Orsay, France
G. Nuzillat
LCR, Thomson-CSF, F-Orsay, France

1 - INTRODUCTION

Ultra-high speed GaAs digital and analog integrated circuits are being developed to meet high throughput requirements for very fast signal processing in telecommunications or military systems as well as for commercial very high speed computers. To achieve high system performance, the "on-chip" delay must be minimized. Also a LSI (VLSI) level must be reached (2.000 to 10.000 gates/chip) for digital IC's. The GaAs technology is then confronted to LSI problems such as parameters uniformity and yield. On the other hand, analog circuitry needs precision operators and high stability.

This paper will attempt to bring out the major technological trends in GaAs IC's and to show in what amount today's state of the art can meet the requirements. An extrapolation towards next generation problems will also be attempted.

Fig. 1 recalls the most general features of a very fast signal processing system and shows the various levels where GaAs can be present. Table 1 summarizes the expected performances for the GaAs key elements of such a system [1] [2]. Fundamental requirements for devices and chip parameters are deducible from these data. The most important are summarized in table 2.

These data, considered in the light of devices and interconnect physical behavior, give some guidelines for the understanding of the trends and efforts in GaAs IC's technology.

2 - ON-CHIP SWITCHING SPEED AND SLEW RATE

In digital applications, one has to look at the total "on-chip gate delay t_{pd}". This delay is derived from the rate of charge sharing between the active part of the transistors and the loads.

The maximum current delivered by a transistor is $I_{max} = Q_t/t_t$ with Q_t being the amplitude of internal charge controlled by the device and t_t its total transit time. Each gate introduces its own out-put parasitic load charge Q_p, so that when one gate is driving N identical gates through interconnection lines, each carrying a charge Q_l, the "on-chip switching time" is approximately

$$t_{pd} = (Q_p + N\ (Q_l+Q_t))/Imax.$$

Let us recall that in digital IC's, the max available current, I_{max}, is limited by the maximum power dissipation, P_{max}, allowed to the gate which is a function of the chip complexity. For free air cooling, the total allowed power dissipation is $2W/cm^2$ [2]. For high speed BFL (D MESFET) technology,

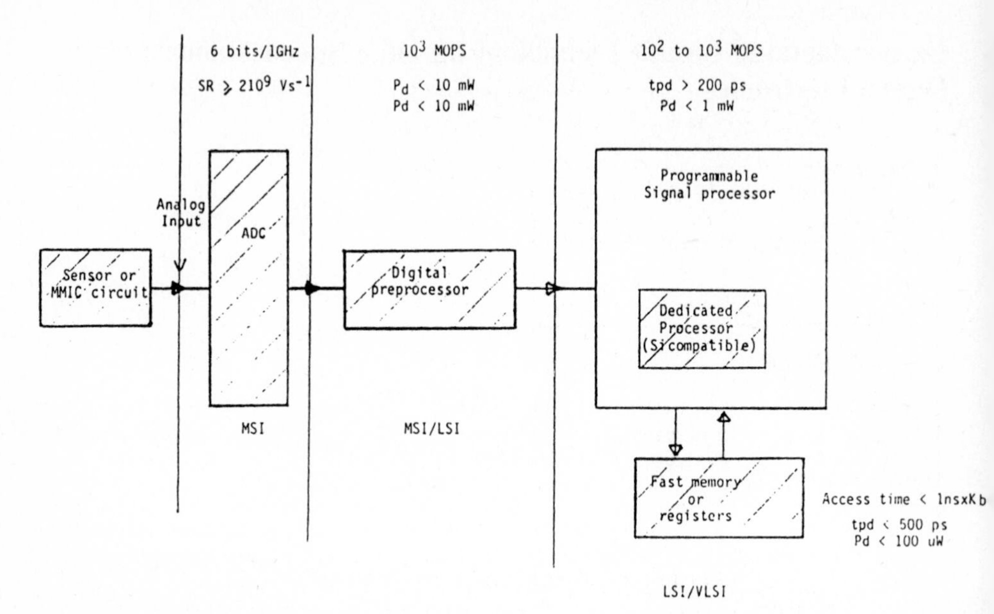

Fig.1. Various stages of a very fast signal processing system (VSFP)
▨ GaAs can be used

Table 1. Summary of the expected performances, integration level, for GaAs elements in fast signal processor

SUB. SYSTEMS	ANALOGS A/D INTERFACE	DIGITAL PREPROCESSORS	DEDICATED PROCESSORS	FAST DATA STORAGE
EXAMPLES	ADC DAC Regeneration circuit	MUX DEMUX DECODER	ALU MULTIPLIER FFT Butterfly Filters, correlators Switching matrix	4k to 16k bit SRAM ROM FIFO
REQUIRED COMPLEXITY Gates/chip		MSI/LSI	LSI 10^3	LSI/VLSI 10^3 ---> 10^4
PERFORMANCES (Typical)	6 bits/1GHz ADC	10^3 MOPS	10^2 to 10^3 MOPS	Access time<1nsxKbit
Required Speed & power performance on-chip (air cooling)	SR > 2 10^9 V/s	20 <tpd <50 ps 10 mW > Pd > 1 mW	100ps <tpd <200ps 1 mW > Pd > 100 uW	200 <tpd <50 ps 100 μW > Pd > 20 μW

the optimum P_{max} is around 20 mW/gate, which leads to $I_{max} \simeq 2{,}5$ mA. On the other hand, for high density DCFL (E/D MESFET) technology (10^4 gates/cm^2), the maximum allowed power would be 200 μW and $I_{max} \simeq 200$ μA.

Table 2. Requirements at chip level for GaAs technology

DIGITAL / ANALOG	"On-chip switching time"	Functionality
Max. Slew Rate (SR)	Device current capability/vs power dissipation Connection and parasitics loading	
Accuracy (in signal acquisition and processing)		V_T uniformity Offset uniformity Stability 1/f noise

Using E FET instead of D FET allows a substantial reduction in bias voltage, and also reduces the number of transistors in a gate. Large scale integration is then possible without drastic switching speed degradation [3]. The corresponding speed criterion for analog circuits is the slewing rate (maximum rate of voltage increase at a critical node in the circuit). The slew rate is also determined by the maximum current available and by the line, parasitic and input capacitances. A similar formulation can be used for the slew rate SR :

$$SR^{-1} = (Q_p + N(Q_L+Q_t))/(I_{max}\Delta V_0),$$

in which ΔV_0 is the voltage swing at the determining mode.

To improve either the switching speed or the SR, one can first increase the transistors width along the critical path in the circuit, keeping other parameters a constant. Q_t is then made larger. This is used, for instance, in the design of bus drivers but cannot be applied in LSI chip design due to the increase in power dissipation.

Speed improvement can also be obtained through the so-called geometrical scaling down [3] [4]. Bias voltage and logical swing being kept at constant values, Q_p Q_l and Q_t are simultaneously reduced by the scaling factor λ ,and therefore one must simultaneously either increase the electron density inside the control region of the device (to recover a high Q_t), or reduce the transit time t_t.

In submicron gate GaAs MESFET's, owing to the overshoot of the electron velocity in the channel, the overall transit time scales down generally faster than the gate length lg, so that the ratio t_t/Q_t decreases with lg in the expressions for t_{pd} or SR^{-1}. On the other hand, short channel effects [5], leading to an elevation of the output conductance g_d and then to a degradation in gain and noise immunity, set a limit to the gate length reduction which is around .5 um.

The increase of the electron density in the control region of the devices can be obtained by elevating the channel doping concentration [6] . On the other hand, a high doping level means a reduction in the electron mobility by Coulomb scattering. In MISFET's and heterojunction TEGFET's, it is possible to confine a high density electron gas along an interfacial barrier in a very low doped material. Conditions are then met for high density, high velocity electron flow.

However, the MISFET is not a high mobility transistor because of electron scattering by the interfacial charges fluctuations in the insulating material [7]. In a TEGFET structure, this is not the case and high electron mobilities are achieved due to the high quality of the interface [8].

3 - FUNCTIONALITY AND ACCURACY

In digital as well as analog IC's, one of the most sensitive parameters to control is the FET threshold voltage V_T [9].

To ensure the functionality of thousand gates digital processors, a 10% long-range uniformity of V_T is desirable. Yet more stringent is the requirement for analog circuits. For instance, in a 4 bit flash ADC, a 2% short-range precision on V_T is required mainly in order to minimize the offset dispersion in differential pairs. Furthermore, such a stringent condition makes analogs very sensitive to parasitic functional drift of V_T. Any of these requirements would entail using very high quality SI substrates with low chromium compensation, to control the active layer thickness at a high precision level and to achieve very good control of the lithography, especially for the gate length lg.

Even so, three important causes of functionality "sickness" would remain :

- the back and side gating effects, which at first analysis can be considered as a gating of the channel by the interfacial (active layer to substrate) barrier
- the surface-stress orientation effects, which make the current of the transistors depend on their orientation on the chip
- the surface depletion layer effect, which deserves special comment.

The first generation GaAs MESFET's was spoiled by the existence of a depletion layer under the free surface of the gate-to-source or gate-to-drain spacings. This is due to the pinning of the surface Fermi level at mid-gap. A high source-to-gate access resistance R_{sg} resulted, leading to the well-known series feedback effect (typically 4 Ω xmm). The available controlled charge Q_t was then significantly reduced. A reduction factor of over 10% was currently observed. Furthermore, the fixed depletion layer in the access zones had approximately the same depth as the zero bias Schottky gate depletion layer. The drastic consequence of that was the impossibility to operate the MESFET with a direct gate bias, i.e. to obtain E MESFET operation.

In addition, surface traps are also responsible for a lagging phenomenon of the gate control. [10].

The recessed gate process provided a means to reduce these effects, but a true planar process permitting high doping levels under the surface of the access regions -in order to limit the depletion layer penetration- was needed. Notwithstanding recent technological advances, long-term surface passivation remains a problem to be solved in GaAs IC technology.

Low-frequency (f<1 MHz) noise in MESFET's is also a severe limitation to the implementation of wide band (direct-coupled) analog devices [11].

When the device operates with open channel, the 1/f noise can be attributed to bulk traps, whereas, when the channel is pinched-off, the noise is due to interfacial traps. Yet, high quality material (epilayers and substrates) as well as clean process are a must.

4 - THE '80-'85 TECHNOLOGICAL EVOLUTION

During the 1980-85 period, GaAs technology has followed a steep learning curve [12].

Perhaps the more significant efforts have been those of various companies to elaborate high quality semi-insulating materials. Liquid encapsulated crystal pulling (LEC) has been generalized. The decrease in the residual impurities concentration ($<10^{15}cm^{-3}$) allows high ($10^{8}\Omega xcm$) resistivity without (or with low) added Chromium concentration. Electron mobility is improved ($\mu_n>5000cm^2V^{-1}s^{-1}$). Use of In-alloyed crystals also reduces the dislocation density [13].

Owing to the higher stability of the material during the annealing process, the direct ion implantation of the active layer into the SI substrate has generalized. The implantation energies were lowered (60 to 80 keV against 150 to 300 keV). Flash capless annealing process was also developed [14].

These improvements resulted in a shrinking of the long - range V_T dispersion [13] (σV_T typically 25 mV over 3" wafers), as well as that of the short-range (typically 10 mV over 1 mm^2), [15].

The offset dispersion of differential pairs was also improved by using EMESFET (typically mean offset of 10 mV with standard deviation 30 mV) [10].It must be noticed that an important part of this dispersion is due to the FET gate length lithographic dispersion. Direct implant into the substrate also permitted the self-alignment of N^+ source and drain access region giving a significant reduction of the overall surface depletion layer effects. EMESFET technologies with low R_{sg} (typically .3 ΩXmm) and high g_m (typically 200 mS/mm) have then developed at a fast pace.

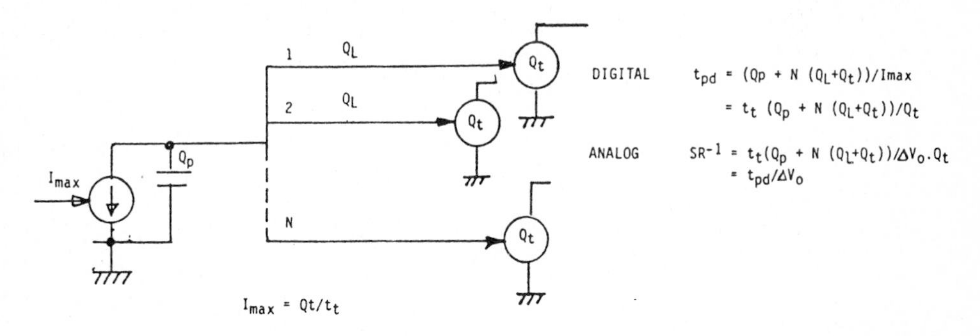

Thus, an important breakthrough was achieved which made E.D. MESFET (DCFL) technology a credible choice for very fast LSI circuits development [15]. Various dedicated operators [16] [17], memories [18] [19] [20] ,and gate arrays have been demonstrated with full functionality and more than 6% yield. Fig. 3 to 5 and table 3 give a short panorama of the increase in circuit complexity and throughput rate.

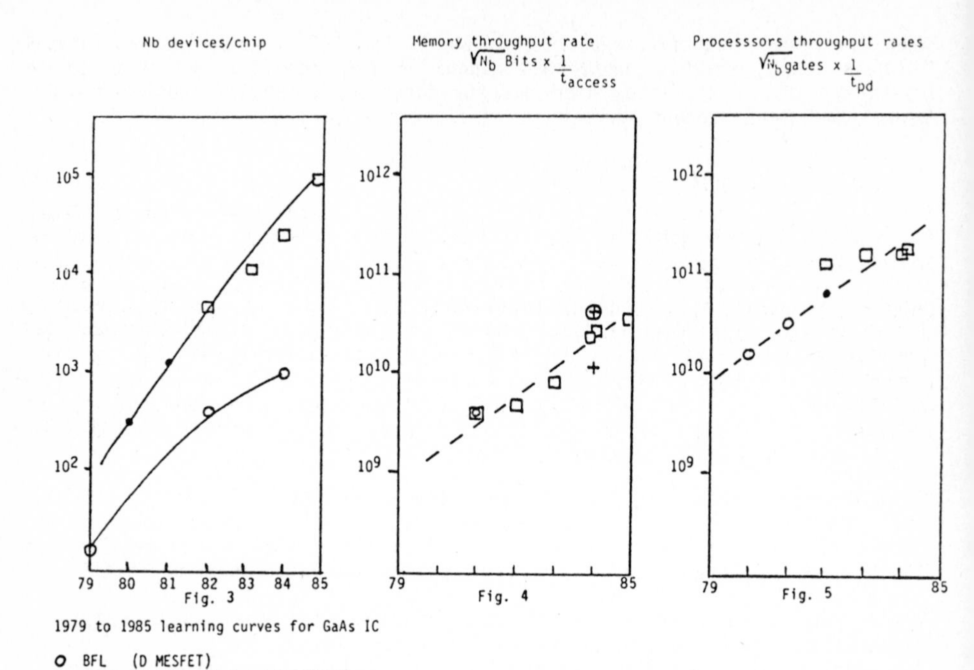

Fig. 3 Fig. 4 Fig. 5

1979 to 1985 learning curves for GaAs IC

- O BFL (D MESFET)
- • SDFL (Schotty diodes + D MESFET)
- □ DCFL (E MESFET)
- ⊠ LPFL (Low pinch off voltage D MESFET)
- + HEMT GaAl As/GaAs

During the same period, fine epitaxial methods (MOCVD and more especially MBE) have permitted the development of the very promising so-called HEMT. After demonstration of very high speed and low power dissipation capability [22] [23], GaAl As/GaAs devices (TEGFET, MODFET, HEMT, SDHT etc..) have continued to develop. In Japan, HEMT 4kb memories [24] have been realized and tested. In the US, MODFET's [8] [25] and SDHT's [27] are reaching MSI or LSI circuit level. In Europe, TEGFET's will enter in a pre-development stage, exhibiting very high electron-mobility at liquid nitrogen temperature ($\mu_n > 10^5$ cm^2/Vs). These devices and circuits are today the best candidates for cryogenic IC's.

Also a contending technology is the GaAl As/GaAs heterojunction Bipolar which has demonstrated its capability to enter the traditional bipolar silicon structures (CML and I_2 L) with a large improvement in speed power product. As for silicon ECL, this structure should be very well suited to drive arrays of logical gates. A 1kbit gates array chip with 400 ps on chip mean propagation delay and power dissipation 1mW/gate has been demonstrated. [35]

Analog IC's are now considered a challenging point for GaAs. Wide band monolithic op amps have been experimented [11] with a "real estate efficiency" of about 110 dB/mm^2 between .1 and 4.5 GHz. Ultra-high speed comparators [28] [29] [30] are under development in many companies. Today's state of the art gives a sensitivity of about 50 mV at 1GHz sampling frequency (LSB=100 mV). 3 and 4 bit flash ADC has been demonstrated operating at 1 GHz [29]. Decision circuits exhibiting 100 ps rise and fall times and high dynamic sensitivity (65 mV) over a 1 GHz bandwidth have been fabricated with a significant yield [31].

GaAs can also be used for other functionalities. For example, a four phase CCD has recently been reported with up to 4 GHz operation and .8 transfer efficiency for 3 x 100 µm cell size [32]. A monolithic 7 tap programmable transversal filter using SAW device has also been successfully tested [33].

5 - THE NEXT GENERATION PROBLEMS

It is an accepted fact that GaAs devices and circuits have now demonstrated their capability to drive interconnection patterns at high speed. A further improvement in device capabilities along with a reduction in load capacitance should bring circuits performances close to device intrinsic speed limits. Chip delay around 10 ps should then be obtained. Such switching time falls in the range of propagation time of the electromagnetic wave along the distributed microstrip interconnection lines. Considering that in digital circuitry, an impedance desadaptation is necessary for saving power, signal degradation by multiple reflection and interconnection cross-talk is foreseeable. Fig. 6 is a simulation result showing that 30% cross-talk

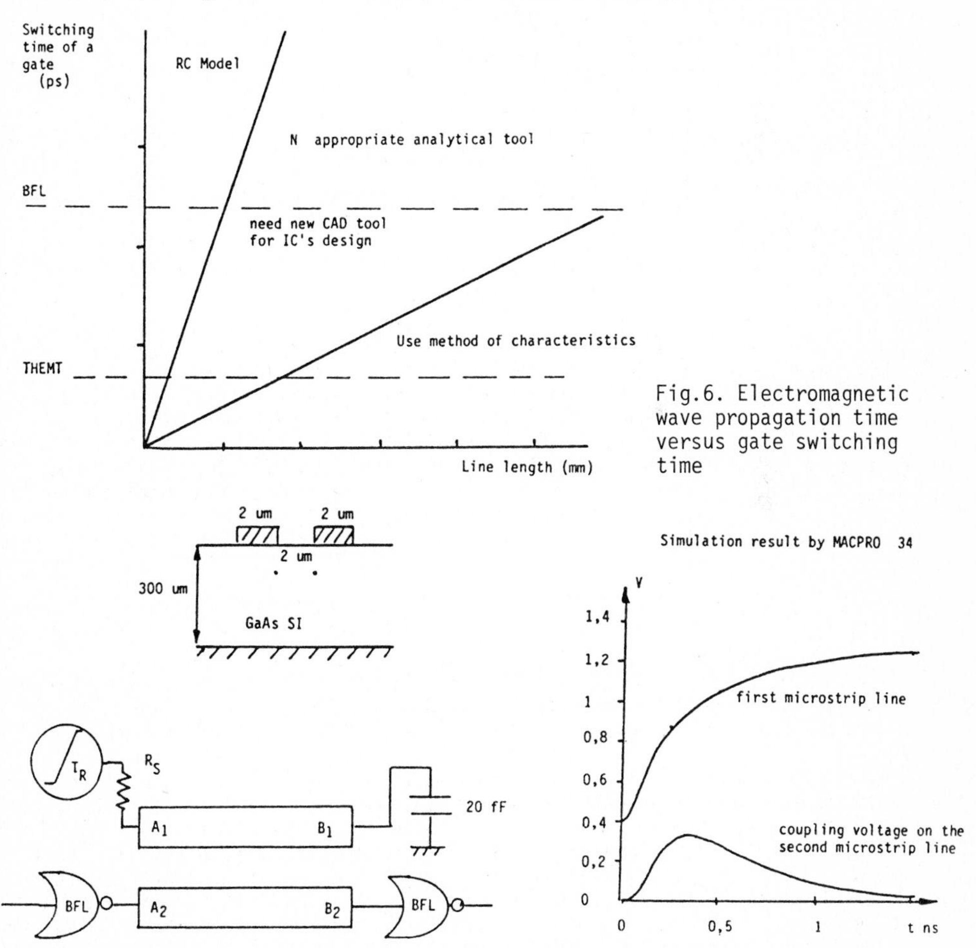

Fig.6. Electromagnetic wave propagation time versus gate switching time

Fig.7. Coupling and cross talk noise between parallel microstrip lines - Simulation results

noise can be expected along parallel 2 μm distant bus lines [34]. Also expectable is a coupling noise along power distribution lines.

The interconnection problem will probably become of importance in LSI, VLSI architectures, using 10 to 20 ps on-chip switching times. Circuit architecture could then be the arena of next generation problems, where today's usual quest for circuit densification will give place to new principles (pipelined or systolic architectures, for instance) and push ahead the design of new CAD tools.

Table 3. Some significant results in analog GaAs circuit

Type of circuit	Performances	Technology	Year
Regenerator (Decision circuit)	Clock frequency 105 GHz	DFET	81
DAC	3 bits 1 GHz	DFET	83
Wide band Amp	42 dB .1 to 2 GHz		
Sample hold	500 MHz	DFET	83
ADC	1 GHz 1% precision	DFET	83
ADC	3 bits flash LSB 200 mV Settling time 600 ps		83
Comparator	1.8 GHz	DFET	83
Comparator	1 GHz 50 mV offset < 30 mV	EFET	84
Regenerator Decision circuit	3 GHz	DFET	84
Sample hold	100 MHz 400 MHz	DFET	84

Table 4. Some significative results for gate arrays

Year	Technology	Nb gates	chip dim mm	mean t_{pd} ps	power/gate mW
1982	SDFL	320	1.05 x 1.23	200	
1983	SDFL	432		250	3
1983	DCFL	1224	4.6 x 4.3	100	0.25
1984	DCFL	1050	3.15 x 3.75	100	0.2
1984	HJBT	1024	3.8 x 3.5	400	1

OTHER ANALOG FUNCTIONALITIES

Four phase CCD	4 GHz 3x100 μm cell size 0.8 transfer efficiency	1984
Monolithic 7 tap transversal (SAW) programmable filter	50 dB dynamic 20 ns delay on off ratio 18 dB	1984

1 B. DUNBRIDGE, "Digital Signal Processing, Applications and Technology" High speed digital technologies conference, San Diego, 1981.
2 G. NUZILLAT and Coll., IEEE J. of SSC, Vol. SC 17, n°3, p. 568, 1982.
See also
G. NUZILLAT and Coll., IEE proceedings I, Solid state and elect. Dev., Vol. 127, n° 5, 1980.
See also
MASAYUKI ABE and Coll., IEEE trans ED 29, n° 7, July 82.
3 P.T. GRELLING, MSN, p. 96, Nov. 84.
4 A. BARNA, C.A. LIECHTI, IEEE, SC 14, n° 4, Aug. 79.
5 J.F. PONE and Coll., IEEE, Ed. 29, n° 8, p. 1244, 1982.
6 H. DAMBKES and Coll. GaAs IC Symposium Digest, p. 153.
7 M. MOUIS and Coll. SSE.
8 D.M. SOLOMON, H. MORKOC, IEEE, Ed. 31, n° 8, p. 1015, Aug. 84.
9 G. NUZILLAT and Coll., IEEE, Ed. 27, n° 6, 1980.
10 M. ROCCHI, "Status of the Surface and Bulk Parasitic Effects Limiting the Performances of GaAs IC's", ESSDERC 1984, paper.
11 V. PAUKER and Coll. Proc. of 14th Europ. Microwave Conf., p.773, Liège, Sept. 84.
12 Sven A. ROOSILD, GaAs IC Symposium Digest, p. 155, 1984.
13 R.E. LEE and Coll., GaAs IC Symposium Digest, p. 45, 1984.
14 R.C. CLARKE, IEEE, Ed. 31, n° 8, Août 84.
15 H.J. HELIX and Coll. GaAs IC Symposium Digest, p. 163, 1984.
16 Y. NAKAVAMA and Coll., IEEE, J. of SSC, SC 18, p. 599, Oct.83.
17 P.G. FLAHIVE and Coll., GaAs IC Symposium, p. 7, 1984.
18 M. HIRAYAMA and Coll., ISSCC Dig. Tchn. Pap., p.46, Feb. 84.
19 Y. ISHII and Coll., GaAs IC Symposium, p. 121, 1984.
20 N. YOKOYAMA and Coll., IEEE, J. SSC, SC 18, n° 5, p. 520, 1983.
21 T. HAYASHI and Coll., GaAs IC Symposium, p. 111, 1984.
22 P.N. TONG and Coll., Electron Letters 18, p. 517, 1982.
23 T. MIMURA and Coll., J. JAP 19, p. 225, 1980.
24 S. KURODA, T. MIMURA and Coll., GaAs IC Symposium, p. 125. 1984.
25 N.C. CIRILLO and Coll., GaAs IC Symposium, p. 167, 1984.
26 NISHIUCHI and Coll., ISSCC, Dig Tech Pap, p. 48, 1984
27 S.S. PEI and Coll., GaAs IC Symposium, p.129, 1984
28 T. DUCOURANT and Coll., 11th Int. Conf. on GaAs and Related Compound, Biarritz, Sept. 84
29 L.C. UPADHYAYULA and Coll., IEEE, Ed. 30, p. 2, 1983.
30 P. GREILLING and Coll., GaAs IC Symposium, p. 31, 1984.
31 M. PELTIER and Coll., IEEE, MTT-S, Washington, 1980.
32 E.A. SOVERO and Coll., GaAs IC Symposiun, 1984.
33 J.Y. DUQUESNOY and Coll., IEEE, 1984. Dallas
34 L. CHUSSEAU, P. CROSAT : MACPRO, a digital circuit macro simulator, including interconnection propagation and coupling effects. Int. Report. IEF, Univ. de Paris-Sud, ORSAY.
35 H.T. YUAN and Coll., ISSCC, Digest Tech Pap, p. 42, 1984

Heterojunction Bipolar Transistor Technology for High-Speed Integrated Circuits

P.M. Asbeck
Rockwell International Corporation, 1049 Camino Dos Rios,
Thousand Oaks, CA 91360, USA

1. Introduction

A bipolar structure provides many potential benefits for picosecond electronic circuits: 1) current flow is vertical, through epitaxial layers whose total thickness is only a few tenths of a micron, decreasing carrier transit times; 2) large amounts of current can be carried by modest size transistors, enabling rapid charging of interconnect capacitances; and 3) threshold voltages are tied to built-in potentials of p-n junctions, which are easy to reproduce device-to-device, enabling accurate logic voltages to be established. Added advantages result from using III-V heterojunctions rather than Si [1]: a) high electron mobility; b) pronounced transient electron velocity overshoot effects; c) ability to incorporate quasi-electric fields into the devices by semiconductor composition variations; and d) ability to insure adequate emitter efficiency with wide gap emitters, and therefore produce devices with heavily doped, low resistance base regions together with lightly doped, low capacitance emitter regions. As a result of these advantages, HBTs offer prospects for switching time delays in the sub 10 ps regime, with a robust, manufacturable fabrication process.

2. HBT Structure

A band diagram for GaAs(GaAl)As single heterojunction HBTs is shown in Fig. 1. Representative layer structures are described in Table 1. An abrupt emitter-base heterojunction gives rise to a conduction-band potential discontinuity that may act as a "ballistic launching ramp" to inject high velocity electrons into the base [2]. With a graded emitter-base heterojunction, this spike disappears and the full bandgap difference contributes to reducing hole injection into the emitter. High gain, high speed devices have so far been made with both approaches. Grading the Al composition, and hence bandgap, across the base can provide a quasi-electric field to drive electrons to the collector more rapidly than by

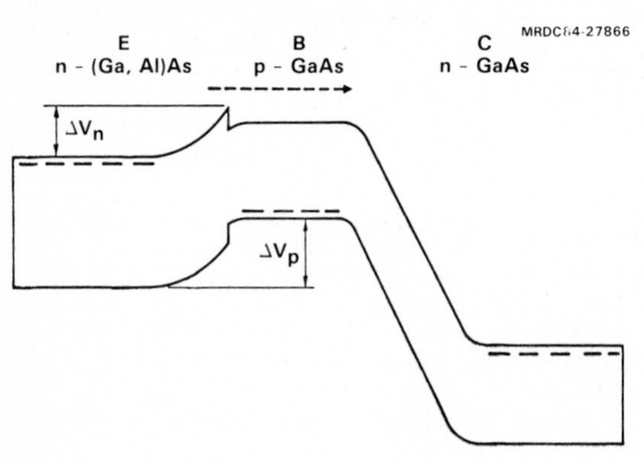

Fig. 1 Schematic band diagram of HBT

Table 1
Typical Epitaxial Structures for HBTs

Layer	Thickness (μm)	Type	Doping (cm^{-3})	AlAs Fraction	
8	0.075	n^+	1×10^{19}	0	Cap
7	0.125	n	5×10^{17}	0	Cap
6	0.03	n	5×10^{17}	.30-0	Grading
5	0.22	n	5×10^{17}	.30	Emitter
4	0.03	n	5×10^{17}	0-.30	Grading
3	0.1	p^+	5×10^{18}	0	Base
2	0.5	n^-	3×10^{16}	0	Collector
1	0.6	n^+	4×10^{18}	0	Subcollector
Substrate		S.I. undoped		0	

diffusion alone [3]. It has been shown using picosecond optical pulse measurement techniques that electron velocities on the order of 2×10^7 cm/s may be obtained in graded bases [4]. Other possibilities include the incorporation of a wide bandgap collector [1] (to suppress hole charge storage during transistor saturation), as well as the incorporation of superlattice regions in the emitter and collector (to overcome GaAlAs material deficiencies) [5]. An alternative structure makes use of a buried emitter and a collector on the wafer surface (inverted device); this has been used to implement I^2L digital circuits [6].

3. Estimates of Performance Potential

Numerical simulations of HBT performance have been carried out at various levels of refinement. Figure 2 shows calculated cutoff frequency, f_t, vs collector current density, as determined from a one-dimensional simulation which accounts for velocity overshoot in an ad-hoc way only [7]. The result illustrates the prospects for f_t above 100 GHz (transit times below 1.6 ps), as well as for high operating current density up to 10^5 A/cm^2. Monte Carlo calculations which account more properly for the hot electron effects have yielded even higher f_t projections; up to 150 GHz for the

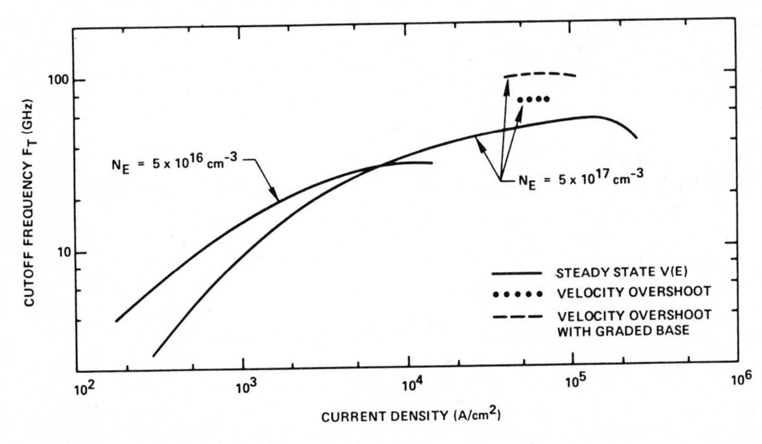

Fig. 2 Calculated cutoff frequency vs collector current density

intrinsic device [8]. Estimates of the maximum frequency of oscillation, f_{max}, attainable are strongly dependent on the lateral dimensions of the device and the values of parasitic resistances, and thus are closely tied to process technology; f_{max} values up to 200 GHz appear to be feasible with submicron lateral feature size. An important factor that influences both f_t and f_{max} is the collector depletion layer thickness, W_c. Small values of W_c reduce the electron transit time and tend to increase f_t; large values tend to reduce base-collector capacitance and favor f_{max}. An optimal choice for W_c hinges on the application (broad band vs narrow band) and the base-collector voltage that must be supported. In digital circuits, simulations suggest that propagation delays under 10 ps may be obtained for power dissipation per gate near 1 mW with submicron emitter widths [9].

4. Fabrication Technology

To implement GaAs/(GaAl)As HBTs at Rockwell International, MBE structures on semi-insulating GaAs substrates have been utilized. Devices have been defined with a quasi-planar process, which uses ion implantation to contact the base, and to provide isolation between devices, as shown in Fig. 3. Rapid thermal annealing is used to avoid excessive dopant diffusion [10]. Oxygen implants are used to produce a compensated, partially insulating spacer layer between extrinsic base and collector to cut down on capacitance [11]. Emitter widths as low as 1.2 μm (governed by the contact width) have been obtained. High performance HBTs have also been demonstrated using MOCVD epitaxial layer growth [12].

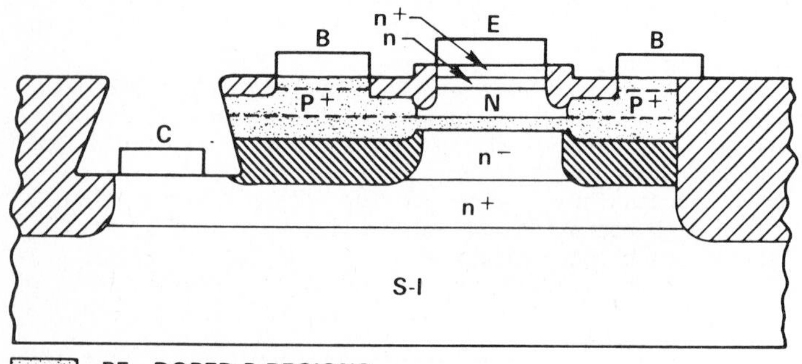

Fig. 3 Schematic cross-section of HBTs fabricated at Rockwell

5. Device Characteristics

Experimental characteristics of HBTs fabricated at Rockwell are illustrated in Fig. 4. Devices have current gain adequate for digital ICs (30-50 in devices with 1.6 μm emitters, higher in larger devices), high current handling capability (up to 10^5 A/cm^2), high transconductance (6000 mS/mm), and high threshold voltage uniformity (standard deviation of 4 mV across a wafer). Data of current gain vs frequency calculated from S parameter measurements are displayed in Fig. 5, indicating a cutoff frequency f_t of 40 GHz. The value of f_{max} associated with this device is on the order of 30 GHz. The devices also have appreciable microwave power handling capability: a 1 dB compressed power of 2.5 W per millimeter of emitter length has been measured at 12 GHz in class A operation (albeit in a small device with 20 μm total emitter length).

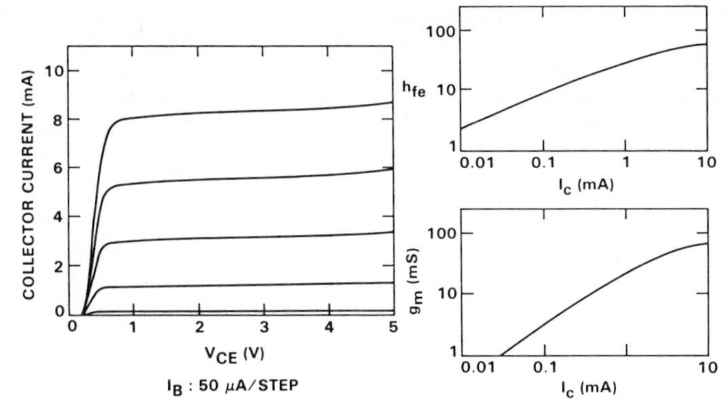

Fig. 4 Experimental HBT I-V characteristics

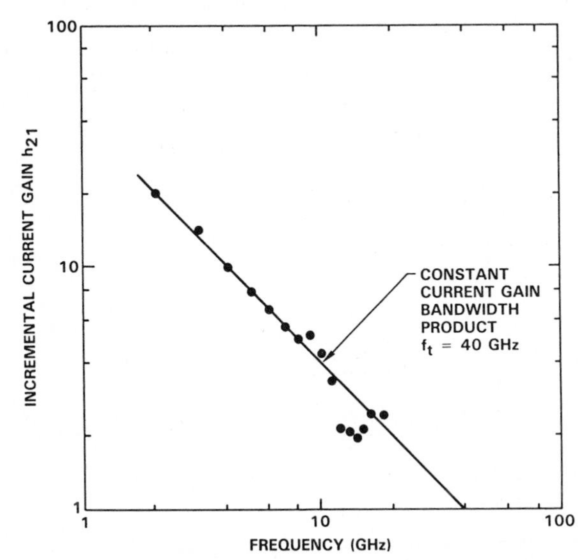

Fig. 5 Current gain, h_{21}, vs frequency from S parameter measurements

6. Circuit Results

Prototype digital circuits used to demonstrate the capabilities of HBT technology include ring oscillators and frequency dividers. The ring oscillators are of the nonthreshold logic (NTL) type, as well as of the current-mode logic (CML) and classical emitter-coupled logic (ECL) types. The NTL gates, while not adapted for use in complex digital systems because of limited fan-out, are the simplest, and the ones whose characteristics are most directly tied to the transistor characteristics. NTL ring oscillators with 17 stages were made and propagation delay times per gate of 29.3 ps (at a power of 4 mW per gate) were measured. Relatively low power operation was also observed (60 ps at 400 μW for a power delay product of about 25 fJ).

Frequency dividers were configured from master/slave D flip-flops implemented with series-gated CML circuits. These classical Si-like circuits allow operation with a single-ended clock at frequencies up to $1/2\ \tau_d$, where τ_d is the propagation delay per (complex) gate. Divide by four circuits of this type have been demonstrated which operate at input clock rates up to 8.6 GHz.

Results obtained with the frequency dividers provide a preliminary indication of circuit yield, which appears to be excellent. The yield of four circuits was 90% on the wafer tested. Moreover, this yield result corresponds to rf functionality (with a test frequency input of 7 GHz) using fixed power supply voltages. The dividers are small-scale circuits with 35 transistors in about 15 equivalent gates.

For both ring oscillators and frequency dividers, the circuit operation was found to be in good accord with SPICE simulations based on calculated and measured characteristics of the transistors. Parameters corresponding to a representative transistor for use in ICs are shown in Table 2.

Table 2
HBT Model Parameters (SPICE)

Emitter Dimensions	1.6 μm x 5 μm
R_E	28 Ω
R_B	390 Ω
R_C	65 Ω
τ_F	2.5 ps
C_{jeo}	11 fF
C_{jco}	17 fF

7. Future Outlook

Applications of HBTs may be envisioned in very high speed digital ICs at various levels of complexity, gate arrays and A/D converters, as well as in microwave and millimeter wave discrete devices. For microwave applications, HBTs benefit from having lower 1/f noise than FETs.

A variety of improvements in HBT technology may be anticipated in the future. These include size reduction and improved dimensional control through the use of projection lithography and electron beam lithography, reduction of parasitic resistances and capacitances through the introduction of self-aligned fabrication techniques, and improved current gain and cutoff frequency through improved understanding and optimization of the device structure. The use of new material systems may provide additional performance benefits. Of particular interest is the introduction of low bandgap semiconductor for the base region to reduce power dissipation in VLSI circuits. Notable work has already been demonstrated using (In,Ga)As and (In,Ga)(As,P) lattice-matched to InP [13]. Integration of optical and electronic devices in the same chip is an additional interesting possibility; laser transistor structures have already been produced [14].

8. Acknowledgements

The Rockwell work reviewed here is the result of the efforts of many people, including D. Miller, R. Anderson, F. Eisen, A. Gupta and others. We are indebted to H. Kroemer for numerous illuminating discussions.

9. References

1. H. Kroemer, Proc. IEEE 70, 13 (1982).
2. D. Ankri and L.F. Eastman, Electron. Lett. 18, 750 (1982).
3. D.L. Miller, P.M. Asbeck, R.J. Anderson and F.H. Eisen, Electron. Lett. 19, 367 (1983).
4. B.F. Levine, W.T. Tsang, C.G. Bethea and F. Capasso, Appl. Phys. Lett. 41, 470 (1982).
5. S.L. Su, R. Fischer, W.G. Lyons, O. Tejayadi, D. Arnold, J. Klem and H. Morkoc, J. Appl. Phys. 54, 6725 (1983).
6. W.V. McLevige, H.T. Yuan, W.M. Duncan, W.R. Frensley, F.H. Doerbeck, H. Morkoc and T.J. Drummond, IEEE Elect. Dev. Lett. 3, 43 (1982).
7. P.M. Asbeck, D.L. Miller, R. Asatourian and C.G. Kirkpatrick, IEEE Elect. Dev. Lett. 3, 403 (1982).
8. K. Tomizawa, Y. Awano and N. Hashizume, IEEE Elect. Dev. Lett. 5, 362 (1984).
9. T. Sugeta (unpublished).
10. P.M. Asbeck, D.L. Miller, E.J. Babcock and C.G. Kirkpatrick, IEEE Elect. Dev. Lett. 4, (1983).
11. P.M. Asbeck, D.L. Miller, R.J. Anderson and F.H. Eisen, IEEE Elect. Dev. Lett. 5, 310 (1984).
12. C. Dubon, R. Azoulay, P. Desrousseaux, J. Dangla, A.M. Duchenois, M. Hountondji and D. Ankri, 1983 IEDM Tech. Dig., p. 689.
13. See, for example, A.N.M Masum Choudhury, K. Tabatabaie-Alavi and C.G. Fonstad, IEEE Elect. Dev. Lett. 5, 251 (1984); R.J. Malik, J.R. Hayes, F. Capasso, K. Alavi and A.Y. Cho., IEEE Elect. Dev. Lett. 4, 383 (1983).
14. J. Katz, N. Bar-Chaim, P.C. Chen, S. Margalit, I. Ury, D. Witt, M. Yust and A. Yariv, Appl. Phys. Lett. 37, 211 (1980).

Permeable Base Transistor*

R.A. Murphy
Lincoln Laboratory, Massachusetts Institute of Technology,
Lexington, MA 02173, USA

1. Introduction

Many important applications exist for a three-terminal device capable of amplifying and generating power in the EHF frequency range and of operating at high speeds in a logic circuit. Although sufficiently fast two-terminal devices are presently available, three-terminal devices provide greater fan-out capability in a logic circuit and greater isolation of input and output signals in an analog circuit. The permeable base transistor(PBT) was conceived at Lincoln Laboratory in 1979 [1,2] as a device capable of being used in these applications [3,4,5]. It is presently one of several devices being developed for high frequency and high speed applications. Other devices include the conventional field-effect transistor (FET) [6,7], the high electron mobility transistor (HEMT) [8,9], and the heterojunction bipolar transistor (HJBT) [10].

This paper will discuss the characteristics and advantages of the PBT, describe the experimental techniques by which we have fabricated GaAs and Si PBTs, and report the results we have obtained. It will also attempt to provide some insight into the properties necessary for high frequency operation and the means by which such devices are characterized.

2. Device Description

A cross-sectional drawing of a GaAs PBT is shown in Fig. 1. The PBT is a vertical device, much like the solid-state realization of a vacuum-tube triode. The emitter and collector regions of the device are separated by a parallel array of metallic stripes which are connected to an external base terminal. Voltage applied to this terminal controls the current flow from the collector terminal to the emitter terminal. In the GaAs PBT the metallic stripes are imbedded in the GaAs by an epitaxial overgrowth process. In the Si PBT anisotropic etching techniques are used to achieve the vertical topology. The emitter, base, and collector nomenclature is used because the control mechanism at low-to-moderate currents is the modulation of a potential barrier, rather than the modulation of channel thickness [3]. Minority carriers do not participate in the current flow in a PBT.

The advantages of a PBT for high frequency and high speed operation can be discussed in terms of the simple equivalent circuit of Fig. 2. In this equivalent circuit, the transconductance (g_m) and the output resistance (R_o) are intrinsic measures of the control of the current by the base voltage and the dependence of the current on the collector voltage,

*This work was sponsored by the Defense Advanced Research Projects Agency and the Department of the Air Force.

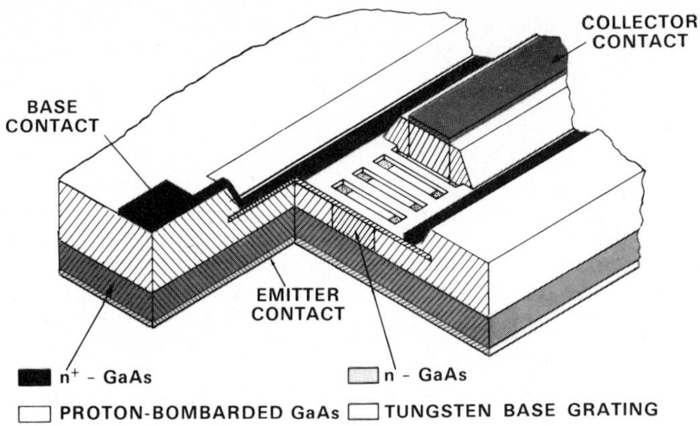

Figure 1. Cross-sectional drawing of a GaAs PBT

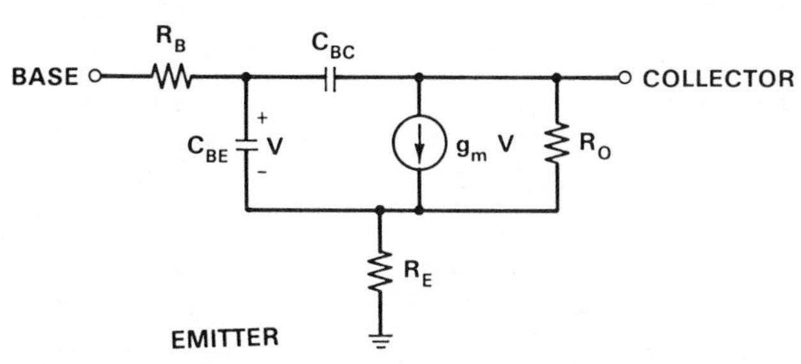

Figure 2. Simplified equivalent circuit of a PBT

respectively. The base resistance (R_B) arises from the finite resistivity of the base metallization, and the emitter resistance (R_E) is due to the resistance of the material between the control region and the emitter contact. The base-emitter capacitance (C_{BE}) and the base-collector capacitance (C_{BC}) are due to the capacitive coupling across the depletion widths surrounding the Schottky barrier base electrode.

High frequency transistors are usually characterized by gains which describe their capabilities in an amplifier circuit [11]. The short-circuit current gain (h_{21}) is the current gain with the output short circuited, and is particularly important for digital circuits, in which the ability to drive successive logic stages is important. The frequency at which this gain equals one is the unity short-circuit current gain frequency (f_T). The maximum available gain (MAG) is the power gain when both input and output ports are impedance matched. This is only possible when the amplifier is unconditionally stable. If not, the amplifier can be described by the maximum stable gain (MSG), which is the most power gain that can be achieved as simultaneous matching is approached. The unilateral gain (U) is that which would exist if the amplifier were neutralized, i.e. if the output-to-input feedback were cancelled out, and both the input and output ports are matched. The maximum frequency of oscillation (f_{max}) is the frequency at which both MAG and U [12] become unity. It is a particularly important figure of merit for analog applications, because it indicates the highest frequency at which the device can provide power gain.

It can be shown that f_T depends on the transit of the charge through the region in which the control electrode exerts influence over the charge [13], namely,

$$f_T = \frac{g_m}{2\pi(C_{BE} + C_{BC})} = \frac{\langle v \rangle}{2\pi L_{eff}} \tag{1}$$

In Eq. (1), the quantity $\langle v \rangle$ is an appropriate average velocity of the electrons transitting the control region, and the quantity L_{eff} is the effective length of the control region. Even though these two quantities are not precisely defined, it is clear from this formula that higher f_T's can be obtained by either increasing the average velocity of the electrons or by decreasing the length of the control region. The former appears to be the primary advantage of HEMT devices, in which the electron velocity is increased by decreasing their collisional scattering. The latter is the primary advantage of vertical devices such as the PBT because short control regions can be obtained more easily.

Other characteristics of the PBT which favor high frequency operation can be understood by considering an expression for f_{max} which can be derived for the equivalent circuit of Fig. 2.

$$f_{max} = \frac{f_T}{2\sqrt{\frac{R_B + R_E}{R_O} + \frac{R_B\, g_m\, C_{BC}}{C_{BE} + C_{BC}}}} \tag{2}$$

Equation (2) clearly indicates that a low base resistance (R_B) is necessary for high frequency operation. This value is low in a PBT because the control electrode is metallic. This feature constitutes the PBT's primary advantage over a HJBT. The emitter resistance (R_E) must also be low, which can be achieved in a PBT by control over the emitter layer thickness and the use of a heavily-doped substrate.

High output resistance (R_O) and low collector-to-base capacitance (C_{BC}) are also crucial to high frequency operation. The configuration of the PBT completely avoids substrate leakage and surface depletion, which can lower R_O in planar devices such as FETs or HEMTs. In addition, the vertical disposition of the PBT makes it possible to introduce doping concentration profiles into the collector region. In this manner an optimal compromise between good output characteristics (high R_O, low C_{BC}) and high transconductance can be obtained.

The advantages of short control region, low control electrode resistance, no substrate or surface leakage, and collector region doping profile capability are uniquely combined in the PBT. We have performed extensive numerical simulations which indicate that the PBT should have an f_T in excess of 50 GHz and an f_{max} in excess of 150 GHz for structures we are presently fabricating, and have even higher values for PBT devices which are scaled to smaller dimensions[3]. Our simulations have not explicitly taken into account the enhancement of electron velocity due to a paucity of scattering events (i.e. ballistic transport), but recent Monte Carlo simulations have predicted f_T values for the PBT of approximately 150 GHz [14].

The primary disadvantage of the PBT is that much of the technology for fabricating the devices is new, and has had to be developed specifically for the PBT. The remainder of this paper will describe the technology we have developed for the GaAs and Si PBT, and report the results we have obtained on the devices we have fabricated.

3. GaAs PBT Fabrication Technology

The three-dimensional drawing of a GaAs PBT in Fig. 1 shows the important construction details of the device. The starting material consists of a 0.2-μm-thick n-type epitaxial layer with a concentration of 5×10^{16} cm^{-3} grown upon an n^+ substrate. The wafer is then coated with a 300 -500 Å layer of tungsten. Both evaporated and RF sputtered tungsten layers have been used, but recent wafers have used the latter, which has a sheet resistance of 5-10 ohms per square. A 3200-Å-periodicity grating of stripes is then patterned into the tungsten using x-ray lithography [15] and reactive ion etching (RIE). Base shorting bars, which provide electrical contact to the stripes, are also formed in this step. Tungsten is used for the base metallization because the overgrowth process of imbedding the tungsten stripes requires temperatures greater than 500°C. A scanning electron micrograph of the patterned tungsten film is shown in Fig. 3.

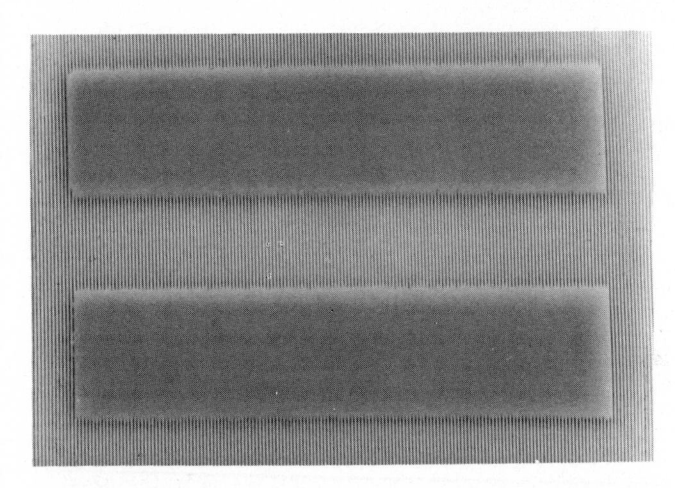

Figure 3. The base metallization pattern on a GaAs PBT immediately prior to epitaxial overgrowth. X-ray lithography is used to pattern 1600 Å lines and spaces

The PBT wafer is cleaned immediately prior to overgrowth with a process designed to remove surface damage and process-induced contaminants without lifting the tungsten stripes [16]. The wafer is then loaded into an epitaxial system and the tungsten stripes are imbedded in the GaAs crystal by growing approximately 2 μm of GaAs over the tungsten stripes. Vapor phase epitaxy (VPE), organometallic chemical vapor deposition (OMCVD), and molecular beam epitaxy (MBE) have been used for this process, as will be discussed below. Emitter, base and collector contacts are then fabricated using conventional photolithography, and individual devices are defined and isolated using proton bombardment. A scanning electron micrograph of a finished device is shown in Fig. 4.

Significant technology developments have been required for all aspects of PBT fabrication. The techniques for base patterning, the base

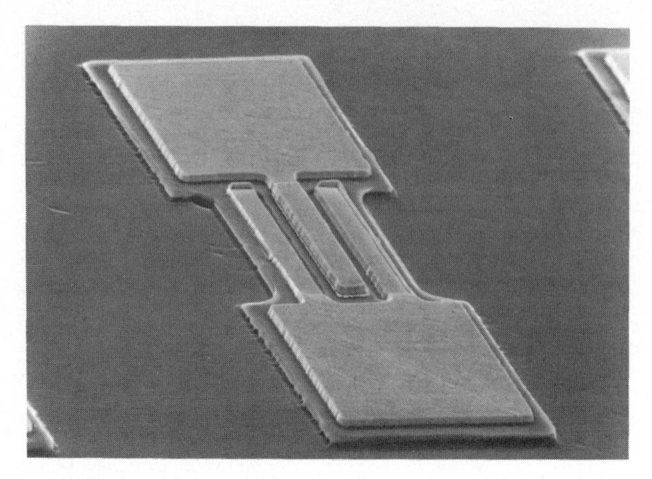

Figure 4. Scanning electron micrograph of a completed GaAs PBT. The single finger forms an 8 x 40 μm collector contact, and the two outside fingers provide base contact

metallization techniques, and the contact fabrication techniques are at present comparatively well understood. However, the achievement of high quality and controllable overgrowths has proven difficult, and remains the primary outstanding problem area connected with the fabrication of GaAs PBTs.

A particularly good PBT wafer was fabricated approximately three years ago using $AsCl_3$-GaAs-H_2 VPE for the overgrowth. This wafer contained a number of devices having an extrapolated f_{max} of approximately 100 GHz [17]. Numerous VPE-overgrown PBT wafers have been fabricated in the years since, but none of the wafers had an f_{max} exceeding 65 GHz. To better understand the problems afflicting VPE overgrowth,a comprehensive set of diagnostic experiments using secondary ion mass spectrometry (SIMS) has been conducted. The most significant result has been the identification of a heavy incorporation of Cl near the base grating, due to the $AsCl_3$-VPE process [18]. These experiments have also shown impurities from the base metallization being incorporated into the overgrown GaAs. As a result, VPE overgrowth has been abandoned and alternate forms of overgrowth, OMCVD [19] and MBE, are being developed.

Recently we have fabricated an OMCVD-overgrown PBT wafer having a MSG of 18.5 dB at 18 GHz [20]. The power and current gains determined by network analyzer measurements are shown in Fig. 5. Extrapolating the MSG data of Fig. 5 to higher frequencies using the 6 dB/octave roll-off predicted by the circuit model of Fig. 2 yields an f_{max} of approximately 150 GHz. A similar extrapolation yields an f_T value of approximately 35 GHz. This is the best result yet obtained with a GaAs PBT.

The other overgrowth technique we are currently concentrating on is MBE. Molecular beam epitaxy has the inherent advantages of cleanliness and superior control over doping profile in the overgrown region. The best f_{max} thus far obtained with a MBE-overgrown PBT is 25 GHz [21]. The performance of MBE-overgrown PBTs has been limited by a uncontrollable dip in carrier concentration at the interface between the emitter layer and the

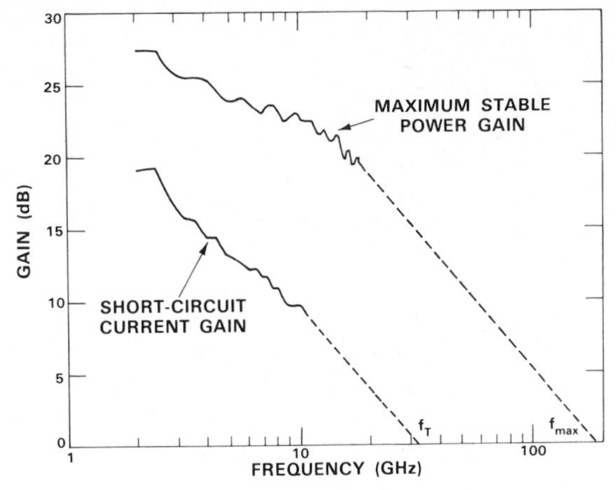

Figure 5. Maximum stable gain and short-circuit current gain for the recent best result. The extropolations are made using a roll-off of 6dB/octave

overgrown MBE layer. This problem results from the difficulty in cleaning the GaAs surface upon which the MBE overgrowth is to be initiated. A number of alternate cleaning and metallization techniques are currently being investigated.

The results which have been obtained with the PBT compare favorably with those obtained with other high frequency devices. Values for f_{max} of approximately 100 GHz have been reported for conventional FETs, with f_T's approximately that of PBTs [6] . Much higher f_T's have been obtained with HEMT devices, on the order of 70 GHz [8,9], and are indicative of the higher carrier velocities in the devices. However, the values of f_{max} obtained have been limited by their output characteristics to values less than 70 GHz. Heterojunction bipolar transistors have also achieved f_T's of approximately 40 GHz, but the f_{max} has been limited by the comparatively high base resistance.

4. Fabrication of Silicon PBTs

A cross-sectional drawing of a silicon PBT is given in Fig. 6. Epitaxial growth is carried out at a much higher temperature in silicon, making the

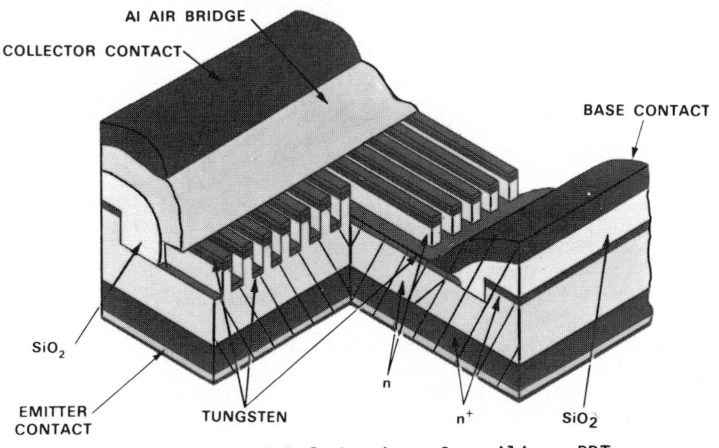

Figure 6. Cross-sectional drawing of a silicon PBT

metal interaction problem during overgrowth more severe. However, since surface depletion can be controlled in silicon, a different structure which does not require overgrowth is possible. Grooves are etched into the n layer using RIE, and the metal is deposited on the top of the ridges and the bottom of the grooves, forming the collector and base metallization respectively. Since proton bombardment is not available as a fabrication technique in silicon, devices are isolated using thick oxide layers. The best f_{max} thus far obtained with the silicon PBT is 20 GHz [22], which is comparable with that obtained by the best bipolar transistors.

5. Summary

A number of devices are under development for high frequency and high speed applications. Each of these devices has its own particular merits. Perhaps the strongest attribute of the PBT is that it combines all of the properties needed for high speed operation, although individual characteristics of the PBT are exceeded by one or the other of its competitors. It is possible that the optimal high speed device will in some way combine that best features of several of the devices currently under development. Nevertheless, the experimental results we have obtained with the PBT have demonstrated the high speed capability of the PBT, and simulations indicate that even higher frequency performance is possible. It is also important to appreciate that the measurement of these devices is a difficult problem. Electronic instrumentation capable of measuring the switching times of those devices, which are less than 10 ps, does not presently exist. It is quite likely that the optical techniques which have been reported by other researchers in this volume may provide the best way of obtaining accurate measurements on these high speed devices.

Acknowledgment

The intensive development of understanding and technology which has occurred in the past five years has required the efforts and dedication of a large number of researchers. The author especially wishes to recognize C. O. Bozler, A. R. Calawa, C. L. Chen, M. A. Hollis, K. B. Nichols and D. D. Rathman for their continuing efforts.

References

1. C. O. Bozler, G. D. Alley, R. A. Murphy, D. C. Flanders, and W. T. Lindley: in Proc. 7th Bien. Cornell Conf. on Active Microwave Devices, 33 (1979)
2. G. D. Alley, C. O. Bozler, R. A. Murphy, W. T. Lindley: in Proc. 7th Bien. Conf. on Active Microwave Devices, 43 (1979)
3. C. O. Bozler, G. D. Alley: IEEE Trans. Electron Devices ED-27, 1128 (1980)
4. C. O. Bozler, G. D. Alley: Proc. IEEE 79, 46 (1982)
5. G. D. Alley: IEEE Trans. Electron Devices ED-30, 52 (1980).
6. E. T. Watkins, H. Yamasaki, J. M. Schellenberg: in ISSCC Tech. Dig., 198 (1982)
7. E. T. Watkins, J. M. Schellenberg, L. H. Hackett, H. Yamasaki, M. Feng: in IEEE Int. Microwave Symp. Dig., 145 (1983)
8. L. Camnitz, P. Tasker, H. Lee, D. ver Merwe, L. Eastman: in IEDM Tech. Dig., 360 (1984)
9. U. K. Mishra, S. C. Palmateer, P. C. Chao, P. M. Smith, J. C. M. Hwang: IEEE Electron Device Lett. EDL-6, 142 (1985)
10. P. M. Asbeck, D. L. Miller, R. J. Anderson, R. N. Deming, R. T. Chen, C. A. Liechti, F. H. Eisen: in GaAs IC Symposium Tech. Dig., 133 (1984)

11. G. D. Vendelin: Design of Amplifiers and Oscillators by the S-parameter Method (Wiley-Interscience, John Wiley and Sons, New York 1982)
12. R. Spence: Linear Active Networks (Wiley-Interscience, John Wiley and Sons, London 1970)
13. E. O. Johnson, A. Rose: Proc. IRE, 407 (1959)
14. Y. Awano, K. Tomizawa, N. Hashizume: to be published in GaAs and Related Compounds (Inst. Phys., London, 1984)
15. D. C. Flanders: J. Vac. Sci. Technol. 6, 1615 (1979)
16. C. O. Bozler, M. A. Hollis, S. W. Pang, R. W. McClelland, K. B. Nichols: submitted for publication
17. G. D. Alley, et al.: IEEE Trans. Electron Devices ED-29, 1708 (1982)
18. M. A. Hollis, K. B. Nichols, C. O. Bozler, A. R. Calawa, M. J. Manfra: in Prog. 1984 Electronic Materials Conf. (1984).
19. K. B. Nichols, R. P. Gale, M. A. Hollis, G. A. Lincoln, C. O. Bozler: IEEE Trans. Electron Devices ED-31, 1969 (1984).
20. C. O. Bozler, M. A. Hollis, K. B. Nichols, S. Rabe, A. Vera, C. L. Chen: submitted for publication
21. A. R. Calawa, M. J. Manfra, G. A. Lincoln: presented at Int. Conf. on Molecular Beam Epitaxy, paper Q-3, 1984
22. D. D. Rathman, B. A. Vojak, D. C. Flanders, N. P. Economou: in Ext. Abs. 16th Conf. on Solid State Device and Materials, 305 (1984)

Two Dimensional E-Field Mapping with Subpicosecond Resolution

K.E. Meyer and G.A. Mourou
Laboratory for Laser Energetics, University of Rochester, 250 East River Road, Rochester, NY 14623, USA

1. Introduction

With the development of very high speed semiconductor devices which respond in the picosecond regime, such as the heterojunction bipolar transistor, permeable base transistor, GaAs MESFET and TEGFET, the need has arisen for new techniques to directly measure device response on this time scale. The continuing advancement of VLSI circuits also requires the development of new characterization techniques. The recent development of photoconductive [1] and electro-optic [2-7] sampling techniques has advanced the art of device characterization to meet some of these needs. The latter has been used to measure electrical transients as fast as 0.46 ps with submillivolt sensitivity [5-7].

One of the advantages of the electro-optic technique is its capability to probe transient electric fields where they occur on devices or along transmission lines. This feature is important for in-situ integrated circuit characterization. In this report we demonstrate the measurement of transient E-fields propagating on microwave coplanar transmission lines using two different sampling geometries. In the first case the striplines have been fabricated on an electro-optic substrate and the dispersion characteristics of the stripline have been measured. In the second case an electro-optic superstrate has been placed in proximity to the stripline (fabricated on a non-electro-optic substrate) to sample the fringing fields. For both geometries subpicosecond temporal resolution as well as good (~10 μm) spatial resolution have been demonstrated.

2. 2D Mapping Using an Electro-Optic Substrate

A schematic of the coplanar stripline, $LiTaO_3$ substrate, and Cr:GaAs photoconductive switch arrangement is shown in Fig. 1. The Cr:GaAs and $LiTaO_3$ crystals were mounted side by side on a glass plate and were subsequently ground and polished together in order to present a continuous surface on which to fabricate the electrodes. For ease of fabrication the electrode widths and separation were chosen to be 50 μm. The optical axis of the lithium tantalate was parallel to the direction of the electric field between the electrodes and the probe beam was perpendicular to the electrode plane. 100 fs optical pulses were used to generate and probe the electrical pulse [5-7].

Figure 2 shows the different electrical waveforms obtained on the stripline as the probe beam was moved progressively further away from the photoconductive switch. The 30 mV input signal had a 10%-90% rise time of 0.46 ps. This is the fastest electrical signal that has been directly measured to date. As the signal propagates it experiences frequency dispersion due to higher-order propagation modes in the stripline. The higher frequency components move slower than the lower frequencies and hence the rise time degrades rapidly as the propagation distance increases. The waveform degradation is quite similar to frequency dispersion on microstrip transmission lines calculated by Hassein, et al. [8], and measured by Valdmanis at LLE.

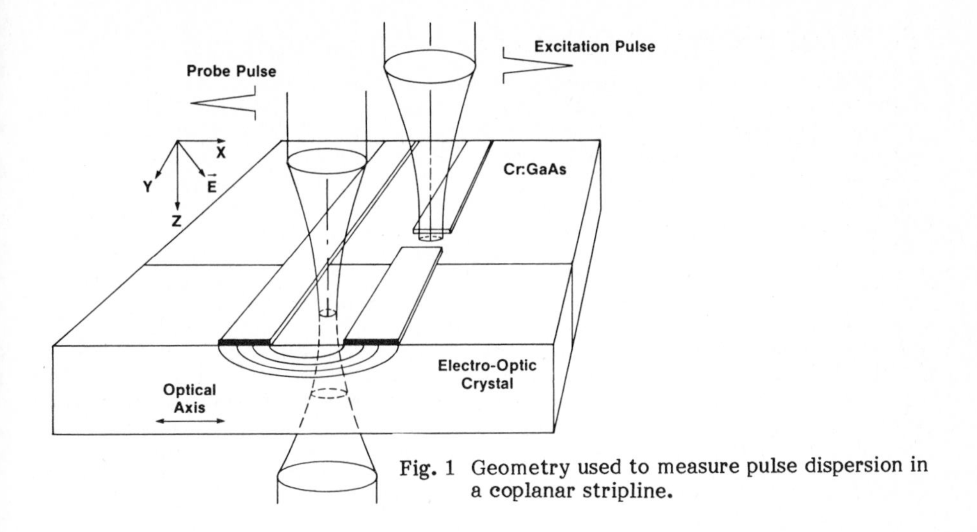

Fig. 1 Geometry used to measure pulse dispersion in a coplanar stripline.

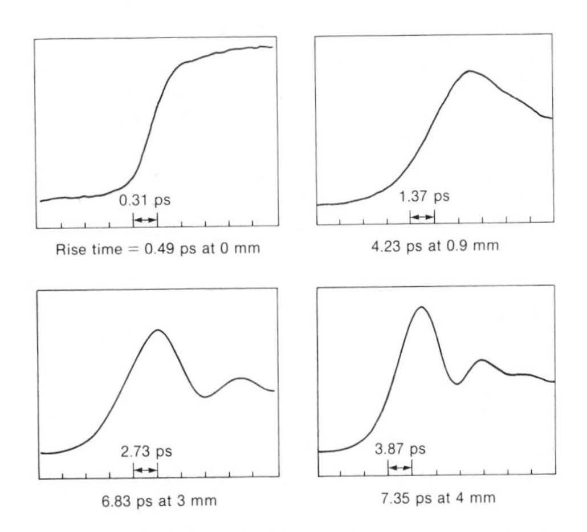

Fig. 2 Electrical waveforms measured as a function of propagation distance along the coplanar stripline. The input pulse had a 10%–90% rise time of 0.46 ps

An obvious extension of this technique takes advantage of the electro-optic properties inherent in GaAs [9]. In this mode the GaAs device being characterized also serves as the sampling medium; this practically eliminates any interference with the device operation. This technique requires a probe pulse with a wavelength greater than 900 nm in order to avoid band-to-band absorption. Short pulses in the IR regime can be obtained by starting with 620 nm pulses created by a synchronously pumped anti-resonant ring laser [10], which are then amplified with a 1 KHz dye amplifier chain pumped by a regenerative amplifier [11]. The remaining 100 ps 1.06 μm pulses from the regenerative amplifier are then mixed in a Kerr cell with the amplified 0.18 ps pulses. Using this technique we have obtained 1.06 μm pulses with a FWHM of 0.56 ps. This new GaAs-based sampling technique is currently under investigation.

3. 2D Mapping Using the Reflection Mode

The reflection mode sampling configuration is shown in Fig. 3. The test circuit, consisting of a photoconductive switch and coplanar stripline fabricated on a Cr:GaAs substrate, has identical dimensions as that used in the previous experiment. The sampler consists of a slab of $LiTaO_3$ which has been coated on one surface with a high reflection dielectric coating. This slab is placed over the test circuit with the HR coating adjacent to the circuit. When the electrical signal propagates down the stripline some of the field lines above the stripline penetrate into the sampling crystal and induce a small amount of transient birefringence. A probe beam is focused into the area of interest, passes through the region of induced birefringence, and is reflected back out to the detection electronics. The induced birefringence is detected as a change in the polarization of the probe beam. Thus the transient electric fields can be mapped out in two dimensions simply by scanning the position of the probe beam over the desired range. Note that it is not necessary for the sampling crystal to be in contact with the circuit, although closer proximity will result in improved sensitivity.

Figure 4 shows the experimental results when 100 fs optical pulses are used to generate and probe the electrical pulses. The amplitude of the electrical signal is

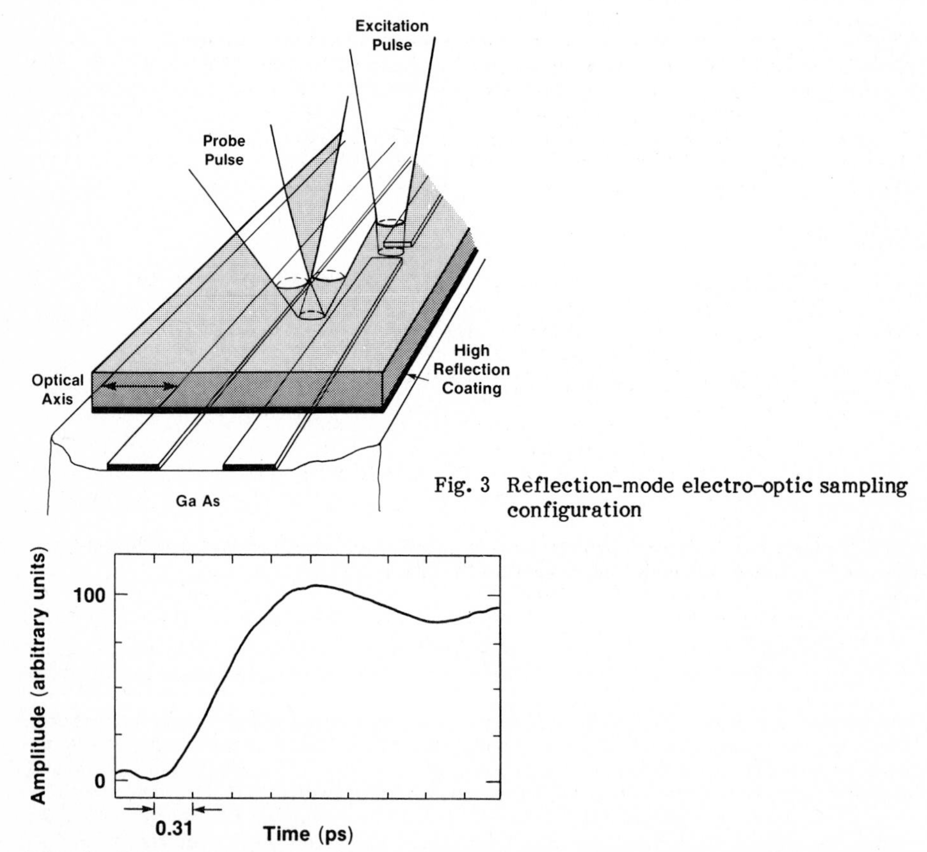

Fig. 3 Reflection-mode electro-optic sampling configuration

Fig. 4 Temporal response of the reflection-mode sampler with coplanar stripline dimensions of 50 μm. The 10%-90% rise time is 0.74 picoseconds.

30 mV. The 10%-90% rise time of the waveform is 0.74 ps. The probe beam was focused to a diameter of 15 μm; better spatial resolution may be obtained by simply focusing the beam more tightly. The good signal/noise ratio achieved indicates that the reflection-mode geometry has a sensitivity comparable to that obtained with the balanced stripline [2-4] or coplanar sampler [5-7].

The temporal resolution of the sampler is determined by four parameters: the intrinsic electrical pulse rise time τ_e, the transit time of the electrical signal across the optical beam waist τ_W, the transit time τ_O of the optical pulse through the region of induced birefringence, and the laser pulsewidth τ_L. The dominating factor in this geometry is τ_O because the probe pulse passes through the region of induced birefringence twice, once before and once after reflection. The depth of penetration of the field into the crystal decreases monotonically with the separation of the coplanar electrodes, hence the temporal resolution of the experiment can be improved by reducing the dimensions of the stripline.

Acknowledgments

The authors wish to thank Princeton Applied Research for supplying the signal averager and lock-in amplifier. They would also like to acknowledge the technical expertise of Arne Lindquist and Herb Graf for crystal fabrication and Lynn Fuller for his advise and access to photolithographic fabrication facilities. This work was supported by the Laser Fusion Feasibility Project at the Laboratory for Laser Energetics which has the following sponsors: Empire State Electric Energy Research Corporation, General Electric Company, New York State Energy Research and Development Authority, Northeast Utilities Service Company, Ontario Hydro, Southern California Edison Company, The Standard Oil Company, and the University of Rochester. Additional support was provided by AFOSR Grant #84-0318. Such support does not imply endorsement of the content by any of the above parties.

References

1. D.H. Auston, A.M. Johnson, P.R. Smith, and J.C. Bean: Appl. Phys. Lett. 37, 371 (1980).
2. J.A. Valdmanis, G.A. Mourou, and C.W. Gabel : Appl. Phys. Lett. 41, 211 (1982).
3. J.A. Valdmanis, G.A. Mourou and C.W. Gabel, SPIE 439, 142 (1983).
4. J.A. Valdmanis, G.A. Mourou, and C.W. Gabel, IEEE J. Quantum Electron. QE-17, 664 (1982).
5. G.A. Mourou, and K.E. Meyer : Appl. Phys. Lett. 45, 492 (1984).
6. K.E. Meyer and G.A. Mourou, Proceedings of the Fourth International Conference on Ultrafast Phenomena, Monterey, California, 1984, pp. 406-408.
7. K.E. Meyer and G.A. Mourou, Proceedings of the Conference on Lasers and Electro-optics, Anaheim, California, 1984, p. 70.
8. G. Hasnein, G. Arjavalingam, A. Dienes, and J.R. Whinnery, SPIE 439, 27 (1983).
9. B.H. Kolner and D.M. Bloom, Elect. Lett. 20, 818 (1984).
10. T. Norris, T. Sizer and G.A. Mourou, J. Opt. Soc. Am. B, to be published.
11. I. Duling, T. Norris, T. Sizer, P. Bado, and G.A. Mourou, J. Opt. Soc. Am. B, to be published.

Picosecond Electrooptic Sampling and Harmonic Mixing in GaAs

B.H. Kolner, K.J. Weingarten, M.J.W. Rodwell, and D.M. Bloom
Edward L. Ginzton Laboratory, Stanford University,
Stanford, CA 94305, USA

Electro-optic sampling is a powerful technique for exploiting the capabilities of modern mode-locked laser systems to make high speed electronic measurements. Using ultra-short light pulses to probe the electric fields of microstrip transmission lines deposited on $LiNbO_3$ and $LiTaO_3$, VALDMANIS et al. [1,2] demonstrated an electro-optic sampling system capable of resolving picosecond and subpicosecond rise-time photoconductive switches. KOLNER et al. [4,5] utilized a similar system to characterize photodiodes exhibiting bandwidths of 100 GHz. In both cases, a hybrid connection between the device under test and the electro-optic transmission line was required. VALDMANIS et al. [6] and MEYER and MOUROU [7] have shown that by placing an electro-optic crystal in contact with the circuit under test, picosecond waveforms could be measured without a hybrid connection. Although these techniques have demonstrated impressive results, they potentially compromise the true device response by reactive loading of the transmission line systems. This occurs due to 1) the fundamental mode mismatch between similar transmission lines on different dielectrics, 2) parasitic reactances associated with the bonding wires between the two transmission line systems or 3) capacitive loading of a transmission line by close proximity to the sampling crystal.

In this paper we report on a new approach to electro-optic sampling of high speed GaAs devices that overcomes these potential limitations. Our system relies on the fact that GaAs is electro-optic and devices and circuits fabricated in this material can be probed directly using picosecond infrared pulses to yield time and frequency domain measurements of a truly noninvasive nature. The circuits can be excited either by on-board photodetectors for impulse response measurements or by external signal generators, phaselocked to the laser pulse train, for analog swept frequency or synchronized digital measurements. In the latter case, the pulse timing stability of the laser becomes an important factor in making accurate measurements. By operating the sampler as a wideband harmonic mixer, we have been able to characterize the timing jitter of the laser and establish the limits it would impose on measurements made with external signal sources.

Of the various possible geometries for electro-optic modulation in GaAs, the longitudinal case illustrated in Fig. 1a is the most attractive. In this configuration, the optical sampling beam passes through the wafer at a point adjacent to the upper conductor of a microstrip transmission line and is reflected back by the ground plane below. For $\langle 100 \rangle$ cut GaAs (the most common orientation for integrated circuits), the electric field lines along the $\langle 100 \rangle$ axis induce birefringent axes along the $\langle 011 \rangle$ and $\langle 01\bar{1} \rangle$ directions. This birefringence is converted to an amplitude modulation of the sampling beam with a polarizer. For a given voltage on the transmission line, the electric field (and hence

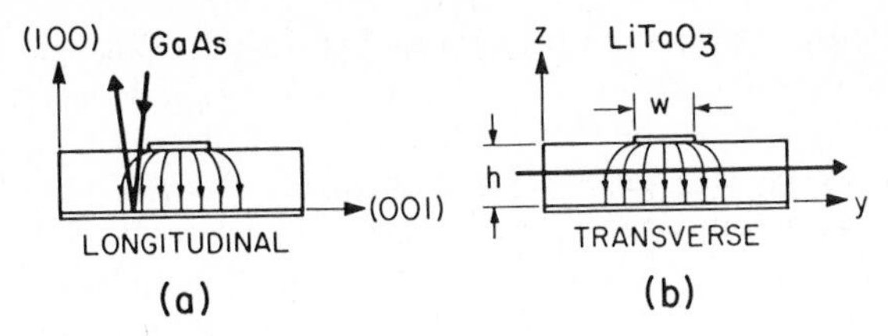

Fig. 1. Microstrip sampling geometries with indicated crystallographic axes.

the birefringence) varies inversely with the substrate thickness. However, since the net phase retardation is proportional to the product of the birefringence and the substrate thickness, the thickness cancels out. The sensitivity, or minimum detectable voltage, is therefore independent of the characteristic impedance of the transmission line and for a 50 Ω microstrip system is nearly ten times better than that of the transverse $LiTaO_3$ sampler (Fig. 1b). Previous approaches to electro-optic sampling relied on a transverse field geometry [1-7] and thus were sensitive to the dimensions of the transmission lines. With this new longitudinal sampling configuration, we can make absolute voltage measurements, independent of transmission line impedances and device geometries.

To make impulse response measurements with on-chip photodetectors, a dual wavelength picosecond source is needed. A Nd:YAG laser is ideal for this application. The 1.06 μm wavelength is well below the absorption edge of GaAs and can be used as the sampling beam. The second harmonic, obtained by frequency doubling to .532 μm, yields an ideally synchronized source for exciting photodetectors. While reliable cw mode-locked Nd:YAG lasers are commercially available, the minimum pulsewidth is limited to 50-100 ps and is too long to be used for high speed sampling. However, recent work on fiber-grating pulse compressors has resulted in the efficient compression of Nd:YAG pulses to less than 5 ps [8]. As a first step toward sampling monolithic GaAs integrated circuits, we used a packaged GaAs photodiode [9] connected to a GaAs microstrip transmission line. We excited the photodiode with 5 picosecond pulses and electro-optically sampled the microstrip line to yield the photodiode impulse response. Although a hybrid arrangement, the initial results demonstrated the effectiveness of the longitudinal sampling geometry. A more complete description of this experiment can be found in [10].

Because an electro-optic modulator produces a photocurrent that is proportional to the product of the optical intensity and the modulating signal, it can be viewed as a mixer. In the frequency domain, any signal at ν_0 propagating on the transmission line will mix with all of the harmonics of the fundamental sampling rate, f_0. Sidebands due to the convolution of these two spectra will appear at frequencies $nf_0 \pm \nu_0$. In particular, if the transmission line is driven with a pure microwave signal, a replica of the nearest harmonic will appear between DC and $f_0/2$, where it can be conveniently viewed on a spectrum analyzer or other receiver. This permits frequency domain measurements to be made throughout the microwave spectrum by driving the circuit under test with an external signal generator and measuring the magnitude of the mixer products at baseband. The phases of the microwave signals are also preserved and any fluctuations or phase noise is transferred to the baseband signal. If the driving signal is a clean sinusoid, the phase noise of the down-converted harmonic is readily apparent and provides a way to quantify the jitter in the laser pulse train. A typical harmonic spectral component contains a delta function at nf_0 and a phase noise pedestal arising from the pulse-to-pulse timing jitter of the laser. For small phase fluctuations, the relative phase noise power at a given offset from the carrier can be shown to vary as the square of the harmonic number, n [11]. Figure 2 shows a series of harmonic spectra mixed down to about 10 MHz. These spectra were obtained by applying signals up to 16 GHz ($n = 199$) to the transmission line with an HP 8340 microwave synthesizer phase-locked to the laser mode-locker driver (HP 3325). The growth of the phase noise sidebands was found to be in excellent agreement with the predicted square-law dependence.

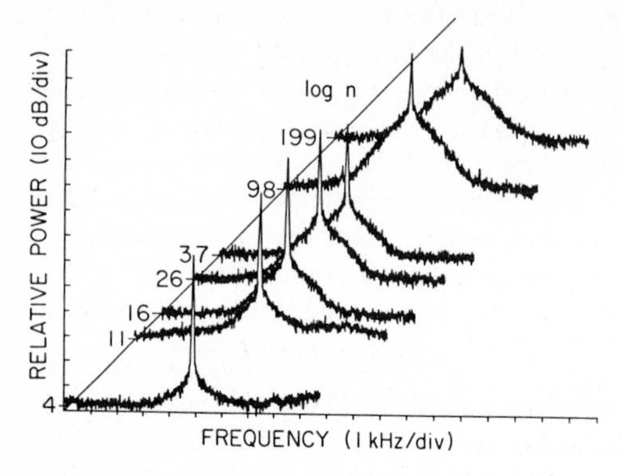

Fig. 2. Laser envelope harmonic spectral components converted to 10 MHz by harmonic mixing in the electro-optic sampler. Center frequency of each component equals $n \times 82$ MHz, where n is the harmonic number.

The power in the phase noise sidebands, P_{DSB}, can be shown to be related to the r.m.s. timing jitter [11,12]. To calculate the total double sideband power, the phase noise spectrum is integrated from some low frequency f_1 near the carrier (nf_0) to some higher frequency f_2 where the phase noise power falls to the level of the AM and Johnson noise. Since the apparent width of the carrier component depends on the resolution bandwidth of the spectrum analyzer, using a narrower bandwidth allows lower frequency phase fluctuations to contribute to the total sideband power. Thus, any calculation of timing jitter using this method must specify the low frequency cutoff, f_1. The expression relating the r.m.s. timing jitter to the carrier power P_a and the phase noise power is

$$\Delta t_{r.m.s.} = \frac{T}{2\pi n}\sqrt{\frac{P_{DSB}}{P_a}} \qquad \text{where} \qquad P_{DSB} = 2\int_{f_1}^{f_2} \frac{P_b(f)}{B}\,df \qquad \text{and } T = 1/f_0$$

Using a spectrum analyzer with a resolution bandwidth of 10 Hz, we determined that $\Delta t_{r.m.s.} \leq 11$ ps for $f_1 \geq 10$ Hz [13]. This suggests that if this pulse train is used to sample a microwave signal and produce less than, say, 10 degrees phase uncertainty, the microwave frequency must be below 2.5 GHz.

In spite of this limitation, we used this pulse train to sample an active monolithic microwave integrated circuit (*mmic*) at a single frequency to observe electronic distortion. We drove a four stage GaAs FET traveling wave amplifier [14] with a microwave synthesizer phase-locked to the laser mode-locker driver. We chose an operating frequency that was an exact multiple of the fundamental sampling rate plus one hertz. Thus, the sampling pulses "walked through" the driving sinusoid at a rate of one hertz. By pulse modulating the synthesizer at 10 MHz, a narrowband receiver could be used for signal-to-noise enhancement. Since the spectrum analyzer we used as the 10

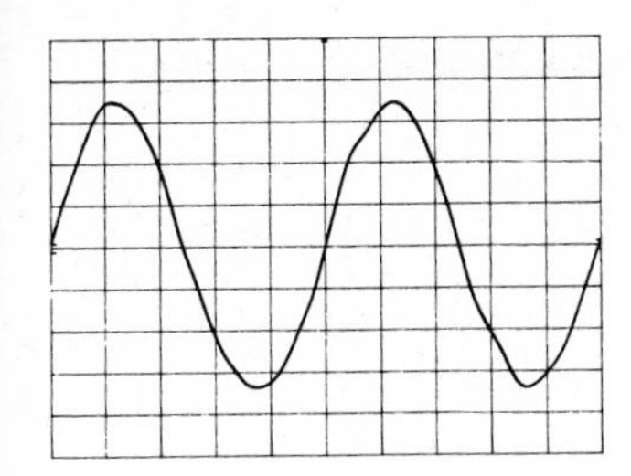

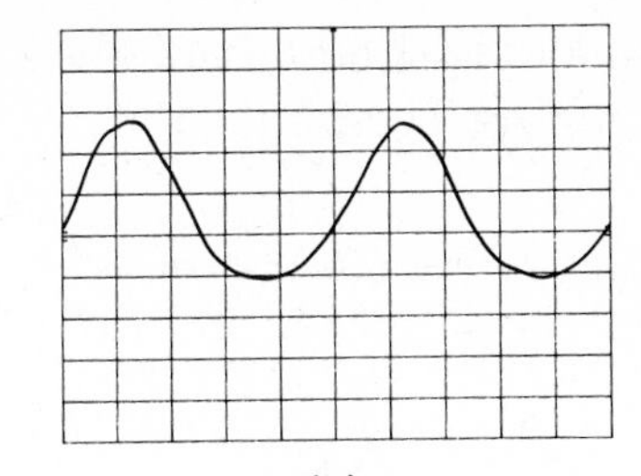

(a) (b)

Fig. 3. Voltage output of a four stage GaAs FET traveling wave amplifier measured by electro-optic sampling in the GaAs substrate. (**a**) Normal drain-source biasing. (**b**) Reduced drain-source biasing demonstrating soft clipping distortion. Frequency ≈ 3 GHz. (Horizontal ≈ 65 ps/div. Vertical ≈ .2 volts/div.)

MHz photodiode receiver displayed only the r.m.s. value of the sampled waveform, we injected a small amount of the 10 MHz chopping signal into the input so that it would sum vectorially with the photodiode signal and produce a true bipolar waveform.

With the synthesizer tuned to approximately 3 GHz, and the TWA biased normally, we measured the waveform shown in Fig. 3a by electro-optically sampling the TWA at the output of its last stage. Then, we reduced the drain-to-source voltage (V_{DS}) from +4 volts to +1.5 volts so that the TWA was operating in the "triode region". Figure 3b shows the soft clipping on the negative peaks as well as the reduction in gain that resulted from this bias condition.

Acknowledgements

The authors wish to thank George Zdasiuk of Varian Associates for supplying the GaAs FET TWA. They also wish to acknowledge the partial support of the Air Force Office of Scientific Research and the Joint Services Electronics Program.

References

1. J.A. Valdmanis, G. Mourou, and C.W. Gabel: *Appl. Phys. Lett.*, **41**, 211 (1982)
2. J.A. Valdmanis, G. Mourou, and C.W. Gabel: *IEEE J. Quant. Elect.*, **19**, 664 (1983)
3. G.A. Mourou and K.E. Meyer: *Appl. Phys. Lett.*, **45**, 492 (1984)
4. B.H. Kolner, D.M. Bloom, and P.S. Cross: *Elect. Lett.*, **19**, 574 (1983)
5. B.H. Kolner, D.M. Bloom, and P.S. Cross: *SPIE Proceedings*, **439**, 149 (1983)
6. J.A. Valdmanis, G. Mourou, and C.W. Gabel: *SPIE Proceedings*, **439**, 142 (1983)
7. K.E. Meyer and G.A. Mourou: unpublished
8. J.D. Kafka, B.H. Kolner, T.M. Baer, and D.M. Bloom: *Optics Lett.*, **9**, 505 (1984)
9. S.Y. Wang, D.M. Bloom, and D.M. Collins: *Appl. Phys. Lett.*, **42**, 190 (1983)
10. B.H. Kolner and D.M. Bloom: *Elect. Lett.*, **20**, 818 (1984)
11. J. Kluge: *Ph.D. Thesis*, Universistät Essen, Fed. Rep. of Germany (1984)
12. W.P. Robins: "Phase Noise in Signal Sources", Peter Peregrinus Ltd., London(1982)
13. B.H. Kolner, K.J. Weingarten, and D.M. Bloom: *SPIE Proc.*, **539**, (1985) to be published
14. Laboratory prototype, courtesy of Varian Associates

Characterization of TEGFETs and MESFETs Using the Electrooptic Sampling Technique

K.E. Meyer, D.R. Dykaar, and G.A. Mourou
Laboratory for Laser Energetics, Univesity of Rochester, 250 East River Road, Rochester, NY 14623, USA

1. Introduction

Recent advances in GaAs technology have resulted in several new classes of devices, all of which have very high speed response [1,2]. These devices have been shown, indirectly, to have response times of tens of picoseconds [1]. Indirect measurements are necessary because available sampling oscilloscopes have a limiting response of 25 picoseconds and jitter of a few picoseconds. Hence, most ultrafast devices are characterized in a ring oscillator configuration of several devices and individual device response is averaged. In the frequency domain most RF measurements are connector limited to the range of 18 GHz.

The electro-optic sampling system, developed at LLE [3], has demonstrated the capability of measuring subpicosecond electrical transients with submillivolt sensitivity [4]. Most recently, a coplanar sampling geometry yielded an improved temporal resolution of 0.46 ps [5]. The sampling system has been fully described elsewhere [3-5]. The electrical signal of interest is triggered by a short (0.1 ps) optical pulse provided by a colliding pulse mode-locked laser. The signal is coupled into a high-bandwidth (coplanar or microstrip) transmission line fabricated on a lithium tantalate substrate. The electrical signal induces a transient birefringence in the lithium tantalate, which is detected as a change in polarization of a short (0.1 ps) optical probe pulse synchronized with the excitation pulse. By sweeping the delay between the excitation and probe beams the electrical signal as a function of time is obtained. In this report we describe the application of this technique to the measurement of the transient response of GaAs TEGFETs and MESFETs [6].

2. Experimental Results

Two different sampling geometries, shown in Fig. 1, have been used in these experiments. Each offered particular advantages and point out the variety of configurations available. In both cases a Cr:GaAs photoconductive switch was used to generate a fast (5 ps rise time) electrical step, superimposed on a d.c. bias, which was the input to the gate of the device. With the source grounded the response of the FET is a current pulse out of the drain. The drain signal is coupled into the electro-optic sampler and then into a load.

TEGFETs, provided by Thompson CSF, were tested using microstrip input switch and sampler geometries (Fig. 1a). All propagation distances were kept to a minimum and connections were made with short wire bonds. Output transients for two different bias conditions are shown in Fig. 2. For both curves the drain bias was +1.0 V. The lower curve shows the result with the gate bias V_g = -0.5 V, i.e. the device is nearly pinched off. The initial negative signal indicates that some of the negative input pulse is being coupled through the device; the positive signal is indicative of gain. For the upper curve the gate voltage was increased to 0.0V and increased gain was observed. The measured 10% - 90% rise time was 16 ps. This is in good agreement with the calculated cutoff frequency of 23 GHz.

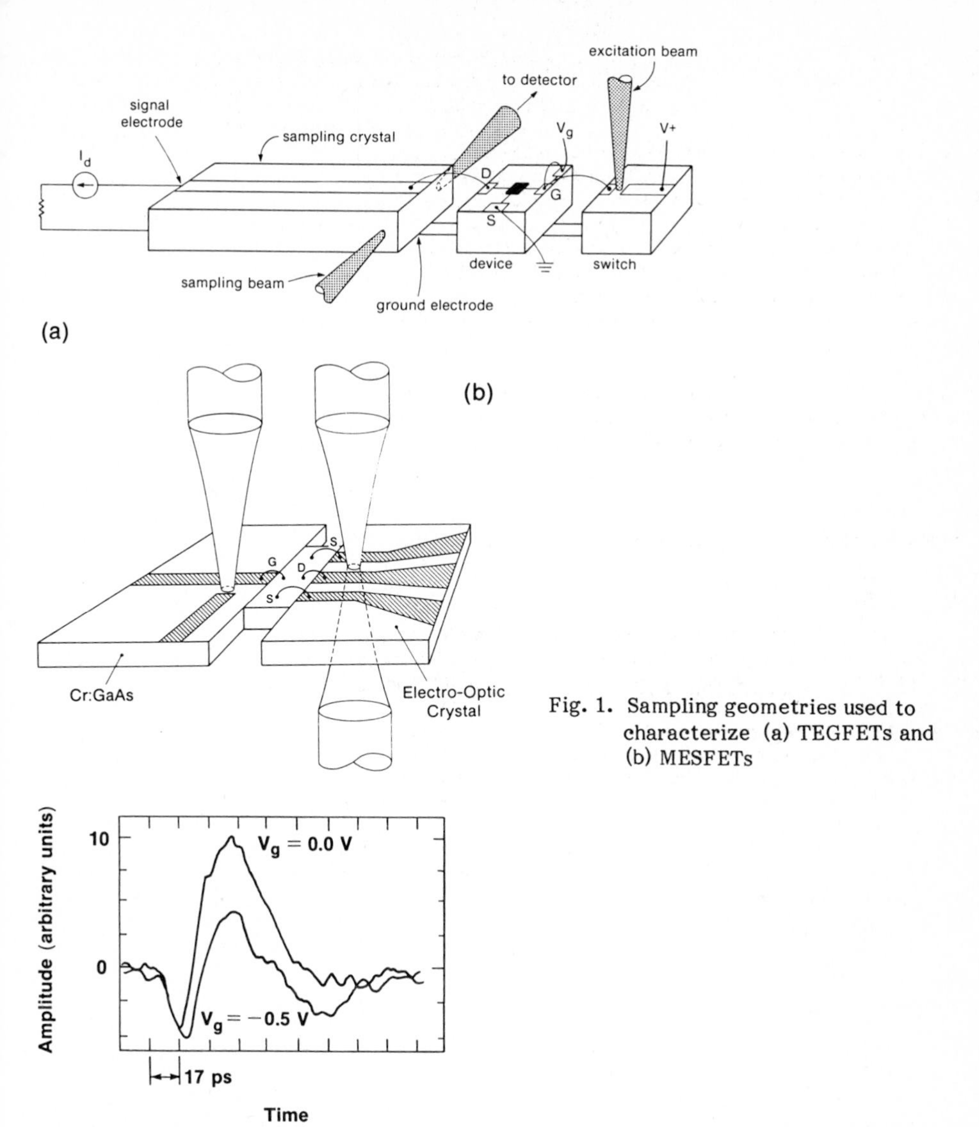

Fig. 1. Sampling geometries used to characterize (a) TEGFETs and (b) MESFETs

Fig. 2. 16 ps TEGFET response with the gate on (upper curve) and the gate pinched off (lower curve)

The GaAs MESFETs were fabricated at the Department of Electrical Engineering of Cornell University. For their characterization, a microstrip input switch and coplanar waveguide sampler were used (Fig. 1b). Experimental results are shown in Fig. 3. In order to measure the input pulse the sampling crystal was butted up against the GaAs switch with no device in between. The switch was connected to the sampler using a small amount of silver paint. The sampler was calibrated by applying an a.c. voltage set at the lock-in frequency directly to the coplanar waveguide and recording the resulting d.c. signal out of the lock-in. With a switch bias of -20V the amplitude of the input signal was -53 mV. The 10%-90% rise time was 5.4 ps.

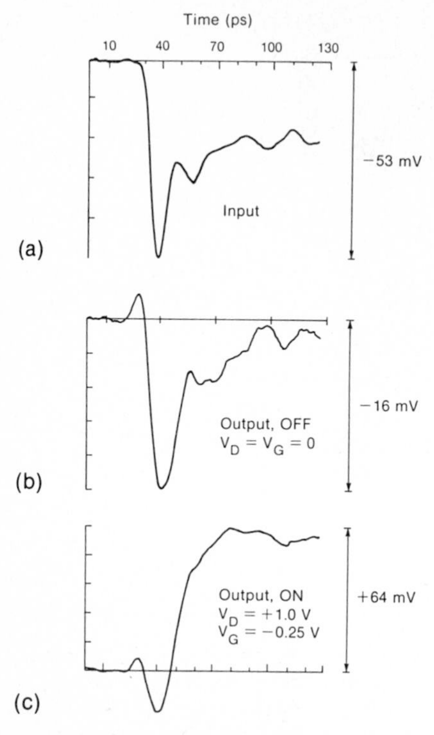

Fig. 3. MESFET response: (a) input pulse (τ_r = 5.4 ps); (b) output pulse with device biased "off" (τ_r = 8 ps); (c) output pulse with device biased "on" (τ_r = 25 ps)

With the device in place connections were made with short wire bonds. Output transients are shown in Figs. 3(b) and 3(c). With $V_D = V_G = 0$ the device is fully "off" and part of the input signal is passively coupled through. The pulse rise time slows slightly (τ_r = 8 ps) as it passes through the device. When the biases are increased to V_G = +0.25V and V_D = +1.0V the MESFET turns "on", the signal is inverted, and some gain is observed. As in the case of the TEGFET, the initial part of the transient is negative. The 10%-90% rise time in the "on" condition is 25 ps, which corresponds approximately to the 20 GHz current-gain cutoff obtained with S-parameter measurements.

3. Device Modeling

The MESFET operation has been simulated using the circuit configuration, shown in Fig. 4(a), suggested by the Cornell group and implemented with SPICE 2G.1. In order to model the experiment as closely as possible the actual measured input in Fig. 3(a) was digitized and used as the input to the model. To represent the "off" condition V_G, V_D, and the transconductance g were set to zero. The other parameters were varied, starting with "reasonable" values, to obtain a good qualitative fit with the experimental curve of Fig. 3(b). Inductors were added at the gate, source, and drain to mimic the effect of the bond wires, but no significant change resulted. To model the "on" condition we set V_G = +0.25V, V_D = +1.0V, and varied the transconductance to optimize the fit with the curve of Fig. 3(c). The resulting good fit is shown in Fig. 4(b). The model parameters for this fit are: $R_G = 16\,\Omega$, C_{GS} = 0.3 pF, C_{GD} = 1.2 pF, $R_s = 2\,\Omega$, $R_{DS} = 20\,\Omega$, and g = 45 mS. Further modeling will be done and better devices characterized with the goal of studying transit time effects in this class of devices.

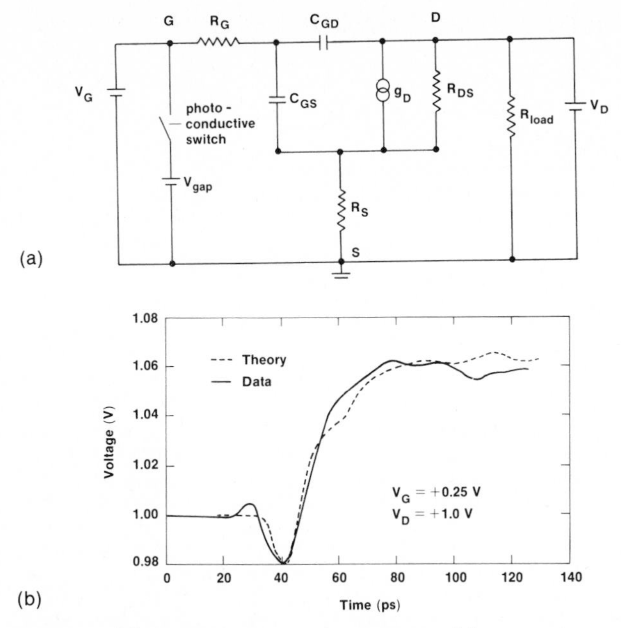

Fig. 4. (a) Circuit model of MESFET; (b) comparison of SPICE simulation with measured device response

4. Acknowledgment

This work was supported by the Laser Fusion Feasibility Project at the Laboratory for Laser Energetics which has the following sponsors: Empire State Electric Energy Research Corporation, General Electric Company, New York State Energy Research and Development Authority, Northeast Utilities Service Company, Ontario Hydro, Southern California Edison Company, The Standard Oil Company, and the University of Rochester. Such support does not imply endorsement of the content by any of the above parties. Additional support was provided by NSF Grant #ECS-8306607 and AFOSR Grant #84-0318. One of us (D.R.D.) has an IBM Pre-Doctoral Fellowship.

5. References

1. H. Morkoc and P.M. Solomon: IEEE Spectrum 21, No. 2, 28 (1984).
2. H. Dämbkes and K. Heime: Two-Dimensional Systems, Heterostructures, and Superlattices, edited by G. Bauer, F. Kuchor, and H. Heinrich, Springer-Verlag, New York, p.125 (1984).
3. J.A. Valdmanis, G.A. Mourou, and C.W. Gabel, Appl. Phys. Lett. 41, 211 (1982).
4. J.A. Valdmanis, G.A. Mourou, and C.W. Gabel: IEEE J. Quantum Electron. QE-17, 664 (1983).
5. G.A. Mourou and K.E. Meyer: Appl. Phys. Lett. 45, 492 (1984).
6. P.R. Smith, D.H. Auston, and W.M. Augustyniak, Appl. Phys. Lett. 39, 739 (1981).

Picosecond Electro-Electron Optic Oscilloscope

S. Williamson and G.A. Mourou
Laboratory for Laser Energetics, University of Rochester,
250 East River Road, Rochester, NY 14623, USA

1. Introduction

The technique of electro-optic sampling [1-3] presently offers the only means by which an electrical waveform can be time-resolved with subpicosecond resolution. With a sensitivity of ~1 mV and the capability of sampling in a contactless configuration, this technique has become a valuable tool for the characterization of ultrafast electronic components. The contactless mode of sampling can also be scaled up to allow sampling on a plane surface permitting the evaluation of any number of discrete components within an integrated circuit. In spite of these attractive features, the electro-optic sampling technique has been adopted by only a few large laboratories, the major drawback being the requirement of a short pulse laser system. The complexity of such a laser results in a sampling oscilloscope that is delicate, maintenance intensive, and expensive, precluding its development in industry and many universities.

Recently, we have developed a conceptually new picosecond sampling oscilloscope that maintains the salient features of the electro-optic sampler while eliminating the need for a short pulse laser. The new oscilloscope, called the electro-electron optic oscilloscope, represents the converse approach to electro-optic sampling.

2. Principle Of Operation

In both systems the process of sensing an E-field induced change in birefringence with polarized light is the same. What has changed (Fig. 1) is the replacement of the short pulse laser with a CW laser and the slow response detector with a streak camera.

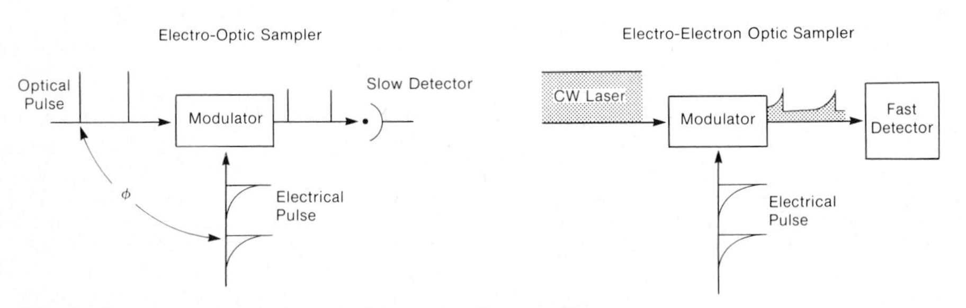

Fig. 1 Converse approaches to electro-optic sampling

A streak camera is comprised of an electron image converting tube, that via the photoelectric effect, converts an optical pulse to an electron pulse replica. This electron pulse can then be rapidly deflected, or streaked in a time-varying electric field. An electron pulse streaked in this manner has its time axis mapped along the direction of deflection onto a phosphor screen, where it can then be recorded. With this arrangement, the temporal resolution of the oscilloscope is determined, not by the shortness of an optical pulse, but, by the resolution of the streak camera. Presently, streak cameras with a few picoseconds resolution are commercially available. As depicted in Fig. 1, by using a CW laser each electrical signal is first converted to an optical replica then to an electron replica within the streak camera. Consequently, each electrical waveform is sampled in its entirety with every shot.

3. Experiment And Results

The experimental setup is shown in Fig. 2. The modulator is a Lasermetrics Pockels cell having a response time of ~40 ps. The Pockels cell and Soleil compensator are placed between crossed polarizers. A 1 mW HeNe laser is used as the CW probe. The device being characterized is an HP step recovery diode model 33002-A, that is driven at 85 MHz along with the deflection plates of the synchroscan type streak camera. In this configuration, a simple phase shift between the two sine waves is all that is required for synchronizing the diode to the streak camera.

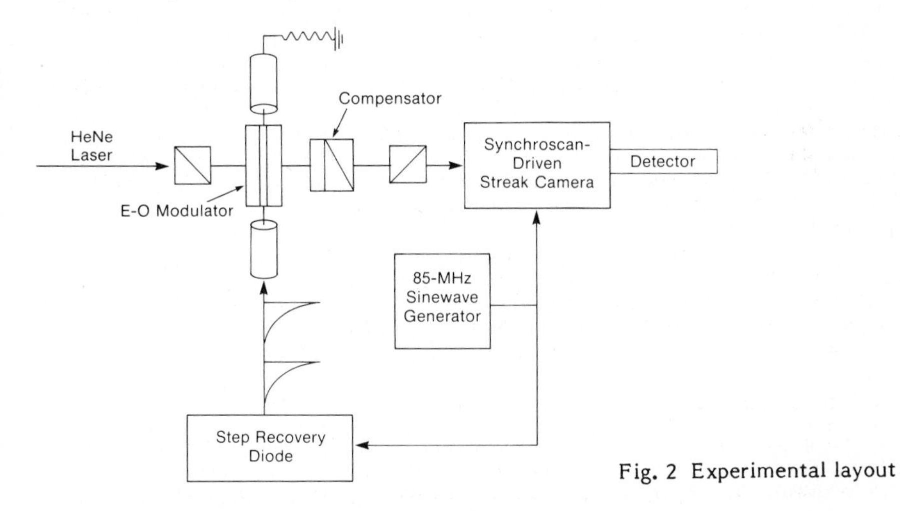

Fig. 2 Experimental layout

For the initial experiment the compensator was adjusted to 50% transmission for maximum sensitivity of modulation, dT/dV. In order to operate at this point it is necessary to attenuate the transmitted beam by 5 OD. This assures us of an average current density from the photocathode safely below the damage threshold ($\sim 10\ \mu A/cm^2$). Attenuating the transmitted signal by 10^5, though, also attenuates the modulated signal by the same factor. As a consequence, no signal could be found even after several tens of seconds of integration time. However, we discovered that if we adjusted the optical compensator off the 50% transmission point down to a transmission factor of 10^{-5} and then removed the 5 OD optical attenuation, a signal large enough to be observed even in real time appeared on the streak camera monitor. An explanation for the enhancement of the modulated signal is offered from Figs. 3 and 4. In Fig. 3 we have plotted both the transmission function, T for a modulator between crossed polarizers as well as the corresponding sensitivity curve dT/dV. We see that the sensitivity indeed has its maximum at $V_{\pi/2}$, that is, at the 50% transmission point. However, when the transmitted background signal can only be a small portion of the incident signal, an enhancement in the depth of modulation relative to the background can nevertheless be achieved. By adjusting the compensator (or equivalently, the DC bias voltage), the efficiency of modulation, (dT/dV)/T goes as $\cot [(\pi/2) V/V_{\pi}]$ (Fig. 4). In this way, the number of modulated photons can be increased, in our case, by a factor of $\sqrt{10^5}$ over the conventional

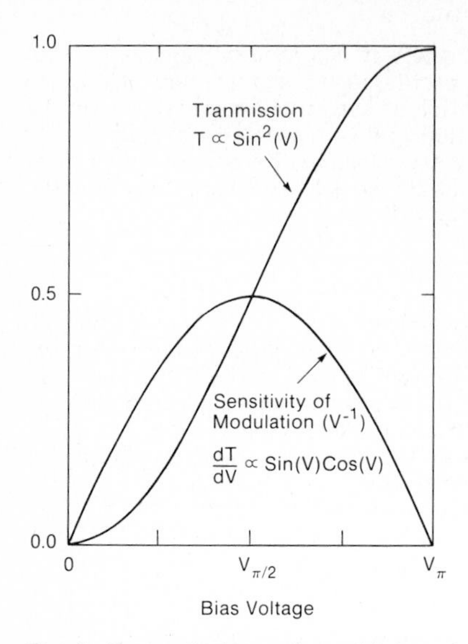

Fig. 3 Transmission and sensitivity of modulation curves for an electro-optic modulator

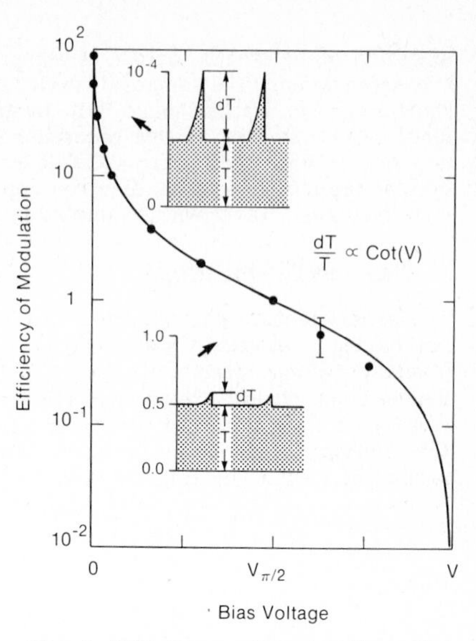

Fig. 4 Efficiency of modulation. Note difference in scales for the inserted transmission curves

approach of biasing to 50% and then attenuating by 5 OD. Though each optical replica consists of just a few tens of photons, the synchroscan streak camera upon accumulating 85 x 10^6 shots/sec reconstructs the waveform within seconds. It is important to bias slightly off the zero transmission point so that a signal possessing both polarities is not rectified. The issue of nonlinearity of response for the modulator is not a serious one since a maximum excursion of a few volts along a transmission curve whose V_π is 5,000 volts results in a deviation of less than 1% from a straight line. Figure 5 displays the signal of the step recovery diode. The response agrees well with a measurement made using a Tektronix sampling oscilloscope.

In conclusion, we have developed a novel picosecond oscilloscope that is based on the electro-optic effect but eliminates the need for a short pulse laser system. The electro-electron optic oscilloscope maintains the attractive features found in the electro-optic sampler but is relatively simple to operate, less expensive, and is

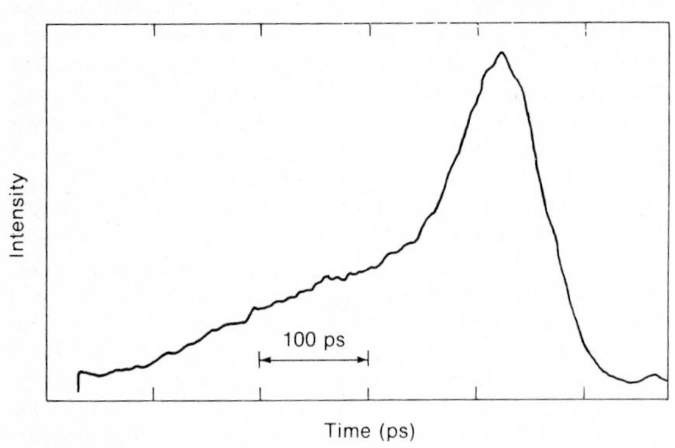

Fig. 5 Time-resolved response for step recovery diode

portable. In addition, the sampling frequency of this oscilloscope can be adjusted to accommodate the optimal operating frequency of the circuit in question. Any CW laser having a wavelength between 250 nm and 1000 nm can be chosen as the probe. As an example, a semiconductor laser can be used, allowing the probing of GaAs which happens also to be an electro-optic material. Finally, because the electro-electron optic sampler samples each waveform in its entirety, the system can potentially be used in single shot mode.

4. Acknowledgement

This work was supported by the Laser Fusion Feasibility Project at the Laboratory for Laser Energetics which has the following sponsors: Empire State Electric Energy Research Corporation, General Electric Company, New York State Energy Research and Development Authority, Northeast Utilities Service Company, Ontario Hydro, Southern California Edison Company, The Standard Oil Company, and the University of Rochester. Such support does not imply endorsement of the content by any of the above parties. The authors would like to thank Norman Schiller and coworkers of the Hamamatsu Corp. for their willingness in offering both a streak camera and assistance making the final result of this preliminary experiment possible.

5. References

1. J.A. Valdmanis, G.A. Mourou, and C.W. Gabel, Appl. Phys. Lett. **41**, 211 (1982).
2. J.A. Valdmanis, G.A. Mourou, and C.W. Gabel, IEEE J. Quantum Electron. **QE-17**, 664 (1983).
3. G.A. Mourou and K.E. Meyer, Appl. Phys. Lett. **45**, 492 (1984).

Picosecond Optoelectronic Diagnostics of Field Effect Transistors

Donald E. Cooper and Steven C. Moss
The Aerospace Corporation, Chemistry and Physics Laboratory, P.O. Box 92957, MS: M2-253, Los Angeles, CA 90009, USA

INTRODUCTION

Picosecond optoelectronics provides the capability to measure the frequency response of solid state devices with much greater bandwidth than conventional techniques. Laser-triggered switches acted as pulse generators and samplers to measure the impulse response of a device with a temporal resolution of a few picoseconds. Fourier analysis of the impulse response functions yields scattering matrix parameters with a frequency bandwidth of 40 GHz or more. In comparison, current frequency-domain technology is limited to a 26 GHz bandwidth. (Higher frequencies can be covered with frequency mixing techniques, at the cost of additional noise and experimental complexity.)

The electrical properties of a linear device can be completely characterized by time-domain measurements of the impulse response. These data can be Fourier transformed to yield frequency-domain information. For a 2-port device, the Fourier transform of the impulse response is the device transfer function, which is equivalent to the S_{21} scattering parameter. Similarly, the response at the input port when the output port is pulsed yields the S_{12} parameter, and the reflection of a pulse off the input or the output port transforms into the S_{11} or S_{22} parameters. This data is equivalent to that obtained from a network analyzer, with the advantage of greater frequency bandwidth. In addition, the de-embedding process is considerably simplified because the pulses are generated and sampled within a few millimeters of the device being characterized. The only intervening elements (other than the bond wires) are short sections of easily characterized transmission line.[1] Windowing of the time-domain response can also enhance the de-bedding process by separating the device response from artifacts due to the test fixture. In this paper we apply picosecond optoelectronic techniques to the characterization of an unpackaged GaAs FET.

EXPERIMENTAL

Optoelectronic pulse generators and samplers were fabricated in microstrip transmission lines on silicon-on-sapphire (SOS) substrates.[2] Two test fixture designs were used (Fig. 1). In the split fixture (Fig. 1(a)) two separate SOS wafers were connected by a gold-plated Kovar strip attached to each ground plane with conducting epoxy. The device under study was epoxied to the Kovar strip in the ~ 1 mm gap between wafers and wirebonded to the microstrips and the ground plane. A second test fixture design was used to simplify fixture fabrication and reduce the inductance of the source bond wires. This design (Fig. 1 (b)) was fabricated on a single SOS wafer, with the source connection to two large, relatively low-impedance pads. The microstrip impedance in both test fixtures was approximately $50\,\Omega$.

The optoelectronic pulse generation and sampling switches were triggered by 4 picosecond optical pulses from a synchronously pumped dye laser. A 6-7 picosecond electrical pulse from a biased switch entered one port and was sampled at the same or the other port by triggering a sampling

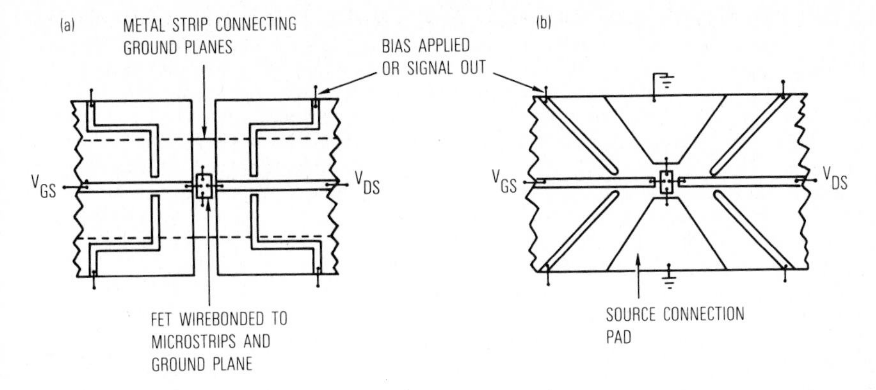

Figure 1. Experimental test fixtures.

switch with a second time-delayed optical pulse. The current conducted across the sampling switch was measured as a function of time delay to produce the impulse response function. The operating point of the device under test was controlled by dc voltages applied to the gate and drain microstrips.

RESULTS AND DISCUSSION

The impulse response of an Avantek AT-8041 in the split fixture is shown in Figure 2. Figure 2(a) and 2(d) show the result of reflecting an electrical pulse off the gate and the drain, respectively. For these measurements the pulses are generated at one port of the device and sampling is done at the same port immediately opposite the pulse generation switch.

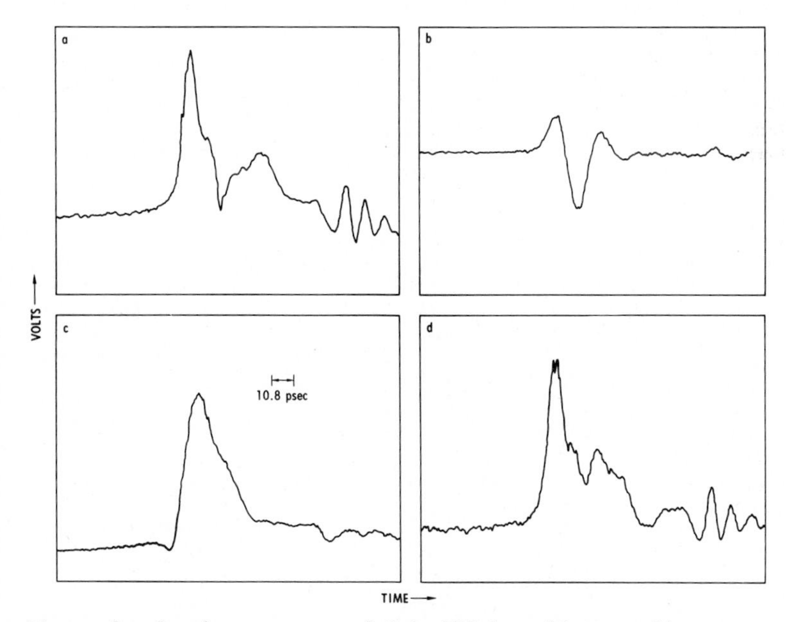

Figure 2. Impulse response of GaAs FET in split test fixture.

Thus the large initial peak represents the profile of the pulse entering the device, the shoulder is the result of the reflection at the wirebond to the microstrip, and the broad peak at later times is the reflection from the device itself. The rapid oscillations at the right of the trace are an experimental artifact due to reflections in the microstrip circuit. Because of the congested nature of these data, the actual pulse reflections are difficult to extract, and the second test fixture was designed to minimize these problems. Figure 2(b) shows the gate response when the drain is pulsed, and Figure 2(c) shows the drain response when the gate is pulsed. These data are quite clean and easy to interpret. Figure 2(c) shows a 10-90% risetime of 8.2 picoseconds when the transistor is used as an amplifier, with a longer tail indicating considerable phase dispersion within the FET. This very rapid rise is close to the temporal resolution of the test fixture. This data was digitized and a fast Fourier transform was performed on a microcomputer. The pulse shape and sampling aperture were deconvoluted by dividing the transformed spectrum by the spectrum of an optoelectronic "autocorrelation" peak. The result is the S_{21} spectrum displayed in Figure 3. The fit to the expected f^{-1} dependence at high frequencies is reasonably good. The switch sensitivities were not calibrated for these measurements, so the absolute gain is not known, but the data is well-behaved out to a 40 GHz bandwidth.

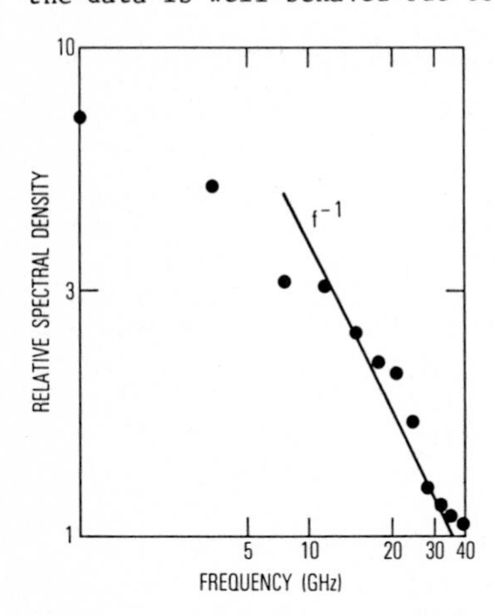

Figure 3. Insertion gain ($|S_{21}|$) vs. frequency, split test fixture.

The planar test fixture produced impulse response data that was significantly different from the split fixture. The measurements involving reflected pulses produced much cleaner data, from which S_{11} and S_{22} could be extracted. The pulse amplification experiment yielded a pulse shape with considerably longer 10-90% risetime (15 picoseconds) and much less asymmetry. The second test fixture used shorter source bond wires, reducing source inductance, but the low-frequency impedance of the transmission line formed by the source pads was about 25 Ω , considerably raising source resistance. These changes can be expected to have a major affect on the measured S parameters.

The data was digitized and transformed to the frequency domain, and a polar plot of S_{21} is shown in Figure 4, with the manufacturer's S_{21} specifications shown as open circles. The input and output amplitudes were measured to yield absolute S_{21} magnitudes. This plot also shows the phase components of S_{21} , which depend upon the designated time origin of the pulse profile. The propagation delay of the short lengths of microstrip between the switches and wirebonds on either side of the FET were mathemati-

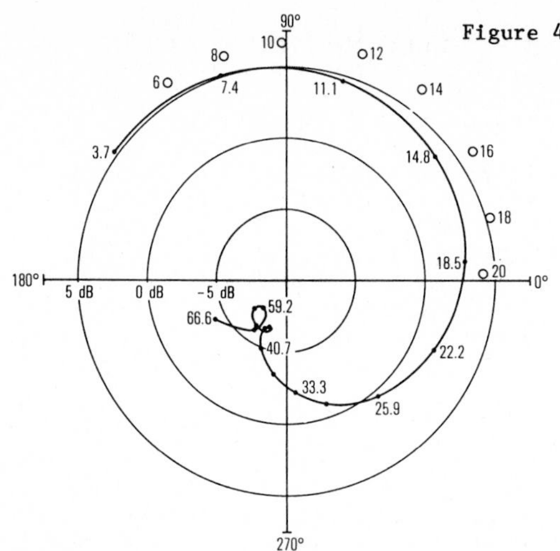

Figure 4. Polar plot of S_{21}, planar test fixture.

cally removed (de-embedded) to bring the reference plane up to the wirebond/microstrip junction. (The microstrip propagation delay was measured in a separate experiment.) This results in a very good fit to the manufacturer's specifications, with the somewhat lower gain probably due to the large source resistance. Beyond 20 GHz the data curve continues in a smooth fashion, with a complicated resonance around 50 GHz due to the two bond wire quarter-wave resonances. The data is well above the noise level to nearly 70 GHz.

SUMMARY

Picosecond optoelectronics provides a useful tool for high-frequency diagnostics of fast solid state devices. Linear scattering parameters are extracted from impulse response measurements, with greater bandwidth than conventional network analyzers and very simple de-embedding procedures. The process was illustrated by the picosecond optoelectronic measurement of the scattering parameters of a GaAs FET.

REFERENCES

1. Donald E. Cooper, "Picosecond Optoelectronic Measurement of Microstrip Dispersion," to be published in Appl Phys. Lett.

2. P. R. Smith, D. H. Auston, A. M. Johnson, and W. M. Augustyniak, Appl. Phys. Lett 38, p. 47 (1981).

Time-Domain Measurements for Silicon Integrated Circuit Testing Using Photoconductors

W.R. Eisenstadt
Electrical Engineering Department, University of Florida,
Gainesville, FL 32611, USA
R.B. Hammond*
Electronics Division, Los Alamos National Laboratory,
Los Alamos, NM 87545, USA
D.R. Bowman* and R.W. Dutton
Integrated Circuits Laboratory, Stanford University,
Stanford, CA 94305, USA

Picosecond time-domain measurements of silicon-integrated circuit interconnects were successfully performed using both bulk silicon photoconductors and polycrystalline (poly-Si) silicon photoconductors integrated on-chip. Standard integrated circuit fabrication techniques, followed by shadow-masked, ion-beam irradiation were used to create the photoconductor/interconnect structures on silicon wafers of 6-ohm-cm to 70-ohm-cm resistivities. A subpicosecond pulsed laser system excited the photoconductors to produce and sample electrical pulses on the Si substrate.

Optoelectronic correlation measurements with photoconductor pulsers and photoconductor sampling gates were performed to characterize both the photoconductors and IC interconnections. Photoconductors processed as fast pulsers produced 50-mV peak magnitude, 5-ps FWHM pulses, and photoconductors processed as large-signal step pulsers produced 1.0-V peak magnitude, 6-ps risetime pulses of 50-ps duration. Photoconductors processed as sampling gates demonstrated 3-dB measurement bandwidths of up to 76-GHz. Due to the virtual absence of noise and jitter, signal delays were measured with subpicosecond precision.

1. Introduction Photoconductor based time-domain measurement techniques for testing Si integrated circuits are described in this paper. Photoconductor circuit elements (PCEs) which were fabricated on a Si IC provided a pulsing and sampling capability that was employed for time-domain measurements. The PCEs were fabricated on the IC substrate, and then ion-beam irradiated to introduce structural damage. This damage permitted picosecond switching times. Here we show the implementation of PCE measurement techniques on an integrated circuit. We present experimental data demonstrating both high-speed-pulsed and step-function time-domain measurements on a Si integrated circuit.

2. Integrated PCEs for Measurements The PCE has been developed and refined to be a reliable picosecond pulsing and sampling device on the Si integrated circuit substrate. PCEs consist of parallelpiped volumes of photoconductive material with metal transmission lines contacting opposing sides. The initial work in developing a Si IC testing capability via PCEs resulted in bulk Si PCEs with characteristic response times as short as

*D. R. Bowman acknowledges partial support from the Fannie and John Hertz Foundation. R. B. Hammond acknowledges support from the United States Department of Energy.

29-ps [1,2]. More recently PCE characteristic response times as short as 12-ps [3] were demonstrated. The Si substrate was utilized for the active regions for the PCEs and deep radiation damage in the substrate was necessary in order to make the PCEs respond at picosecond speeds. Deep damage prevented long-lived carriers, microns below the surface of the substrate, from introducing long-lived currents.

Isolation of the PCE active regions from the Si substrate is desirable for two reasons: 1) deep level substrate damage is not required and a conventional ion-implanter could perform the PCE ion-irradiation and 2) low conductivity PCE active regions could be placed on top of high conductivity Si substrates.

LPCVD poly-Si PCEs were recently built on an insulating oxide layer above the Si substrate. These PCEs are isolated from the Si substrate [4,5]. These poly-Si PCEs exhibit superior switching speeds as well as the desirable isolation properties mentioned above.

Figure 1 displays a test structure implemented on the Si substrate used for both Al interconnect characterization and PCE correlation measurements. The Al lines are isolated from the substrate by a 1-micron-thick layer of SiO_2, and the lines contact the PCE active region in the shaded areas in the figure. These contacts occur at opposite sides of five gaps in the Al interconnect lines. The left-most gap which is located in the horizontal interconnect is a PCE pulser and is biased from the left. When this gap is excited via a laser pulse it generates a pulse traveling to the right on the wide horizontal interconnect line. The other four PCE gaps serve as samplers and measure signals that are launched by the pulser.

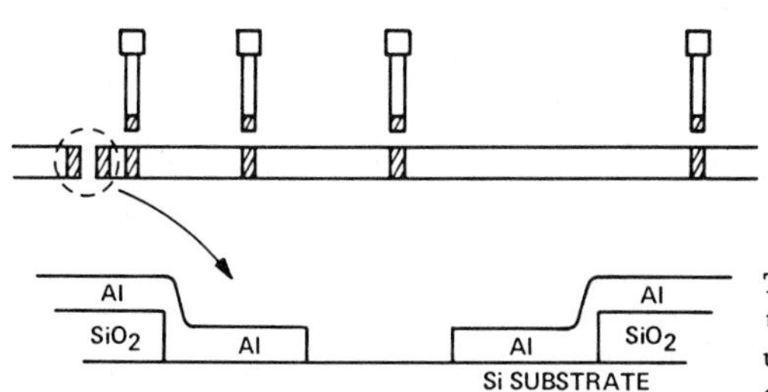

Fig. 1.
Top view and cross-sectional view of the photoconductor test structure used for testing on the Si integrated circuit.

Figure 2 displays a series of poly-Si PCE correlations between an undamaged PCE pulser and 3 PCE samplers. The samplers were damaged with 1.6 MeV Ne to a dose of 10^{15}-cm^{-2}. The measurements were done using the test circuit shown in Fig. 1. The pulsing PCE is the gap in the horizontal transmission line, and the sampling PCEs measured are the three left-most gaps below the vertical metal tabs. The sampling PCEs are located at 100-microns, 500-microns and 1000-microns from the pulsing PCE. The PCE-produced pulses were transmitted on the Al interconnect line and their measurements are superimposed on the same graph in Fig. 2. The risetime of the step-pulse increased consistently as the pulse was sampled along the transmission line (at 100-microns tr = 5.8-ps, at 500-microns tr = 6.0-ps, and at 1000-microns tr = 7.3-ps). The lengthening of the risetime was due to transmission line dispersion because the sampling PCE apertures were constant and of very short duration. The step waveform attenuated as it propagated along the interconnect line as shown by Fig. 2.

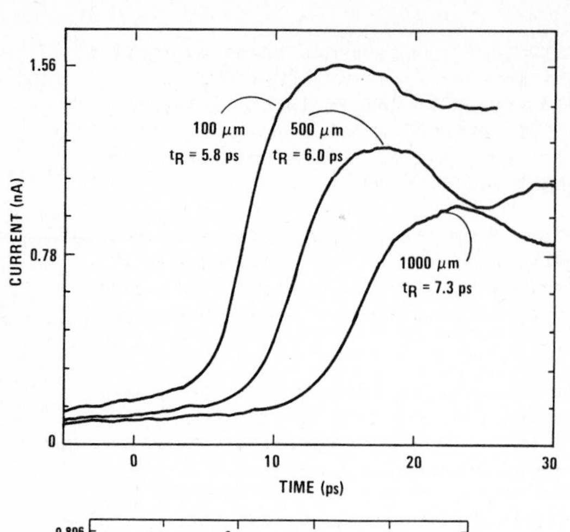

Fig. 2.
Correlation measurements of an undamaged PCE step pulser and PCE samplers damaged with 1.6-MeV Ne ions to 10^{15}-cm^{-2} dose. The test structure is displayed in Fig. 1, and the pulser and sampler separations are 100-microns, 500-microns, and 1000-microns.

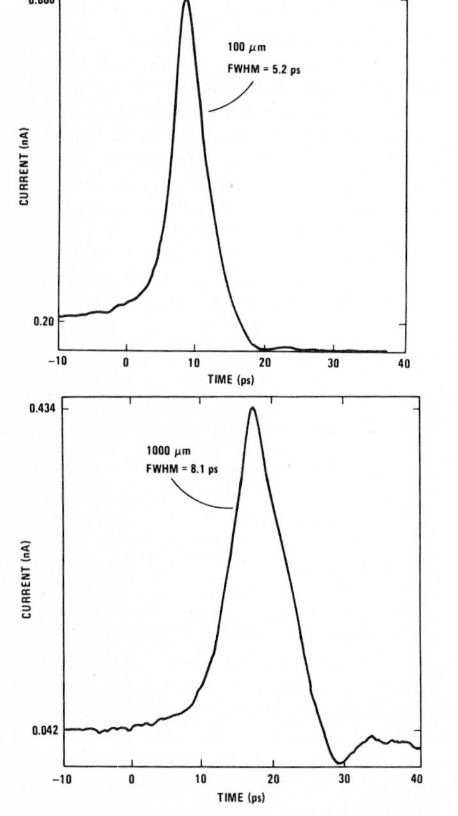

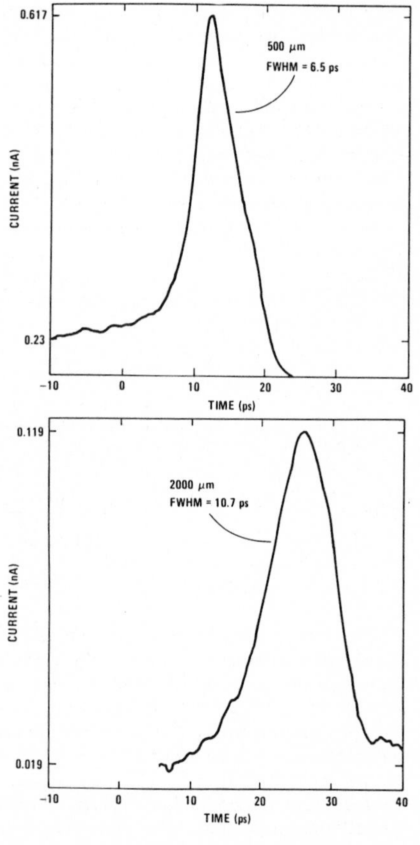

Fig. 3.
Correlations measured using the test structure displayed in Fig. 1. The pulser and sampler PCEs were bombarded by 1.6 MeV Ne ions to 10^{15}-cm^{-2} dose. The measurements displayed are for PCE pulser and sampler separations of 100-microns, 500-microns, 1000-microns, and 2000-microns from the PCE pulser.

Figure 3 shows a series of poly-Si PCE autocorrelation measurements made using the test structure illustrated in Fig. 1. The correlations were made between the pulser PCE and sampling PCEs spaced 100-microns, 500-microns, 1000-microns and 2000-microns from the PCE pulser. The sampler PCEs were bombarded with 1.6 MeV Ne to a dose of 10^{15}-cm^{-2}. The pulser poly-Si PCE measured in Fig. 3 had 50-mV peak magnitudes. The correlation shows very nearly the true shape of the pulse. The pulse length is determined by circuit limits in the PCE pulser [6]. The sampler PCEs had 3-dB measurement bandwidths in excess of 76 GHz. The broadening of the pulse in the sampling measurements was caused by dispersion in the Al interconnect line. Also, the decreasing amplitude indicates loss in the line.

Measurements such as those shown in the pulse measurements of Fig. 3 and the step-pulse measurements of Fig. 2 may be used to calculate the frequency-dependent transmission line properties of the Al interconnect. Active device measurements are also possible using the short pulses and step pulses illustrated in Fig. 2 and Fig. 3. The short pulses may be used to stimulate active devices on the Si integrated circuit and generate their impulse response. If the magnitude of the pulses is small, then small-signal frequency information describing the active device can be found directly using the Fourier transform. The step-pulse PCEs with 6-ps risetimes and 50-ps durations could be used to determine the step responses of very high-speed active devices on the Si integrated circuit. Since the step pulsers have the ability to generate at least 1.0-V pulses, they can be used to examine large-signal transients and non-linearities describing active device switching behavior.

4. Summary The PCE measurement technology has been demonstrated as an effective tool for IC interconnect characterization. The use of the poly-Si PCE for both small-signal and large-signal active device characterization has been discussed and the modeling of the PCE in an impulsive sampling system on the IC substrate has been presented.

REFERENCES

1. W. R. Eisenstadt, R. B. Hammond, and R. W. Dutton, "Integrated Silicon Photoconductors for Picosecond Pulsing and Gating," IEEE Elec. Dev. Letts. EDL-5, 296-299 (Aug. 1984).
2. R. B. Hammond, N. G. Paulter, R. S. Wagner, and W. R. Eisenstadt, "Integrated Picosecond Photoconductors Produced on Bulk Silicon Substrates," Appl. Phys. Letts. 45, 404-405 (1984).
3. W. R. Eisenstadt, R. B. Hammond and R. W. Dutton, "On-Chip Picosecond, Time-Domain Measurements for VLSI and Interconnect Testing Using Photoconductors," IEEE Trans. on Elec. Dev. (Feb. 1984) and IEEE Jour. of Sol. State Cir. SC-20, 284-289 (Feb. 1984).
4. R. B. Hammond and N. M. Johnson, "Impulse Photoconductance of Thin Film Polycrystalline Silicon," to be published in Jour. App. Phys.
5. D. R. Bowman, R. B. Hammond, and R. W. Dutton, "Improved Integrated Photoconductors for Picosecond Pulsing and Gating Using Polycrystalline Silicon," unpublished.
6. D. H. Auston, "Impulse Response of Photoconductors in Transmission Lines," IEEE Journal of Quantum Electronics, Vol QE-19, pp. 639-648, 1983.

Modeling of Picosecond Pulse Propagation on Silicon Integrated Circuits

K.W. Goossen* **and R.B. Hammond**
Electronics Division, Los Alamos National Laboratory,
Los Alamos, NM 87545, USA

We have modeled the dispersion and loss of pulses on microstrip transmission line interconnections on silicon integrated-circuit substrates. Geometric dispersion and conductor linewidth, as well as losses from conductor resistance, conductor skin effect, and substrate conductance are considered over the frequency range from 100 MHz to 100 GHz. Results show the enormous significance of the substrate losses, and demonstrate the need for substrate resistivities >10 Ω-cm for high performance circuits. The results also show the effects of geometric dispersion for frequencies above 10 GHz and the unimportance of conductor skin-effect losses for frequencies up to 100 GHz.

1. Introduction Because of the increasing speed of very large scale integrated circuitry (VLSI) and the development of very high speed integrated circuits (VHSIC), knowledge of pulse dispersion and loss in the subnanosecond regime are becoming important to the Si integrated-circuit designer. HASEGAWA and SEKI [1] have shown that the common circuit-design approach of using lumped-capacitance models for interconnections is not adequate when switching speeds are <100 ps. Reflections at discontinuities, substrate losses, conductor losses, and geometric dispersion are all important. The microstrip transmission line has excellent high-frequency properties that can be easily incorporated into Si ICs. The quasi-TEM analysis is valid to switching speeds of ∿7 ps because cut-off frequencies are above 50 GHz. Although the isolated microstrip transmission line does not exactly represent the integrated circuit interconnection embedded in a complex circuit environment, a thorough understanding of the isolated line is a first step to developing useful high-frequency models for real device interconnections.

In this paper, we model pulse propagation on microstrip transmission lines on Si-IC substrates. Our analyses use equations for the effective dielectric constant, conductor loss, skin-effect loss, and dielectric loss, which are derived in the cited references. We assume quasi-TEM-mode propagation throughout the analysis.

We have developed a computer model and performed time-domain analyses of pulse propagation on Si substrates. We chose microstrip parameters and input waveforms that are of interest because they can be used to perform time-domain studies of high-speed pulse-propagation phenomena on Si ICs [2]. We considered pulses of the form

$$V(t) = \exp(-t/\tau_1)[1 - \exp(-t/\tau_2)] \quad . \tag{1}$$

*Current Address: Department of Electrical Engineering, Princeton University, Princeton, NJ.

2. Analysis We assumed the substrate to be a simple dielectric with relative dielectric constant ε_r at the frequencies of interest. (This assumption is valid in Si up to $\sim 10^{13}$ Hz.) Although the following analysis assumes that the insulator thickness t_0 is zero, the effect of finite insulator thickness can reasonably be incorporated by replacing the insulator with an open circuit at zero frequency and by using a short circuit at all other frequencies. This is possible because the analysis is performed at discrete frequency increments of about 100 MHz. At 100 MHz, the capacitance produced by the insulator introduces negligible impedance as long as $t_o \ll w$.

The effective dielectric constant ε_{eff} and characteristic impedance Z_0 at zero frequency have been calculated by SCHNEIDER [3] (Fig. 1). The maximum relative error in these expressions is less than 2%, but corrections should be made for $t/h > 0.005$ [4].

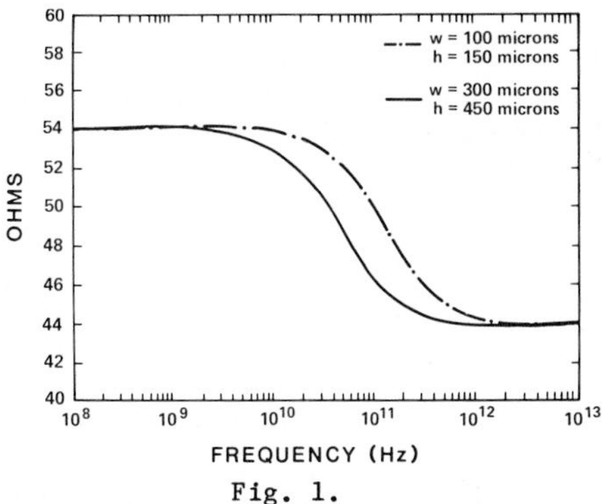

Fig. 1.
Characteristic impedance vs frequency for two microstrip geometries on Si.

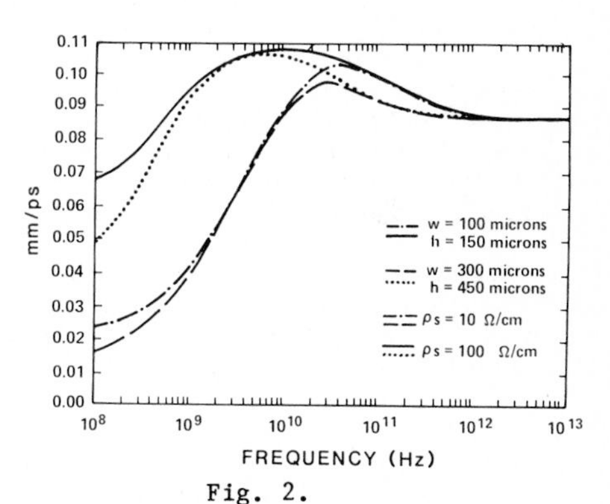

Fig. 2.
Phase velocity vs. frequency for four microstrip cases.

At high frequencies, ε_{eff} must be corrected. YAMASHITA et al. [5] have derived an expression by curve-fitting to a full-wave analysis: This formula is accurate to approximately 1%.

If the material is lossless, the propagation factor β_0 is given by

$$\beta_0 = 2\pi f\sqrt{\varepsilon_{eff}}/c \quad ; \tag{2}$$

however, finite resistivity of the conductor and finite conductivity of the substrate introduce attenuation. For low frequencies, the conductor loss factor α_c is given by

$$\alpha_c = \rho_c/(2wtZ_0) \tag{3}$$

in nepers. Here ρ_c is the resistivity of the metal. This expression, however, assumes a uniform distribution of current in the conductor. At higher frequencies, in the skin-effect regime, the current is not distributed evenly. PUCEL et al. [6] have formulated the conductor loss in this regime.

In practice, the expression for uniform current distribution yields a higher value of attenuation at low frequencies and a lower value at high frequencies than does Pucel's expression for skin-effect loss. For our analyses we used the expression for α_c that yielded the higher attenuation at any given frequency.

An expression for dielectric loss α_d due to nonzero conductivity in the substrate has been derived by WELCH and PRATT [7].

$$\alpha_d = 60\pi\sigma_s(\varepsilon_{eff} - 1)/[(\varepsilon_r - 1)\sqrt{\varepsilon_{eff}}] \tag{4}$$

The time response of the transmission line is then given by

$$V(z,t) = F^{-1}\{\exp[(\alpha + j\beta)z]*F\{V(0,t)\}\} \quad , \tag{5}$$

where F and F^{-1} denote the forward and inverse Fourier transforms. This is similar to the approach used by HASNAIN et al. [8]. α and β are given by an analysis of a transmission line with shunt and series resistances.

4. Results Figure 1 is a plot of Z_0 vs. frequency for the two microstrip geometries we considered in our calculations. The quasi-TEM impedance of the lines varies with frequency due to geometric dispersion. The frequency dependence in Z_0 directly reflects the frequency dependence of the effective dielectric constant for the lines.

Figure 2 shows calculated phase velocity vs. frequency for two substrate thicknesses and two substrate resistivities. Greater velocity occurs for both substrate thicknesses at the higher substrate resistivity,

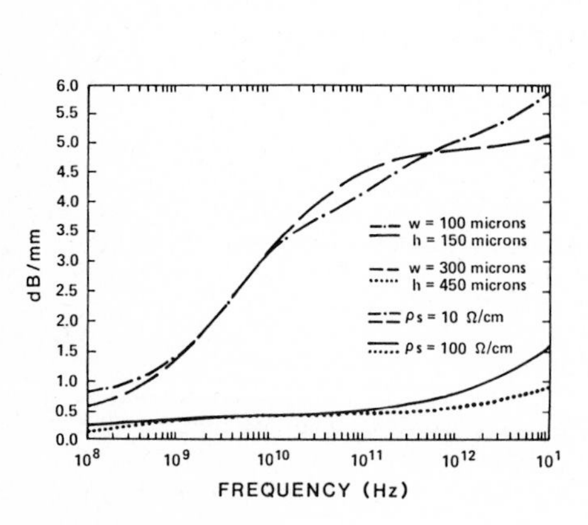

Fig. 3.
Microstrip line loss vs. frequency for the same four cases considered in Fig. 2.

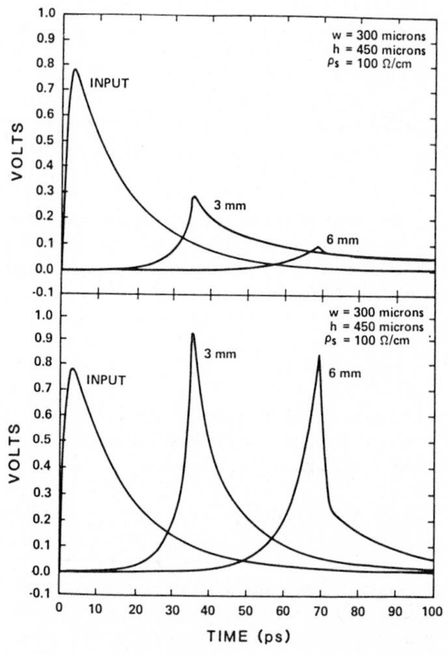

Fig. 4.
Effects of dispersion and loss on the propagation of picosecond pulses for two different cases of microstrip lines on Si substrates.

100 Ω-cm. The decreasing velocity with decreasing frequency is reminiscent of the "slow-wave" on Si. Slow wave is an inductive effect due to the presence of the SiO_2, however, and is not actually present in our purely quasi-TEM analysis.

Figure 3 is a plot of line loss vs. frequency for the same four cases considered in Fig. 3. The loss is dominated by substrate losses except at the extremely high frequencies which are above the cut-off frequencies for the transmission lines, i.e. above ∿50-100 GHz. Losses due to skin-effect in the conductor begin to appear at these very high frequencies.

Figure 4 shows the effects of dispersion and loss on the propagation of picosecond pulses. We considered input pulses of the form shown in Eq. (1). Pulse sharpening due to geometric dispersion can be observed as well as the severe loss due to the 10 Ω-cm substrate compared with the 100 Ω-cm substrate. The propagation lengths, 3 and 6 mm, are realistic for long interconnections on current ICs.

5. Acknowledgements This work was done under internal supporting research at the Los Alamos National Laboratory, which is operated by the University of California for the United States Department of Energy under contract W-7405-ENG-36. Keith Goossen also acknowledges the support of the US Department of Energy Magnetic Fusion Energy Technology Fellowship Program.

REFERENCES

1. H. Hasegawa and S. Seki, "On-chip Pulse Transmission in Very High Speed LSI/VLSI," Technical Digest of the IEEE Microwave and Millimeter-wave Monolithic Circuits Symposium, San Francisco, 1984, 29-33.
2. W. R. Eisenstadt, R. B. Hammond, and R. W. Dutton, "Integrated Silicon Photoconductors for Picosecond Pulsing and Gating," IEEE Electron Device Lett., EDL-5, 296-299 (1984).
3. M. V. Schneider, "Microstrip Lines for Microwave Integrated Circuits, "Bell Syst. Tech. J. 48, 1421 (1969).
4. K. C. Gupta, R. Garg, and R. Chadha, "Computer-Aided Design of Microwave Circuits," (Artech House, Dedham, Massachusetts, 1981), p. 62.
5. E. Yamashita, K. Atsuki, and T. Ueda, "An Approximate Dispersion Formula of Microstrip Lines for Computer-Aided Design of Microwave Integrated Circuits," IEEE Trans. Microwave Theory Tech., MTT-27, 1036 (1979).
6. R. A. Pucel, D. J. Masse, C. P. Hartwig, "Losses in Microstrip," IEEE Trans. Microwave Theory Tech. MTT-16, 342 (1968).
7. J. D. Welch and H. J. Pratt, "Losses in Microstrip Transmission Systems for Integrated Microwave Circuits," NEREM Rec. 8, 100 (1966).
8. G. Hasnain, G. Arjavalingam, A. Dienes, and J. R. Whinnery, "Dispersion of Picosecond Pulses on Microstrip Transmission Lines," Conf. Picosecond Optoelectronics, San Diego, 1983, SPIE 439, 159-163.

Part II

High-Speed Phenomena in Bulk Semiconductors

Picosecond Processes in Carrier Transport Theory

D.K. Ferry
Center for Solid State Electronics Research, Arizona State University, Tempe, AZ 85287, USA

1. Introduction

High electric field transport has been studied for some three decades. In recent years, it has become of much greater interest due to the advent of semiconductor devices on the micron and submicron scale, and to the interaction of these devices with sub-picosecond optical pulses. Theoretically, hot-electron transport (as this high-field transport is usually called) has been traditionally discussed in terms of the Boltzmann Transport Equation (BTE). However, semiconductor transport in these high electric fields is a classical example of carrier response described by far-from-equilibrium thermodynamics. More importantly, on the short time scale of the response of these devices, particularly to the short optical pulses, retardation of the carrier dynamics must be fully considered. Such retardation does not appear in any dynamical description based upon the BTE, and more exact formulations must be adopted.

In general, we are concerned with the evolution of the average velocity and energy ensemble in the presence of the high electric field, and Langevin-like equations have been developed for describing this evolution. Retardation in the relaxation integrals is encountered, an effect which is not predicted from the more usual BTE formalism. Evaluation of the retardation functions leads to a set of kernals which in turn lead to the evaluation of a set of correlation functions. While these correlation functions do not normally appear in BTE approaches, it has been recognized for some time that parameters such as the differential mobility and the diffusion constant are related to these functions. As is known for other far-from-equilibrium systems, the transport parameters of interest can be obtained through a unique set of Green's functions supplemented by a proper knowledge of all relevant moments of the proper nonequilibrium statistical operator. Other approaches to the same end result utilize proper quantum transport formulations, such as the density matrix in the position representation and the Wigner distribution function.

In the present paper, we shall first show in a simple and straightforward manner how the retarded functions are required and arise in the moment equations. We shall then talk of general approaches to obtaining the proper correlation functions and Green's functions for the hot electron problem. An approach based upon the nonequilibrium statistical operator for inhomogeneous systems is discussed. Finally, we shall briefly review full quantum approaches for ultra-small semiconductor devices.

2. Retarded Transport Equations

It is readily apparent to one who has dealt with hot electron transport in large devices, where the BTE may readily be utilized, that we have survived for quite a long time without worrying about correlation functions and retardation. However, we may readily demonstrate that systems, which exhibit the popular "overshoot" phenomenon in their velocity response, must incorporate these effects for discussion of the relevant physics. In fully retarded form, we may describe the momentum balance equation in terms of a retarded Langevin equation, as

$$\partial_t p = eF - \int_0^t \gamma(t-\tau)p(\tau)d\tau + R(t) \ , \tag{1}$$

where ∂_t is a short-hand notation for $d(\cdot)/dt$, $R(t)$ is a random force whose ensemble average is zero, and γ is the relaxation function. In general, where we are in a steady-state or where the velocity response is very slow on the scale of the variation in γ, we may approximate (1) by removing $p(t)$ from the integral, and using the fact that

$$1/\tau = \int_0^\infty \gamma(t)dt \quad . \tag{2}$$

The non-retarded form, with the ensemble average to yield $mv=\langle p\rangle$, is the normal one derived from moments of the BTE. The more general form we have adopted here is more correct [1]. To proceed, we Laplace transform (1), so that

$$P(s) - P(0) = eF/s - \Gamma(s)P(s) \quad , \tag{3}$$

where Γ is the transform of γ, and

$$P(s) = eF/\{s[s + \Gamma(s)]\} = (eF/s)X(s) \ , \tag{4}$$

in the case $P(0)=0$, and we have introduced the reduced correlation function $X(s)=1/[s+\Gamma(s)]$. We may now find the time response as

$$p(t) = eF\int_0^t x(t)dt \quad . \tag{5}$$

We may readily demonstrate that $x(t)$ is a correlation function by multiplying (1) by $p(t')$ (retaining the Laplace transform as only over the variable t) and taking the ensemble average. This gives, if we assume that the momentum and the random force are uncorrelated as will be the case for non-interacting electrons,

$$\phi(t',t) = \langle p(t')p(t)\rangle - \langle p(t')\rangle\langle p(t)\rangle \ , \tag{6}$$

so that

$$x(t) = \phi(0,t)/\langle p^2(0)\rangle \tag{7}$$

is indeed a normalized momentum correlation function. This is no more than a generalized statement relating the dissipation function γ to the fluctuation function of the momentum.

Consider now the case where the correlation function has a simple exponential decay, $\exp(-t/\tau)$, as is the case in equilibrium systems. Then the transform of $x(t)$ becomes

$$X(s) = \tau/(s\tau+1) \quad , \tag{8}$$

and

$$\Gamma(s) = 1/\tau \ . \tag{9}$$

Obviously then, $\gamma(t)$ is just the delta function $\delta(t)/\tau$, and the usual long-time result obtained from the BTE results. Clearly, we may then state that the BTE results are valid only on the long-time scale or in the case where the system is in equilibrium. Certainly, the far-from-equilibrium system, on the short-time scale, is not expected to satisfy the Boltzmann equation. In Fig. 1, the normalized velocity autocorrelation function is shown for a system of non-interacting electrons responding to a step, high electric field. Clearly, the response is not simply a single exponential, but describes a complicated time decay, with elements of anomalous time behavior, in which the decay follows a negative cube-root variation [2]. From (5), we may readily expect that the negative portion of x(t) in Fig. 1 is exactly related to the overshoot of the velocity in its approach to steady-state. Such overshoot is directly related to the non-monotinicity of the response apparent in x(t). We can also show that it is related to the retardation in relaxation by a simple argument. Let us assume that

$$\gamma(t) = (1/\tau^2)\exp(-t/\tau) \quad , \tag{10}$$

which clearly satisfies (2). Then, it is easy to find that

$$x(t) = \exp(-1/2\tau)\cos[(3/4\tau^2)t] \quad , \tag{11}$$

which illustrates the non-monotonic behavior expected in these systems. In actual fact, the true behavior evident in Fig. 1 is more complicated due to the presence of at least two relaxation times in the system [3].

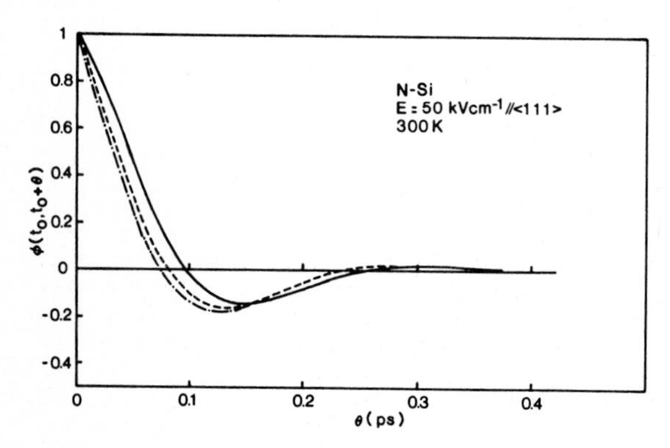

Fig. 1 - Monte Carlo calculations of the velocity correlation function for three initial times: solid curve, t=0; dashed curve, t=0.2 ps; dot-dashed curve, t=0.5 ps. All curves are for silicon with an applied field of 50 kV/cm.

The above discussion introduces us to the concept of having several relaxation times in the system. In a semiconductor, numerous collisions occur,and it is these collisions which provide the mechanism of exchange of energy and momentum. There are collisions between the carriers which randomize the energy and momentum within an ensemble but do not relax either of these quantities for the ensemble as a whole. There are also elastic collisions between the carriers and impurities or acoustic lattice vibrations which relax the momentum but not the energy. Finally, there are inelastic collisions between the carriers and the lattice vibrations which relax both the energy and the momentum. In general, we can identify four generic time scales [4]

$$\tau_c < \tau < \tau_R < \tau_h \quad . \tag{12}$$

Here, we denote the average duration of a collision by τ_c. This is the time required to establish the energy-conserving delta-function that appears in the Fermi golden rule of time-dependent perturbation theory. The average time between collisions, the mean-free time, is denoted by τ. For time scales such that $t<\tau$, the evolution of the system depends strongly on the details of the initial state and on the details of the correlation functions. Generally, $\tau>>\tau_c$, but this is not always the case at high electric fields and the breakdown of this inequality can lead to new transport effects, such as multiple scattering and strong self-energy shifts of the energy levels. The establishment of a non-equilibrium steady-state can be achieved within a few, or a few tens of, τ. The characteristic time associated with this latter process is the relaxation time τ_R. If configuration space gradients exist, the situation becomes more complex. Relaxation in momentum space proceeds on the scale of τ_R and establishes a local equilibrium over regions smaller than a macroscopic scale, perhaps only a few mean free paths in extent. The achievement of a uniform nonequilibrium steady-state requires a longer time, the hydrodynamic time $\tau_h>\tau_R$. Only for times $t>\tau_h$ can the ensemble truly be said to be stationary,and only for times on this scale are the processes even beginning to become ergodic. The above-mentioned cube-root decay is thought to be related to the differences between equilibriation within the ensemble but not to the background.

The above discussion of the multiple scattering times illustrates the need to fully consider the fact that all of the correlation functions are in general two-time correlation functions. This is because the ensemble is not in equilibrium, but in fact is evolving with time. Therefore, the time at which the ensemble average is performed becomes important in the evolution of the correlation function. Only in the stationary, steady-state is the ensemble such that the correlation functions become single time functions. It is only in this stationary, steady-state that the ensemble can be considered as possessing the ergodic properties usually invoked in transport theory. As in other non-equilibrium systems, it is to be remembered that ensemble averages are the important property, and time averages are irrelevant except in the few special cases in which one can be assured that the ergodic nature has been established, i.e. in the stationary, steady-state situation. By summary of these facts, it is obvious that excitation of the semiconductor by intense laser beams, of only a few femto-second duration, will produce transient situations which will never possess ergodic properties during any of this transient time.

3. Correlation Functions

The form that appears in equation (5) is fully equivalent to the well-known Kubo form [5]. We can clearly use this equation to define the large-signal mobility as [6]

$$\mu(t,0) = \int_0^\tau (e/m)x(t')dt' , \tag{13}$$

where m is the effective mass of the carriers. We have introduced the initial time as zero, since we are taking the total response function. If we start from the stationary, steady-state we will be dealing with a different response function x'(t), which is still a non-monotonic function. This latter function is characterized by a different value for the initial condition reflecting the non-equilibrium nature of the system. The non-monotonicity is retained because of the multiple relaxation times in the system, which lead to multiple sources of fluctuations in the system, a point to which we shall return below.

The second important transport parameter in which we are interested is the diffusion coefficient. This is also related to the correlation function, and [6]

$$D(t) = \int_0^t (m^{-2})\phi(t',t)dt' \quad , \tag{14}$$

where we have introduced ϕ earlier. We note here that the integrations in (13) and (14) are different. In the case of the mobility, the integration is over a single function initialized at t=0. However, in (14) the integration is over the initial times of the correlation functions, so that we are summing the contributions of fluctuations, starting at different times, at the time t. Only in the case of the equilibrium system, where the correlation functions are single-time functions and the system has a single relaxation time do these two integrations merge into a single entity. What these equations are telling us is that the idea of fluctuation-dissipation theorem fails generally for non-equilibrium systems [3,7]. This failure has to do with the fluctuation of other quantities (than momentum) affecting the momentum fluctuations. Niez [8] has developed expressions for terms which will provide the corrections, but this work is still somewhat premature.

In general, the momentum fluctuations are composed of many factors. In equilibrium, we need worry only about the specific fluctuations of the velocity about the mean, lattice temperature. In the non-equilibrium state, these fluctuations are about the electron temperature. In addition, there are fluctuations in the momentum due to fluctuations in the temperature itself. Often these two components are out of phase, producing the negative portions of the correlation coefficient. In addition, there may be fluctuations of the momentum due to density fluctuations in a non-homogeneous system. As a general rule, each source of fluctuation is characterized by its own time constant, discussed above, and all of these different time scales are reflected in the correlation coefficient.

Another consideration worth noting at this point is the non-Gaussian nature of the diffusion process itself. The equilibrium case arises from

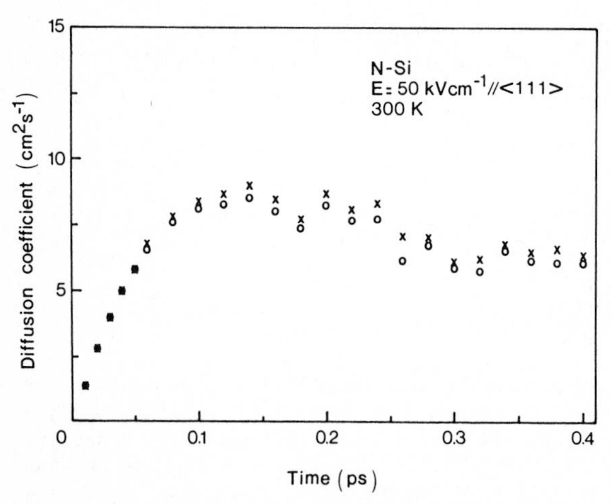

Fig. 2 - A comparison of the direct Monte Carlo calculation of the diffusion coefficient with that obtained from integrations of the correlation function (crosses).

linear, Markovian transport and leads to a Gaussian distribution for the diffusing pulse of photo-excited carriers. Here, we are dealing with a nonlinear, and certainly non-Markovian, system. Both of these factors lead to a general non-Gaussian behavior for this pulse of carriers. Indeed, the spatial distribution develops tails which extend beyond that expected for the Gaussian distribution. Because of the non-Markovian and nonlinear behavior, the governing random-walk equations for the transient diffusion case do not reduce to the normal Fick's law [9]. However, Ensemble Monte Carlo calculations do include the proper retarded transport [10], and may be used to simulate the actual case. In Fig. 2, the transient diffusion coefficient is plotted versus time for silicon. This is compared with results obtained using the correlation function (for non-interacting electrons) and shows the expected agreement. It is clear that the actual value of the diffusion coefficient depends upon time and field, and this is reflected in Fig. 3. The non-Gaussian nature of the diffusive spread is indicated in Fig. 4, where the third and fourth cumulants of the ensemble

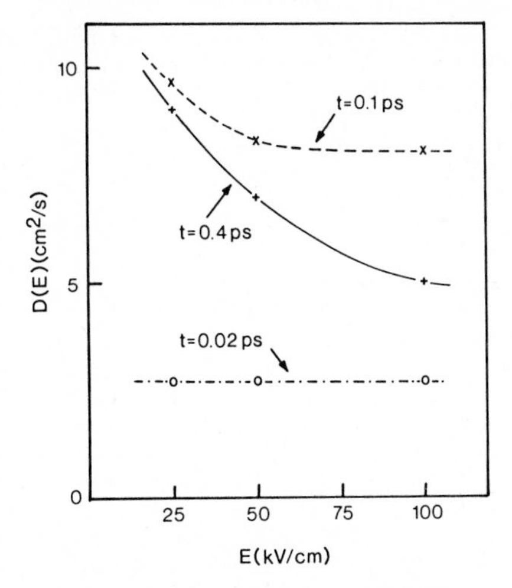

Fig. 3 - The diffusion coefficient as a function of the electric field for three different times, measured from the initial time t=0. This is the transient diffusion coefficient displayed in Fig. 2.

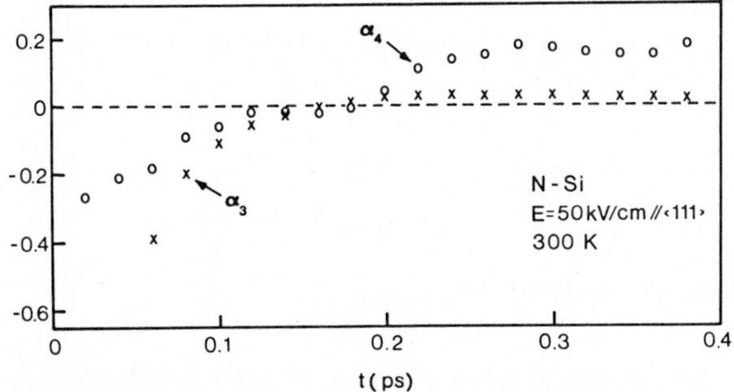

Fig. 4 - The third and fourth cumulants of the spatial distribution, based upon an initial Gaussian function. These should be identically zero for a Gaussian.

are plotted as a function of time. For a Gaussian spread, these quantities are identically zero, but here the difference represented by the fourth cumulant points to enhanced overall spread, which is symmetric in momentum.

If we begin to consider interacting electrons, then the full many-body calculation of the correlation function must be included. In general, this problem has not been addressed effectively, since most of the calculations must be carried out in a quantum formalism. Only in the case of very dense electron-hole systems are major differences in drift and diffusion to be expected, although other many-body corrections (such as energy-gap narrowing) can play some role.

The calculation of the correlation functions is quite straightforward for the non-interacting case, and can be related to the general Green's function propagator of transport theory. We do not have the space here to give a full development, but the quantum approach for second-order perturbation theory is given by Niez et al. [7]. These authors also show that the approach is fully equivalent to that of Zubarev [11], in which a quasi-equilibrium statistical operator formalism is assumed for the density matrix. This latter approach has also been extended to the case of far-from-equilibrium inhomogeneous systems [12]. The quasi-equilibrium statistical operator is the quantum equivalent of the drifted-Maxwellian that has been utilized for many years in high electric field transport.

4. Summary

In this paper, several aspects of the retarded transport that must be considered on the picosecond time scale have been discussed. In particular, the transport is easily treated by consideration of the actual correlation functions for the physical quantities. These correlation functions illustrate the root of the failure of the first fluctuation-dissipation theorem in the non-equilibrium transport. In turn, they may be calculated from Green's functions determined from either classical or quantal transport equations.

The author would like to ȧcknowledge the collaborations of P. Lugli, J. Zimmermann, J.-J. Niez, W. Poetz, and R. O. Grondin, and their enormous contributions to this work. The work was supported in part by the Office of Naval Research.

5. References

1. R. W. Zwanzig: in Lectures in Theoretical Physic, Ed. by W. E. Britton, B. W. Downs, and J. Downs (Interscience, New York, 1961).
2. D. K. Ferry: Phys. Rev. Letters 45, 758 (1980).
3. P. J. Price: in Fluctuation Phenomena in Solids, Ed. by R. E. Burgess (Academic, New York, 1965).
4. G. V. Chester: Rept. Prog. Phys. 26, 41 (1963).
5. R. Kubo: J. Phys. Soc. Jpn. 12, 570 (1957).
6. J. Zimmermann, P. Lugli, and D. K. Ferry: J. Physique Coll. 42(C7), 95 (1981).
7. J. J. Niez, K.-S. Yi, and D. K. Ferry: Phys. Rev. B 28, 1988 (1983).
8. J. J. Niez, private communication.
9. D. K. Ferry and J. R. Barker: J. Appl. Phys. 52, 818 (1981).
10. J. Zimmermann, P. Lugli, and D. K. Ferry: Sol.-State Electron. 26, 233 (1983).
11. D. N. Zubarev: Nonequilibrium Statistical Mechanics (Consultants Bureau, New York, 1974).
12. W. Pötz and D. K. Ferry: in Proc. 17th Intern. Conf. Phys. Semicon., in press.

Carrier-Carrier Interaction and Picosecond Phenomena in Polar Semiconductors

P. Lugli
Dipartimento di Fisica, Università di Modena, Via Campi 213/A, I-41100 Modena, Italy
D.K. Ferry
Center for Solid State Electronics Research, Arizona State University, Tempe, AZ 85287, USA

As the dimension of solid-state devices reaches the submicron limit and sub-picosecond phenomena become relevant, the electron-electron interaction can be of great importance to the device performance. Such interaction defines a characteristic microscopic time scale much shorter than the typical relaxation times of the electron-phonon interaction [1]. In the following, we present a complete treatment of the carrier-carrier interaction, starting with the analysis of the wave-vector- and frequency-dependent dielectric function $\varepsilon(q,\omega)$. An Ensemble Monte Carlo (EMC) technique is then used to study the effect of the interaction on the transport properties of GaAs.

It is well known [2] that the quantity $\mathrm{Im}\{1/\varepsilon(q,\omega)\}$ contains the complete spectrum of the excitation of an electron gas. Furthermore, it is proportional to the scattering strength of a particular mode [3]. We are mainly interested here in the phenomena associated with the long range of the Coulomb forces, since those are usually neglected in the study of transport in semiconductors. The analysis of the dielectric function has shown the importance of plasma effects in such materials, for arbitrary degeneracy, and over a wide range of temperatures and electron densities [1],[4],[5].

If electrons and holes are simultaneously present(for example as a result of the creation of electron-hole (e-h) pairs by laser excitation), the dielectric function of the total system is given by the sum of the electron and hole contribution. Fig.1 shows $\mathrm{Im}\{1/\varepsilon(q,\omega)\}=[\varepsilon_r/(\varepsilon_r+\varepsilon_i)]$ for an e-h plasma in GaAs at 77 K, with a concentration of $5\ 10^{17}\mathrm{cm}^{-3}$. The imaginary part of the dielectric function, ε_i, is given analytically, the real part ε_r is calculated numerically using Kramer-Kronig relations. The wavevectors and frequencies are normalized to the Fermi wave-vector $q_F=(3\pi^2 n)^{1/3}$ and to the longitudinal optical frequency ω_{lo}, respectively. The electron plasmon shows up as a singularity at small wave-vectors, and

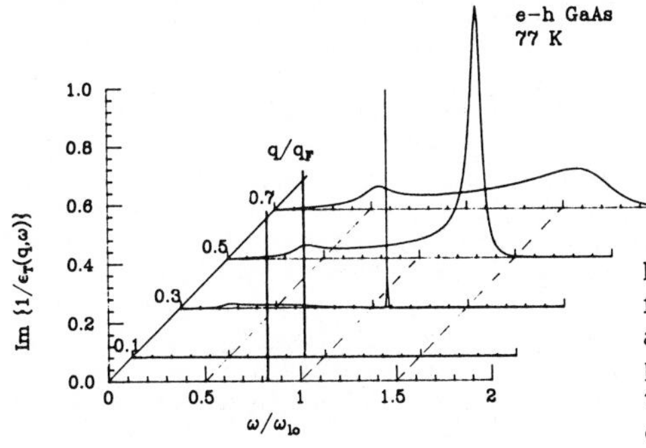

Fig.1 Plot of $\mathrm{Im}\{1/\varepsilon(q,\omega)\}$ as a function of normalized wave-vector q and frequency ω, for an electron-hole plasma. The temperature is 77 K and the carrier concentration is 5×10^{17} cm^{-3}.

is broadened as q increases, as a result of an efficient Landau damping. No collisional damping is included here, corresponding to a low doping condition. At q=0, the electron plasma frequency is $4.5\times10^{13}s^{-1}$. As q increases, the plasma frequency (defined from the peak of $Im\{1/\varepsilon(q,\omega)\}$) follows an almost quadratic dispersion $\omega(q)$. Due to the heavier effective mass, the plasma freqency for holes is a factor 2.5 smaller than that of the electrons. The hole plasmon mode is completely washed out, at small q, by the fact that it occurs at a frequency where the imaginary part of the dielectric function is very large (due to the electron contribution). At higher q, the two modes couple, to give a diffused peak in the excitation spectrum.

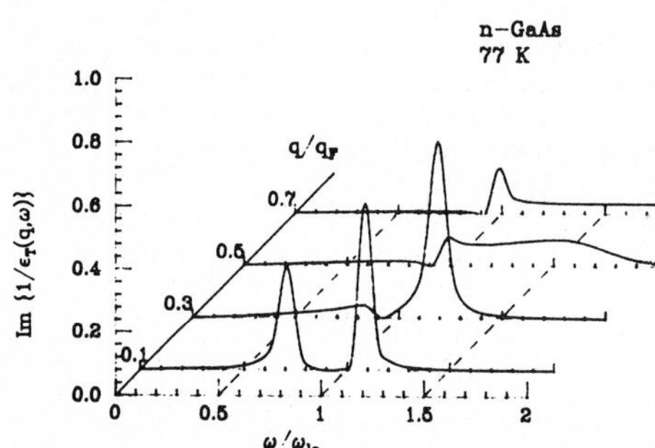

Fig.2 Plot of $Im\{1/\varepsilon(q,\omega)\}$ for an electron plasma, showing the coupling of the plasmon and phonon modes. The temperature is 77 K and the electron concentration is 5×10^{17} cm^{-3}. Collisional damping is included.

Another interesting feature of polar semiconductors is the coupling of plasmon and phonon modes. Fig.2 shows $Im\{1/\varepsilon_T(q,\omega)\}$ for the total system, that includes the electron and lattice contributions. The electron concentration is $5\times10^{17}cm^{-3}$, and the temperature is 77 K. Collisional damping has been included here, by introducing an imaginary component of the frequency, proportional to the inverse momentum relaxation time τ_m [1]. The collisional damping accounts for the effect that electron-impurity and electron-phonon scattering have in disrupting the organized behavior of the electron gas. A value of τ_m equal to 1.6×10^{13} s^{-1} has been used. At small q, two peaks show up in the excitation spectrum. The low frequency one is associated with a plasmon-like mode, the high frequency one with a phonon-like mode. The effect of collisional damping is to broaden the singularity that would show up at small q, where Landau damping is not very effective. As q increases, the plasmon peak (now the high frequency one) shifts towards higher frequencies and is smeared out by the combined effect of the Landau and collisional damping. Nevertheless, plasmon modes are still present for values of q up to half of the Fermi wave-vector, and will modify the transport properties of polar semiconductors at high densities, both by providing an additional scattering mechanism and by shifting the optical phonon frequencies.

An EMC simulation has been performed to study the effect of the complete e-e interaction (including the screened e-e contribution and the electron-plasmon scattering). The parameters used in the simulation have been described elsewhere [4]. Electrons with an initial energy of 0.24 eV are injected into a doped region, with a density of ambient cold electrons of 10^{17} cm^{-3}. No plasmon-phonon coupling occurs at these densities. Fig.3 shows the average velocity as a function of the distance travelled by the injected packet of electrons. The dashed curves are obtained considering only impurity and phonon scatterings, the solid curves include the total e-e interaction. A low field (400 V/cm) is considered. The absence of

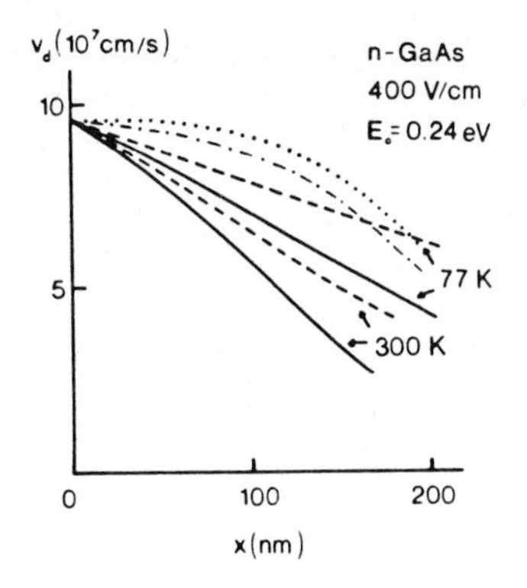

Fig.3 Average drift velocity as a function of distance, with (solid) and without (dashed) the full e-e interaction, for two different lattice temperatures. The upper curves are for an applied field of 10 kV/cm at 77 K, with (dot-dashed) and without e-e scattering.

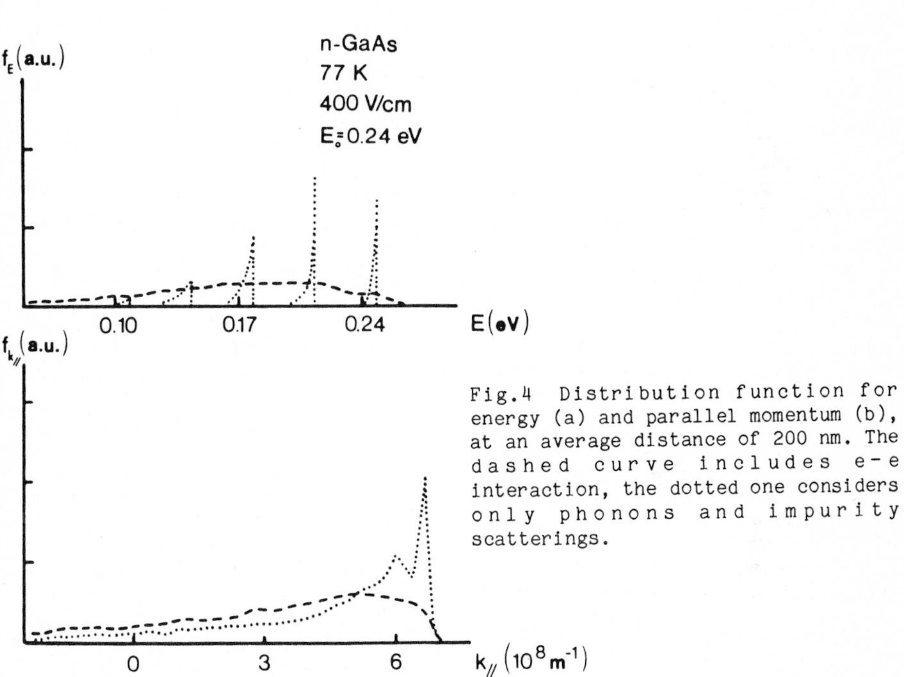

Fig.4 Distribution function for energy (a) and parallel momentum (b), at an average distance of 200 nm. The dashed curve includes e-e interaction, the dotted one considers only phonons and impurity scatterings.

ballistic motion is evident, even at low temperatures (77 K). The polar-optical phonon scattering is efficient enough to reduce the average drift velocity over very short distances. The degradation of the drift velocity is increased when electron-electron and electron-plasmon interactions are considered. At room temperature, the larger availability of phonons causes a much stronger loss of momentum than at lower temperature, while the relative contribution of the total e-e scattering is reduced. If a high electric field is applied to the doped region, it can be expected that the continous supply of energy and momentum from the field would balance the losses due to collisions. The dotted and dash-dotted lines show the results obtained at 77 K with an electric field of 10 kV/cm. The upper curve ignores the e-e interaction, which is included in the lower curve. It is clear that the field is able to sustain high velocities up to a distance of

100 nm. At longer distances though, intervalley transfer starts to occur, as the fastest electrons scatter into the slower upper valleys, leading to a strong loss of average momentum.

The very fast randomization within the injected packet caused by the e-e collisions is evident in Fig.4, that shows the energy (a) and momentum (b) distribution functions, with (dashed curves) and without (dotted curves) inter-particle scattering. The distribution functions are evaluated when the center of the packet is at 200 nm from the point of injection. The distinct peaks in the energy curve with no e-e scattering are related to discrete energy losses via emission of polar optical phonons. The coulomb scattering is extremely effective in setting up a smooth distribution on a sub-picosecond time scale.

In conclusion, we have shown the importance of carrier-carrier interaction and plasma effects in GaAs. The performance of submicron devices, based upon overshoot or quasi ballistic effects, can be strongly affected by them.

Supported in part by the Office of Naval Research.

1. P. Lugli: "Electron-Electron Effects in Semiconductors", Ph.D. Dissertation, Colorado State University, March 1985.
2. D. Pines: Elementary Excitations in Solids (W.A. Benjamin, New York, 1963).
3. M. E. Kim, A. Das, and S. D. Senturia, Phys. Rev., 18, 6890 (1978).
4. P. Lugli and D. K. Ferry, IEEE Electron Dev. Lett., EDL-6, 25 (1985).
5. P. Lugli and D. K. Ferry, Appl. Phys. Lett., to be published.

Subpicosecond Raman Spectroscopy of Electron - LO Phonon Dynamics in GaAs

J.A. Kash and J.C. Tsang
IBM T.J. Watson Research Center, P.O. Box 218,
Yorktown Heights, NY 10598, USA
J.M. Hvam
Fysisk Institut, Odense Universitet, DK-5230 Odense M, Denmark

Highly excited ("hot") conduction band electrons in GaAs lose their energy primarily by the emission of zone center LO phonons[1]. This process is of interest because of the high-speed devices fabricated from GaAs. We have measured with sub-picosecond time resolution the spontaneous Raman spectrum of optically-pumped GaAs. When the optically-injected carrier density is less than 10^{17} cm^{-3}, we temporally resolve the growth of the optically-induced non-equilibrium LO phonon population and show that the electron phonon scattering time is about 165 femtoseconds. This is the first direct measure of the scattering rate. Previous work[2] at 77K and 3 psec time resolution has measured the lifetime of these phonons to be 7 psec. In addition to studying the dynamics of LO phonon emission, we have used sub-picosecond Raman scattering at higher injected carrier densities to study the dynamics of the screening of thermal LO phonons by a non-equilibrium electron plasma. We find that the screening of the phonon by the plasma is not efficient until the carriers have relaxed to the band edges, which takes about 3 psec.

In order to directly observe the electron - LO phonon dynamics, we have performed pump-probe Raman scattering experiments on GaAs using 600 femtosecond laser pulses. The photoexcited carriers and Raman spectra were excited by compressed[3] pulses from a synchronously-pumped Rh6G dye laser operating at 588 nm. The compressed pulses had an autocorrelation full width at half maximum (FWHM) of 900 femtoseconds, average power of about 20 milliwatts at a repetition rate of 80 MHz, and a spectral FWHM of 35 cm^{-1}. The samples, held at room temperature, were GaAs (100) where only the LO phonon is Raman active[4]. The backscattered light from the GaAs surface was analyzed by a triple monochromator and detected by a microchannel plate photomultiplier with a position sensitive resistive anode[5]. This multichannel detector allowed us to observe conveniently the Raman spectrum of GaAs even when the probe power level was less than 1 mW and the excitation linewidth was in excess of 30 cm^{-1}. The detector response time is greater than 5 ns. To perform sub-picosecond time-resolved measurements, we split each compressed laser pulse into a pump pulse and a weak (10-20%), orthogonally-polarized, delayed probe pulse which were focussed to a single spot on the sample. Because the polarizations of the pump and probe were perpendicular, the LO phonon Raman scattering excited by the pump and the probe beams were also orthogonal[2]. By placing a prism polarizer before the entrance slits, only Raman scattering from the probe was detected. The pump beam also produced broad band, unpolarized emission from

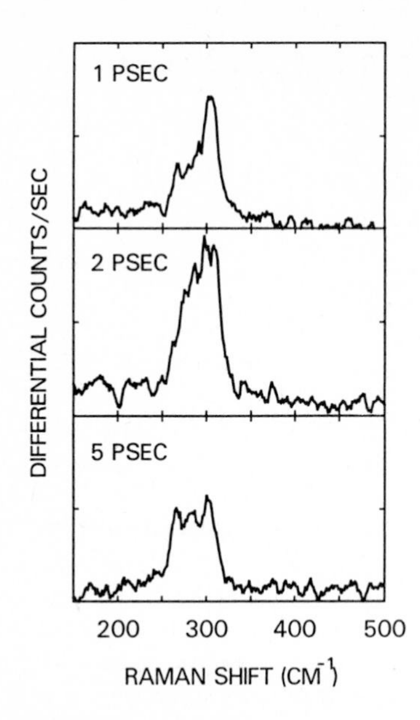

Figure 1. Room temperature time-resolved anti-Stokes Raman difference spectra at 1, 2, and 5 psec, obtained as explained in the text. The instantaneous carrier concentration induced by the pump laser was about 5×10^{16} cm^{-3}. The full-scale count rate is 0.125 counts/second for all spectra

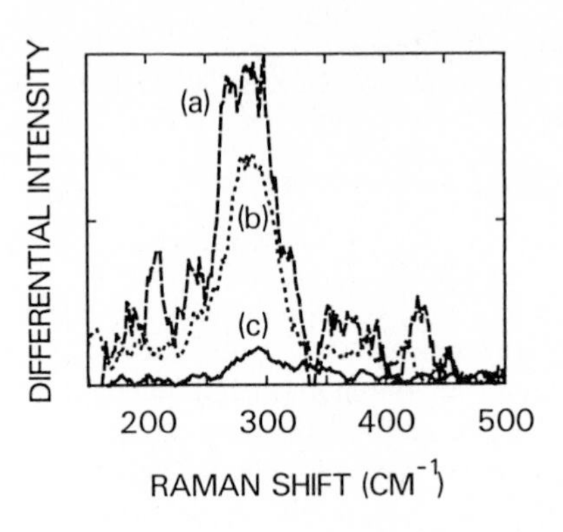

Figure 2. Anti-Stokes Raman difference spectra for probe delay of 3 psec. Instantaneous carrier concentrations are approximately (a) 8×10^{16} cm^{-3}, (b) 3×10^{17} cm^{-3}, and (c) 1×10^{18} cm^{-3}. The curves are scaled so that all should be the same height and shape if the non-equilibrium LO phonon population is linear with pump power and the Raman scattering cross-section is constant

our samples. We measured the Raman spectrum for a negative probe delay and subtracted this background spectrum from the spectrum measured for the delay time of interest. The resulting difference spectrum gives just the additional Raman signal of the probe that is induced by the pump. Anti-Stokes Raman difference spectra at 1, 2, and 5 psec are shown in Fig. 1. The LO phonon frequency of 290 cm^{-1} is at the center of the 35 cm^{-1} wide bands. The width of the Raman peak is due to the width of the excitation. To first order, the spectral weight of these peaks measure the non-equilibrium phonon population. The population is maximum at 2 psec, where the non-equilibrium population is about 25% of the thermal population, and then it decays in 4 psec. (We have also made low temperature measurements which show that the population rise is independent of temperature. The decay time, as expected however, lengthens as the temperature is decreased.) Given the photon energy of 2.11 eV used in these experiments, each optically excited electron creates 16 LO phonons in a cascade before reaching the band edge. (We ignore the ~20% absorption of the light hole band) Only LO phonons with a momentum of $8.4\text{x}10^5$ cm^{-1} are detected in our backscattering geometry. The Fröhlich matrix element which causes the phonon emission favors the emission of the smallest wavevector LO phonons[1]. Therefore, the largest contribution to the Raman active phonons we detect are emitted near the twelfth step in the cascade process. Thus our observed peak in the phonon population occurs at this time. Since the electron-phonon scattering rate is independent of electron energy[1], we deduce that the electron-phonon scattering time is about one-twelfth of our observed maximum, i.e. 165 fsec, consistent with earlier theoretical estimates[1].

Comparing Figs. 2(a) and 2(b), we see that for low pump power densities, the non-equilibrium LO phonon population increases linearly with pump power. However, as seen in Fig. 2(c), this population saturates for optically-injected carrier concentrations $\gtrsim 10^{18}/cm^3$. At these high power densities the spectra are very different, as shown in Fig. 3, which should be compared to Fig. 1. At the shortest delays, the spectrum is now dominated by a broad continuum which varies in a superlinear fashion with pump power. After 3 or 4 picoseconds, we observe a well-defined dip in the pump-induced spectrum at the LO phonon energy. The continuum which dominated the spectrum at shorter times is now much smaller. For times greater than 6 picoseconds, the continuum disappears and the difference spectrum contains just the dip. This negative-going structure, which remains essentially unchanged for delays from 10 psec to at least 30 psec, is a reduction in the strength of the allowed LO phonon scattering.

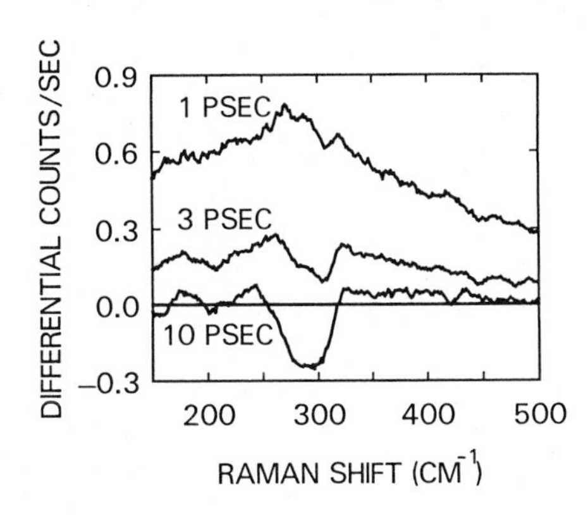

Figure 3. Anti-Stokes Raman difference spectra for probe delays of 1, 3, and 10 psec. Instantaneous carrier concentration for all curves is $\sim 3 \times 10^{18}$ cm^{-3}

The spectroscopic results in Fig. 3 can be correlated with the relaxation of the optically-induced carrier population. SHANK et al.[6] have shown that for times greater than 3 or 4 picoseconds, the optically-injected carriers are almost in thermal equilibrium with the lattice. For carrier concentrations $\gtrsim 10^{18}$ cm^{-3}, the plasmon due to these relaxed carriers couples with and screens the LO phonon[4]. Because of the inhomogeneous nature of our laser excitation, we do not expect to resolve the the coupled modes produced by this mixing. We can however, observe the loss of phonon oscillator strength at the unscreened phonon frequency. Thus the screening causes the negative difference spectrum observed at long times and high carrier concentrations. The screening will persist until the carriers diffuse into the crystal[7]. For delays less than 2 picoseconds, we do not find the dip. Thus, screening due to free carriers does not become fully effective until the carriers have relaxed to the band edges and the distribution reaches quasi-equilibrium. This suggests that the electron-phonon interaction in ballistic electron transport, where the carriers are far from equilibrium with the lattice, may be significantly different from the equilibrium case.

We thank W. Wang and T. Kuech for GaAs samples, J. Custer for technical assistance, and F. Stern, P. Price and Prof. E. Burstein for helpful discussions.

References

1. E. M. Conwell and M. O. Vassel, IEEE Trans. Electron. Devices ED-13, 22 (1966)
2. D. von der Linde, J. Kuhl and H. Klingenburg, Phys. Rev. Lett. 44, 1505 (1980)
3. H. Nakatsuka, D. Grischkowsky and A. C. Balant, Phys. Rev. Lett. 47, 910 (1981)
4. M. V. Klein, in Light Scattering in Solids, ed. by M. Cardona, (Springer-Verlag, Berlin-Heidelberg, 1975), p. 147
5. J. C. Tsang in Dynamics on Surfaces, ed. by B. Pullman, J. Jortner, A. Nitzan and B. Gerber (D. Reidel, Dordrecht, 1984), p. 379
6. C. V. Shank, R. L. Fork, R. F. Leheny and J. Shah, Phys. Rev. Lett. 42, 112 (1979)
7. C. L. Collins and P. Y. Yu, Solid State Commun. 51, 123 (1984)

Acoustic Phonon Generation in the Picosecond Dynamics of Dense Electron-Hole Plasmas in InGaAsP Films

Jay M. Wiesenfeld
AT&T Bell Laboratories, Crawford Hill Laboratory, Holmdel, NJ 07733, USA

In spite of the extensive use of semiconductor alloys of InGaAsP in lasers, LEDs, and photodetectors, relatively little work relating to the ultrafast dynamics of photoexcited carriers has been reported. It will become increasingly important to understand the dynamics of free carriers in these materials when they become a material for integrated optoelectronic devices. In the present contribution, the dynamics of dense electron-hole plasmas created by ultrashort-pulse optical excitation of InGaAsP films is examined by probing transient transmission and reflection. Both the transient transmission and reflection data show an oscillation superimposed on a monotonically changing response. Similar oscillations have been reported by THOMSEN, et al[1] in a study of the dynamics of films of a - As_2Te_3 and cis-polyacetylene photoexcited by ultrashort optical pulses. They have shown that the oscillations are due to coherent acoustic phonons generated in the relaxation of the hot, optically excited carriers. In the present work, the oscillatory behavior is reported and analyzed for films of InGaAsP and GaAs.

Semiconductor films are excited by 0.5 ps pulses at 0.625 μm from a cavity-dumped, passively modelocked cw ring dye laser. Pulse energies up to 2nJ at a rate of 200 kHz are used. Changes in transmission and reflection are monitored by the probe pulse, which is a time-delayed replica of the pump pulse, with less than 0.1 nJ energy. The noncollinear pump and probe beams are focused to a common spot which is either 10μm or 40μm diameter. All experiments are performed at 300 K. The samples are films of $In_{0.70}Ga_{0.30}As_{0.66}P_{0.34}$ (E_g = 0.96 eV) and $In_{0.58}Ga_{0.42}As_{0.93}P_{0.07}$ (E_g = 0.80 eV), grown by LPE on a [100] InP surface and GaAs, thicknesses from 0.2 μm to 1.0 μm. These are the same films used in semiconductor film lasers.[2] Excitation densities are 1.7 mJ/cm^2 or less, which implies initial carrier densities of up to 3 x 10^{20} cm^{-3}.

Transient transmission and reflection data are shown in Fig. 1 for a 0.2 ± 0.02 μm thick film of $In_{0.70}Ga_{0.30}As_{0.66}P_{0.34}$. The transmission curve shows an initial, rapid increase after excitation followed by a rapid decrease which recovers on a long (>600ps) time scale. The reflectivity increases rapidly following photoexcitation and subsequently recovers on a long (>600ps) time scale. The oscillations seen in transmission and reflection have opposite phase and a period of 120 ps. The oscillation period is sensitive only to the thickness of the film: for a 0.6 μm thick film of $In_{0.70}Ga_{0.30}As_{0.66}P_{0.34}$, the oscillation period is 360ps. The oscillation period is insensitive to intensity of the pump beam, as shown in Fig. 2.

Subsequent to ultrashort pulse excitation, the excess kinetic energy of the photoexcited plasma (2.0 eV - bandgap energy) is thermalized producing a temperature rise in the irradiated volume. Thermal expansion within this volume produce stress and strain waves, primarily along the direction normal to the plane of the film, i.e. the [100] direction. That wave which is a standing wave in the film becomes resonantly enhanced and produces a modulation of the bandgap of the film, and hence a modulation of transmission and reflection. A standing wave, rather than a pulse,

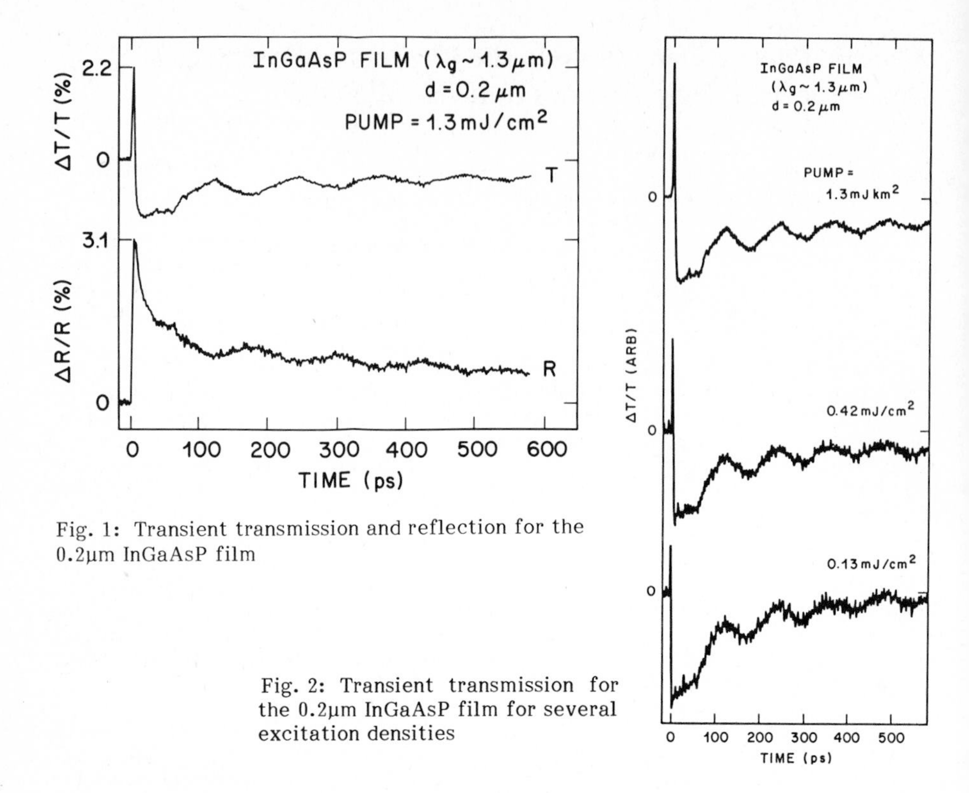

Fig. 1: Transient transmission and reflection for the 0.2μm InGaAsP film

Fig. 2: Transient transmission for the 0.2μm InGaAsP film for several excitation densities

is generated because the photoexcited plasma expands along the film thickness in a time much shorter than 100 ps. The wavelength of the standing wave is twice the film thickness, d, and therefore the velocity of sound for the LA phonon in the [100] direction is

$$v_S = 2d/T \tag{1}$$

where T is the period of oscillation. From the data of Figs. 1 and 2, $v_S = 3.3 \times 10^5$ cm/s. Measured valves of v_S [100] for both InGaAsP stoichiometries and GaAs are listed in Table I. The value for GaAs can be compared to a previously reported experimental value of 4.7×10^5 cm/s.[3] The values for InGaAsP have not been previously measured.

The fractional modulation of transmission due to the oscillations, ΔT/T, is proportional to the energy density E/A deposited in the sample by the pump.[1,4] According to the analysis presented elsewhere [4],

$$\frac{\Delta T}{T} = W \cdot \frac{d\alpha}{dE_g} \cdot \frac{\beta f(1-R)(1-T)}{\rho c} \cdot \frac{E}{A}, \tag{2}$$

where

$$W = 3K \cdot \frac{dE_g}{dp} \quad \text{or} \quad W = \frac{1+\nu}{1-\nu} \cdot \frac{dE_g}{d\eta} \quad . \tag{3}$$

In these equations, $(d\alpha/dE_g)$ is the variation of absorption coefficient with bandgap energy, β is the linear thermal expansion coefficient, f is the fraction of photon energy that exceeds the bandgap energy, R and T are power reflection and absorption coefficients, ρ is the sample mass density, c is the specific heat, K is the bulk modulus, and ν is Poisson's ratio. (dE_g/dp) and $(dE_g/d\eta)$ are the variation of bandgap energy with hydrostatic pressure and with isotropic (hydrostatic-like) strain, respectively. Figure 3 shows a plot of $\Delta T/T$ vs E/A for the 0.2 μm $In_{0.70}Ga_{0.30}As_{0.66}P_{0.34}$ sample. Using values for the needed material constants in Eqs. (2) and (3), [5,6], values are determined for (dE_g/dp) and $(dE_g/d\eta)$. These values are listed in Table I along with values measured for the GaAs film. The value for (dE_g/dp) for GaAs agrees well with the previously measured value of 11.8 meV/kbar.[7] (dE_g/dp) for $In_{0.70}Ga_{0.30}As_{0.66}P_{0.34}$ is determined here for the first time. $(dE_g/d\eta)$ is equal to the difference between hydrostatic-like acoustic deformation potentials of electrons and holes, and is difficult to compare to anything in the literature.

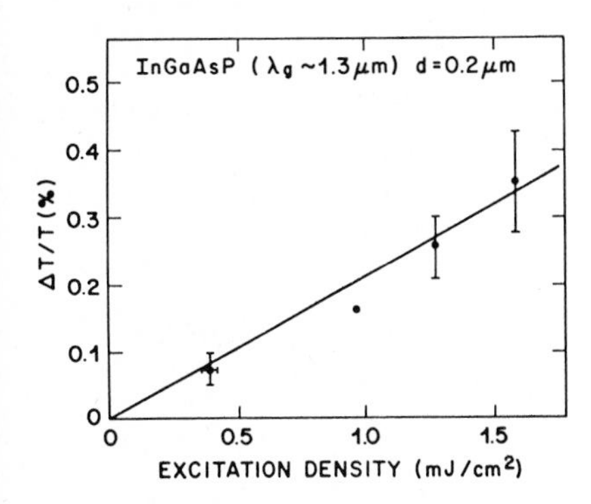

Fig. 3: Magnitude of oscillations vs excitation density

TABLE I: Material properties determined in this work

	$In_{0.70}Ga_{0.30}As_{0.66}P_{0.34}$	GaAs	$In_{0.58}Ga_{0.42}As_{0.93}P_{0.07}$
$v_{LA}[100]$ (10^5cm/sec)	3.3±.4	5.5±1.2	2.8±0.4
dE_g/dp (meV/kbar)	8.9±1.3	9.4±1.9	--
$dE_g/d\eta$ (eV)	9.2±1.4	11.±3	--

In summary, oscillatory behavior has been observed in the transient transmission and reflection response of films of InGaAsP and GaAs excited by subpicosecond optical pulses. Analysis of the oscillations has led to determination of acoustic properties of these materials as listed in Table I.

1. C. Thomsen, J. Strait, Z. Vardeny, H. J. Maris, J. Tauc, and J. J. Hauser, Phys. Rev. Lett., 53, 989 (1984).
2. J. Stone, J. M. Wiesenfeld, A. G. Dentai, T. C. Damen, M. A. Duguay, T. Y. Chang, and E. A. Caridi, Opt. Lett., 6, 534 (1981).
3. J. S. Blakemore, J. Appl. Phys., 53, R-123, (1982).
4. J. M. Wiesenfeld, submitted to Appl. Phys. Lett.
5. S. Adachi, J. Appl. Phys., 53, 8775 (1982).
6. H. Burkhard, H. W. Dinges, and E. Kuphal, J. Appl. Phys., 53, 655 (1982).
7. B. Welber, M. Cardona, C. K. Kim, and S. Rodriguez, Phys. Rev., B12, 5729 (1975).

Picosecond Time-Resolved Photoemission Study of the InP (110) Surface

J. Bokor, R. Haight, J. Stark, and R.H. Storz
AT&T Bell Laboratories, Holmdel, NJ 07733, USA
R.R. Freeman and P.H. Bucksbaum
AT&T Bell Laboratories, Murray Hill, NJ 07974, USA

Angle-resolved ultraviolet photoemission spectroscopy (ARUPS) is a powerful and well established tool for the study of bulk and surface electron states in solids. We report the first demonstration of picosecond time-resolved ARUPS. The technique was used to study the dynamics of electrons photoexcited by a 50 picosecond visible laser pulse on a cleaved InP (110) surface.

On the (110) surface of InP, an unoccupied surface state lies at[1] an energy slightly above the conduction band minimum (CBM). During excitation of carriers into the bulk conduction band, this surface state becomes populated and rapidly comes into equilibrium with the bulk. Using time-resolved ARUPS, we have observed the transient population of this surface state, located its energy minimum at $\bar{\Gamma}$, the center of the surface Brillouin zone, and measured the surface band dispersion along two surface symmetry directions.

Clean InP (110) surfaces, produced by cleavage in ultra-high vacuum, were excited by 50 psec pulses[2] of 532 nm laser radiation at a fluence level of 0.5 mJ/cm^2. The excited surface was probed by measuring the energy and angular distributions of electrons photoemitted by a 50 psec pulse of 118 nm (10.5 eV) laser harmonic radiation. A time-of-flight electron spectrometer with energy resolution of 100 meV, and angular resolution of $\pm 2.5^{\circ}$ was used.

Figure 1 shows the energy spectra for electrons emitted along the surface normal direction obtained in this experiment. The zero of energy is chosen as the valence band maximum (VBM). Figure 1(a) displays a spectrum for the unexcited material where the probe pulse arrives at the sample before the pump pulse (t=-133 psec). When the pump and probe pulses overlap in time as in Fig. 1(b), a new feature appears centered at +1.47 eV. The insets in Figs. 1(b and c) show a magnified view of this feature. If the probe is delayed relative to the pump by 266 psec the spectrum in Fig. 1(c) is observed. Note that the 1.47 eV peak is significantly reduced both in intensity and width. Finally, Fig. 1(d) displays a spectrum obtained with coincident pump and probe pulses following submonolayer coverage of the surface with chemisorbed hydrogen. The -0.2 eV, -1 eV, and +1.47 eV peaks have essentially disappeared. Such sensitivity to hydrogenation is a generally accepted signature of surface states on semiconductor and metal surfaces. We assign the shoulder at -0.2 eV and the broad peak

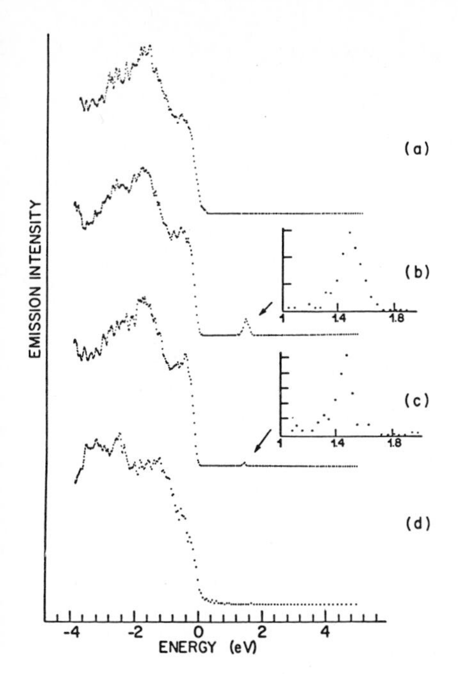

Fig. 1: Pump-probe photoelectron spectra from InP (110) for a) t = -133 psec; b) t = 0; c) t = 266 psec; and d) t = 0 where the surface has been hydrogenated. The insets in b) and c) have the vertical scale magnified by a factor of 10 and 60, respectively.

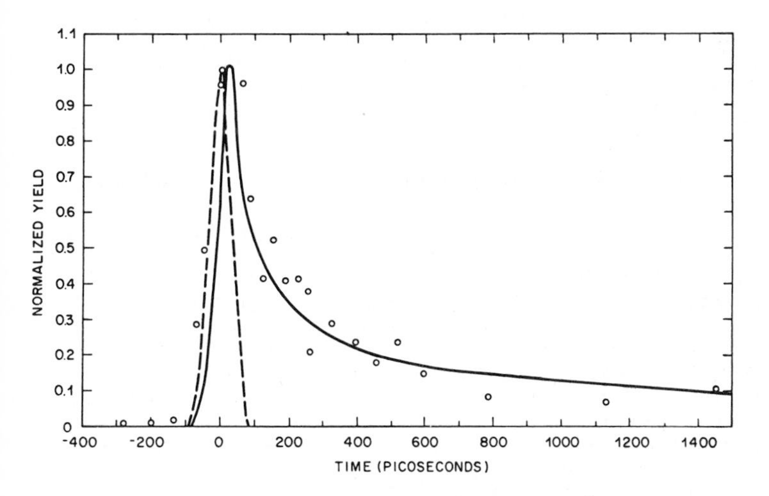

Fig. 2: Integrated transient surface state signal intensity as a function of time delay. The dashed curve represents the system time resolution function. The solid curve is a fit to the data as described in the text.

centered at - 1 eV to occupied surface states. The 1.47 eV peak is attributed to a normally unoccupied surface state.

The time dependence of the 1.47 eV signal is displayed in Fig. 2. The integrated intensity of the 1.47 eV signal is plotted as a function of relative time delay between the pump and probe pulses. A nonexponential decay is observed with the intensity dropping to half maximum within 100 psec. Since the surface state lies so close to the CBM (the band gap in InP is 1.35 eV), a quasi-equilibrium may be established between the surface and the bulk bands. Then the data shown in Fig. 3 represents the time dependence of the bulk carrier density at the surface. The expected evolution of the bulk plasma was calculated using a one-dimensional model including the temporal and spatial dependence of the carrier photoexcitation rate, radiative and Auger recombination, and ambipolar carrier diffusion away from the surface. The results of this calculation are shown as the solid line in Fig. 2. The fact that the surface state signal follows the expected behavior of the bulk carrier density in the near-surface region is taken as evidence that the surface state population is strongly coupled to the bulk plasma.

The insets in Fig. 1 show that the energy width of the surface state peak varies with time. The full width at half maximum (FWHM) drops from its maximum of 154 meV at t=0 to the instrumental limit of 105 meV by 260 psec. Broadening is only observed on the high energy side of the peak. We attribute this result to bandfilling. With this interpretation, and assuming equilibrium between the surface and bulk electrons, we can estimate the peak bulk electron density as 1.8×10^{19} cm^{-3}.

Information about the electron momentum distribution in the 1.47 eV surface state was obtained by measuring the peak intensity as a function of electron exit angle. This data combined with the peak width data has been used to obtain the energy dispersion of the surface state (effective mass) near $\bar{\Gamma}$ along the two surface symmetry directions. We obtain $0.20 \leq m^*/m_e \leq 0.24$ along $\bar{\Gamma}\bar{X}$, and $0.10 \leq m^*/m_e \leq 0.15$ along $\bar{\Gamma}\bar{X}'$, where m_e is the free electron mass. These values differ significantly from the nearly isotropic bulk conduction band effective mass of $0.07\ m_e$, confirming our identification of this peak as a surface state.

In summary, we have used picosecond time-resolved ARUPS to study the energy and momentum of excited electrons on the InP (110) surface. A normally unoccupied surface state was directly observed to be populated via coupling to bulk photoexcited electrons. Measurements of the angular distribution of photoemitted electrons and the energy width of the surface state peak show that the surface band minimum must lie at $\bar{\Gamma}$, and allow a determination of the energy dispersion (effective mass) along two surface symmetry directions.

Reference

1. J. van Laar, A. Huijser, and T. L. van Rooy, J. Vac. Sci. Technol. 14, 894 (1977).

Monte Carlo Investigation of Hot Carriers Generated by Subpicosecond Laser Pulses in Schottky Barrier Diodes

M.A. Osman, U. Ravaioli and D.K. Ferry
Center for Solid State Electronucs Research, Arizona State University, Tempe, AZ 85287, USA

The self-consistent Ensemble Monte Carlo (E.M.C.) method is used to study the dynamics of carriers generated by subpicosecond laser pulses in a Silicon n+/n/metal submicron Schottky diode. The E.M.C. model for the Schottky diode is the same presented in [1], with the inclusion of hole dynamics. In this model the Poisson's equation is solved using an accurate collocation method [2], which gives a very precise solution for the electric field E(0) on the ohmic contact boundary. This allows us to determine the current injected through the ohmic contact using the relation $J=n^{+}\mu_{o}E(0)$, where μ_{o} is the low field mobility in an n^{+} contact region. This is justified if the width of the region is at least a few Debye lengths, so that we consider the carriers at the ohmic contact boundary to be in equilibrium with the lattice. In this way we can model the depletion of carriers and fully account for the non-charge-neutral behavior of the Schottky barrier diode, keeping track of the carriers exiting the device and allocating on the metal side the ones which are not reinjected. Tunnelling current from the semiconductor to the metal is included by a WKB approach as in [1], with an improved formulation of the tunnelling probability, which takes into account nonparabolicity and the effect of finite bandgap [3].

The photoexcitation of carriers is made by simulating a laser pulse with subpicosecond duration. For Silicon the useful wavelengths which we can investigate span between 0.5 and 1.2 μm [4]. The excess energy of the photoexcited electrons has been evaluated including nonparabolicity of the conduction band, and the dependence on whether the generated hole is light or heavy. Parabolic, spherical valence bands are considered to compute the excess energy of the photogenerated holes.

In our approach we neglect the energy spread due to the phonons, so that the transitions between valence and conduction band are treated as direct. This is justified for our purposes since the excess energy of the generated carriers is much greater than the phonon energy. Moreover, we have to consider that soon after generation, the carriers relax to the bottom of the conduction band by emitting successive optical phonons, and during this particular process the equilibrium state of the crystal lattice is basically unaltered.

The transport of holes is fully taken into account in the E.M.C. simulation using the scattering rates given in [5]. The boundary condition for holes at the ohmic contact depends on the rate of recombination which can take place at the interface [6]. The two limiting situations of infinite and zero recombination rates are taken into account using absorbing or reflecting boundary conditions, respectively. Both cases are considered in our simulations. Holes that reach the Schottky barrier interface are absorbed instantly.

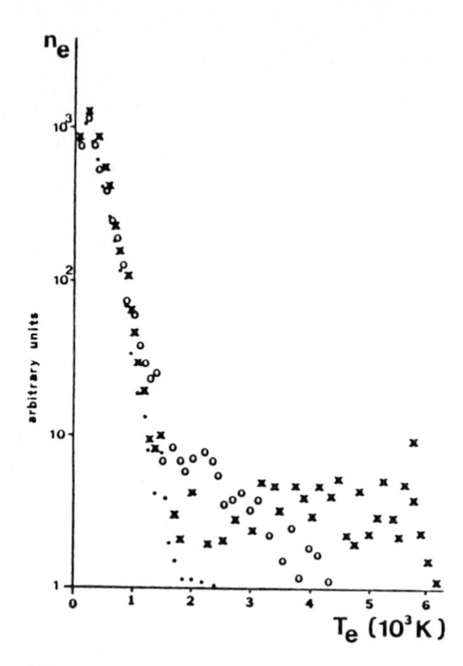

Fig. 1 - Electron Distribution as a Function of Electron Temperature: (.) t=0; (x) t=.25ps; (o) t=.5ps.

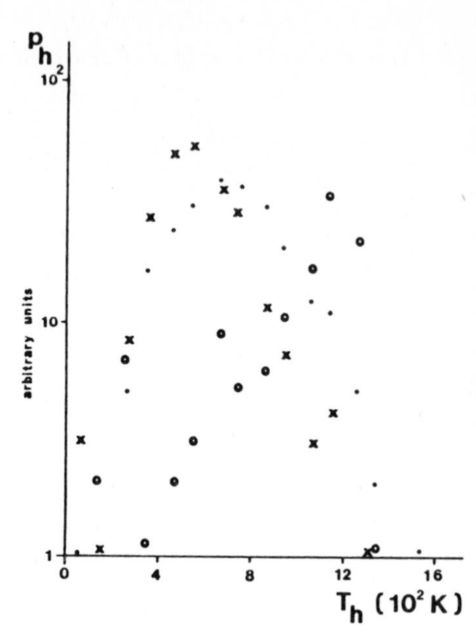

Fig. 2 - Hole Distribution as a Function of Hole Temperature: (o) t=.25ps; (.) t=.5ps; (x)t=2ps.

Using the E.M.C. simulation we do not have to make any simplified assumption on the carrier transport [4], since all scattering and nonlinear transport effects are naturally included in the E.M.C. model. Carrier-carrier interaction, energy-gap narrowing and state filling effects [7], which are effective at high excitation and/or doping levels are neglected in this study.

Energy-dependent electron-hole recombination can also be considered. However, since we simulate a submicron diode, the recombination rate across the device turns out to be very small, for the doping levels in our model, and is therefore neglected in the present approach.

The electron ensemble we simulate consists of background electrons in equilibrium with the lattice and photoexcited electrons. The electron-hole pairs are generated continuously for 0.5 picoseconds by a 2 eV gaussian pulse. The generated electrons relax quickly to the bottom of the conduction band by emitting zero- and first-order LO optical phonons, as shown in Fig. 1. On the other hand, holes relax by emitting zero-order LO optical phonons only, according to our model, which prevents the holes from reaching the equilibrium with the lattice, as shown in Fig. 2. Inclusion of zero- and higher-order inelastic acoustic phonon scattering will allow holes to dissipate more energy and reach the equilibrium with the lattice.

---

* Supported by the Army Research Office and the Office of Naval Research

References

1. P. Lugli, U. Ravaioli and D.K. Ferry: "Monte Carlo simulation of contacts in submicron devices", AIP Conference Procedings n. 122, The Physics of VLSI, p. 162, Palo Alto (1984).
2. U. Ravaioli, P. Lugli, M.A. Osman and D.K. Ferry: "Advantages of collocation methods over finite differences in one-dimensional simulation of submicron devices", II Conference on Numerical Simulation of VLSI Devices, Boston (1984), to be published.
3. H. Flietner: Phys. Stat. Sol. B 54, 201 (1972).
4. R.J. Phelan, D.R. Larson, N.V. Frederick and D.L.Franzen: Proceedings of SPIE, Picosecond Optoelectronics, 439, 207 (1983).
5. M. Costato and L. Reggiani: Phys. Stat. Sol. B 58, 461 (1973).
6. D.L. Scharfetter: Solid-State Electron. 8, 299 (1985).
7. D.K. Ferry: Phys. Rev. B 18, 7033 (1978).

Part III

Quantum Structures and Applications

Properties of GaAlAs/GaAs Quantum Well Heterostructures Grown by Metalorganic Chemical Vapor Deposition

R.D. Burnham, W. Streifer, T.L. Paoli, R.L. Thornton, and D.L. Smith
Xerox Palo Alto Research Center, 3333 Coyote Hill Road,
Palo Alto, CA 94304, USA

The existence of the heterojunction and the recent development of methods for growing a variety of epitaxial ultrathin III–V semiconductor layers (wider-gap or narrower-gap, doped or undoped, abrupt or graded) enables the fabrication of sophisticated quantum-well heterostructures. In the past few years, one of these techniques, metalorganic chemical vapor deposition (MO–CVD), has become increasingly popular as a consequence of its versatility and economy. The chemistry of MO–CVD growth of III–V compounds was pioneered by Manasevit as early as 1968 [1]; however, it was DUPUIS and DAPKUS [2] in 1977 who demonstrated that high quality GaAlAs/GaAs heterostructure lasers can be grown by MO–CVD.

MO–CVD growth is an extremely complex combination of chemical reactions and mass transfer kinetics. The basic mechanism of GaAs growth by combination of single trimethyl gallium (TMGa) and arsine molecules to form GaAs plus three methane molecules [3] is almost certainly made more complex by the importance of intermediate reaction steps.

In addition to understanding the basic physical processes, the impact of materials quality and growth parameters on device quality continues to be a subject for careful investigation [4,5]. The availability of high purity AsH_3 has been identified as particularly important. Experiments [4] indicate that variations in electroluminescent intensity by four orders of magnitude can be related to variability in the purity of the AsH_3 source, with the variation being correlated with cylinder preparation as opposed to AsH_3 source preparation [4]. One of the main causes of this variability is the H_20 and O_2 adsorbed and/or possibly chemisorbed on the inner wall of an AsH_3 cylinder due to variable surface preparation [6,7]. Purity of AsH_3 sources is vital, since desorption of the volatile As from the crystal itself requires that the growth of these compounds occur in an excess of AsH_3 [5]. The critical molar ratio $(AsH_3)/(TMGa + TMAl)$ at which *p*- to *n*-type conversion occurs decreases with increasing growth temperature but increases with an increase of Al fraction [4]. Also $n_D - n_A$ varies by about four orders of magnitude (10^{13} to 10^{17} cm^{-3}) as the growth temperature is increased from 660°C to 800°C [4]. To make matters even more confusing, about an order of magnitude increase in photoluminescent intensity is achieved if $Ga_{.85}Al_{.15}As$ is grown at 850°C instead of 700°C [8]. It is therefore quite remarkable that significant advances in devices have been made in spite of an inadequate understanding of the MO–CVD growth process.

The impact of MO–CVD on device research has perhaps been greatest in the growth of structures utilizing ultrathin layers with thicknesses less than ~500 Å, where the quantum size effect is responsible for producing unique properties [9,10]. There are two techniques which allow for the growth of such ultrathin layers. One technique uses interrupted growth sequences to achieve square wells with interfaces that are smooth to less than ±1 monolayer, even though growth rates of 20 Å/sec are used [11]. The other technique uses slow growth rates on the order of 3.5 Å/sec, high flow velocities and an optimized reactor design to eliminate the need for the interruption of growth [12]. To date, MO–CVD has produced quantum well heterostructure (QWH) lasers with both the lowest threshold current densities and highest output powers measured for diode lasers in the visible and near infrared regions. QWH lasers exhibit long life even at cw output powers of 100 mW [13]. The wavelength tuning range in an external cavity is more than three times that of conventional double heterostructure lasers [10]. By thermally annealing a crystal containing a QWH, it is possible to change the shape of a quantum well from an initially square well of GaAs to a rounded shallower AlGaAs well, thus shifting the confined particle electrons and hole states to higher energy. This is a convenient method for adjusting the wavelength of a QWH laser without significantly increasing the lasing threshold [10]. Perhaps the most exciting capability of QWH's is the variation in energy gap which can be obtained either laterally or vertically via impurity-induced disordering (*n* or *p*) [10,14]. This process has produced a simple form of buried heterostructure laser [10,15] and suggests how QWH's and impurity-induced disordering can serve as the basis for fabricating new forms of integrated electrical and optoelectronic circuits.

To date, MO–CVD has not produced high-electron mobility transistor (HEMT) structures that are of the same quality as those grown by moleculear beam epitaxy (MBE) [16]. However, this is not thought to be a fundamental limitation of the MO–CVD process. Progress has been made on inelastic and resonant tunneling in AlAs/GaAs heterostructures [17]. These structures might be extremely promising as a high speed tunneling transistor [18].

Considering that only about five years have elapsed in which extensive research has been performed on MO–CVD, the achievements thus far are remarkable, and the future potential is excellent.

References

1. H. M. Manasevit: Appl. Phys. Lett. 12, 146 (1968).
2. R. D. Dupuis and P. D. Dapkus: Appl. Phys. Lett. 31, 466 (1977).
3. H. M Manasevit and W. I. Simpson: J. Electrochem. Soc. 118, 647 (1971).
4. T. Nakanisi: J. Crystal Growth 68, 282 (1984).
5. P. D. Dapkus, H. M. Manasevit, K. L. Hess, T. S. Low and G. E. Stillman: J. Crystal Growth 55, 10 (1981).
6. E. E. Wagner, G. Hom and G. B. Stringfellow: J. Electron. Mater. 10, 239 (1981).
7. E. J. Thrush and J. E. A. Whiteaway: Inst. Phys. Conf. Ser. 56, 337 (1981).
8. J. P. André, M. Boulou and A. Micrea-Roussel: J. Crystal Growth 55, 192 (1981).

9. N. Holonyak, Jr., R. M. Kolbas, R. D. Dupuis and P. D. Dapkus: IEEE J. Quantum Electron. QE-16, 170 (1980).
10. R. D. Burnham, W. Streifer, T. L. Paoli and N. Holonyak, Jr.: J. Crystal Growth 68, 370 (1984).
11. R. D. Dupuis, R. C. Miller, and P. M. Petroff: J. Crystal Growth 68, 398 (1984).
12. M. R. Leys, C. van Opdorp, M. P. A. Viegers, and H. J. Talen van der Mheen: J. Crystal Growth 68, 431 (1984).
13. G. L. Harnagel, T. L. Paoli, R. L. Thornton, R. D. Burnham, and D. L. Smith: to be published in Appl. Phys. Lett.
14. K. Meehan, N. Holonyak,Jr., J. M. Brown, M. A. Nixon, P. Gavrilovic and R. D. Burnham: Appl. Phys Lett. 45, 549 (1984).
15 K. Meehan, P. Gavrilovic, N. Holonyak, Jr., R. D. Burnham and R. L. Thornton: to be published in Appl. Phys. Lett.
16. J. P. André, A. Brière, M. Rocchi, and M. Riet, J. Crystal Growth 68, 445 (1984).
17. R. T. Collins, A. R. Bonnefoi, J. Lambe, T. C. McGill and R. D. Burnham: Proceedings of the International Conference on Superlattices, Champaign-Urbana, Illinois, 1984.
18. N. Yokoyame, K. Imamura, T. Ohshima, H. Nishi, S. Muto, K. Kondo, and S. Hiyamizu: Jpn. J. Appl. Phys. 23, L311 (1984).

Molecular Beam Epitaxy Materials for High-Speed Digital Heterostructure Devices

D.L. Miller
Rockwell International Corporation, Microelectronics Research and Development Center, 1049 Camino Dos Rios, Thousand Oaks, CA 91360, USA

1 INTRODUCTION

Molecular beam epitaxy (MBE) has made many important contributions to research and development of III-V compound semiconductor devices. This is particulary true for devices using the lattice-matched GaAs/AlGaAs alloys. This has occurred despite the relatively high cost of MBE apparatus and the reputation of MBE as an esoteric technology.

The successes which have been achieved by the many people using MBE for highspeed device research and development stem in part from the suitability of MBE to growth of the thin layers and abrupt interfaces,which are the heart of highspeed heterostructure devices. This suitability arises from the simplicity and controlability of the process.

The control inherent in the MBE growth process arises from the vacuum environment. Fluxes of gallium, aluminum, arsenic, and dopants are created with resistivity heated ovens, and the fluxes are determined by the temperature of the ovens. In the vacuum of the deposition chamber, simple shutters may be used to interrupt the beams. Inputs to the apparatus elicit a rapid response in the epitaxy growth. This is in contrast to LPE or vapor phase techniques, where the times required to transport material to the substrate through the liquid or gaseous medium can be significant and often can complicate the matter of making abrupt interfaces. Furthermore, the relatively low growth temperatures (550°C -720°C for GaAs/AlGaAs) serve to preserve the abruptness of interfaces by minimizing interdiffusion during growth.

There are also several deficiencies in the MBE technology which detract from its appeal, particularly for application to circuit production. The growth rate is relatively low, so that throughput is low. Reproducibility from wafer-to-wafer is not yet good enough for production. AlGaAs quality must be improved, and MBE appears to be more susceptible to surface defects than other technologies. These topics are discussed in more detail in Sect. 3.

2 MBE Materials in HEMT and HBT Research and Development

Two devices, the HEMT and HBT, are particulary relevant to the illustration of MBE materials for highspeed digital electronics. All of the recent impressive highspeed digital circuit performance demonstrations using HEMT and HBT devices have been accomplished with MBE material.

The HBT is a good example of the use of MBE as a versatile research tool. HBTs fabricated by MBE have demonstrated propagation delays (in ring oscillators) of 29 ps [1], and f_τ (in discrete devices) of 40 GHz [2]. Projections of performance for optimized devices result in a propagation delay of 10 ps, with f_τ approaching 100 GHz. The structure of an HBT designed with the emitter "up" (toward the surface of the epitaxial material) for high-speed ECL circuit applications is shown in Fig. 1. These devices typically contain at least five layers with doping ranging from the low $10^{16}/cm^3$ range for the n-type collector to the upper $10^{18}/cm^3$ range for the p-type base and n-type subcollector and contact cap layers. In MBE, the n-type doping is usually accomplished with a single oven containing silicon. The temperature is raised or lowered to vary the impinging dopant flux; transitions between different doping levels can usually be accomplished in about 30 s, corresponding to the growth of about 100Å of GaAs. If more rapid transistors are desired, a second oven containing silicon may be added.

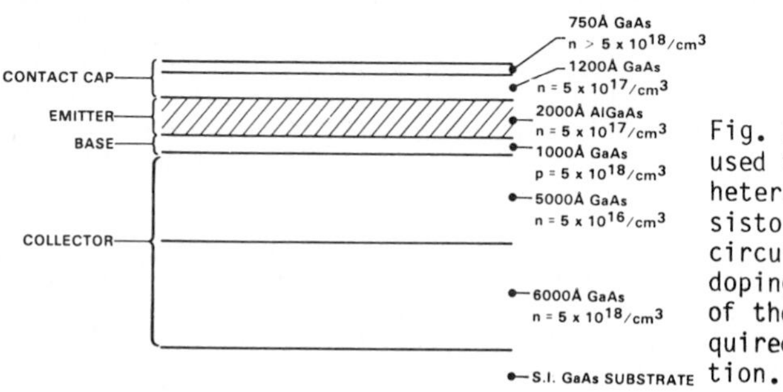

Fig. 1 An epitaxial structure used to fabricate "emitter up" heterojunction bipolar transistors for highspeed ECL circuits. A wide range of doping and precise placement of the heterojunction are required for device optimization.

Dopant placement is just as important as the ability to control doping over a wide range of carrier densities. It can be seen that even a slight displacement of the p-type dopant from the GaAs base into the AlGaAs emitter can create an AlGaAs homojunction instead of the desired AlGaAs-GaAs heterojunction, with an accompanying loss of injection efficiency. The relatively low growth temperature of MBE is helpful in preventing diffusion of dopants during growth, but is not entirely sufficient to guarantee that junctions will remain where they belong. Fortunately, MBE can also be used as a tool to investigate dopant diffusion. The precision of MBE in growing thin layers allows special structures to be grown with sufficient confidence in the initial dopant placement that deviations from the nominal distribution may be attributed to diffusion. Figure 2 shows the Be distribution at an AlGaAs/GaAs heterointerface, illustrating the redistribution which can occur during growth. Investigations of Be diffusion as a function of substrate temperature, As flux, alloy composition, and background doping, has provided guidelines for optimizing HBT structures [3,4].

HBTs also provide the opportunity to improve device performance through "bandgap engineering" beyond the incorporation of a wide gap emitter. An example of this is the creation of a graded bandgap in the HBT base region. AlGaAs is grown with increasing Al content toward the emitter end of the base in order to provide a built-in field to assist the transport of injected electrons across the base region. This is done by ramping the temperature of the Al oven upward during growth of the base in such a way that the desired composition profile is obtained. Computerized control of oven temperature allows this to be done reproducibly. The resulting graded-bandgap HBTs should allow the use of thicker bases (and thus, reduced base resistance) without unduly increasing the base transit time for minority carriers.

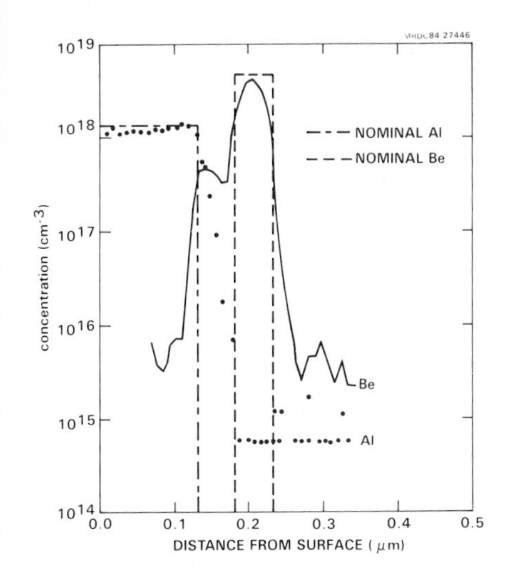

Fig. 2 Be distribution, determined by secondary ion mass spectrometry, near a GaAs/AlGaAs heterointerface grown by using MBE at 610°C. The Be shutter was opened during the growth of 500Å of GaAs, located 500Å from the onset of AlGaAs growth.

The HEMT device structure is an excellent match to the capabilities of MBE. In Fig. 3, a double-heterostructure HEMT is illustrated [5]. All HEMT structures rely on the precise placement of donors in AlGaAs a small distance from a high quality AlGaAs-GaAs interface. For high transconductance devices, the spacing between the interface and the nearest donors is about 20Å. This corresponds to a growth time of about 6 s for MBE. The dopant beam is interrupted by the shutter in a small fraction of a second, and there is neglible bulk diffusion of silicon donors under typical MBE growth conditions. Therefore, it is a simple task to place the donors in the desired position.

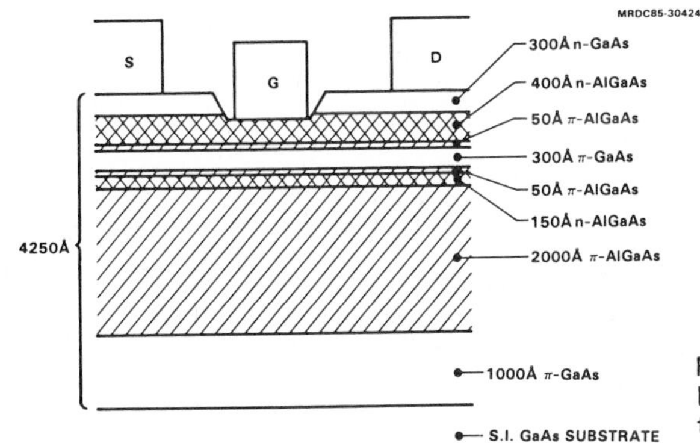

Fig. 3 Double heterostructure high electron mobility transistor epitaxial structure.

To improve current drive capabilities and transconductance, it is desirable to provide as many electrons as possible in the two-dimensional electron gas (2DEG) conducting channel. The number of carriers which can be created in a single 2DEG at the interface between GaAs and AlGaAs is limited by the maximum concentration of Si donors in AlGaAs and by the quantum well density of states to about $1 \times 10^{12}/cm^2$. In order to provide more free carriers, multiple 2DEG structures, such as the double heterostructure HEMT illustrated in Fig. 3, may be grown. The growth by MBE of such a structure containing a so-called "inverted" 2DEG channel is more difficult than "normal" HEMT. The reasons for this are not completely understood; it has been speculated that impurity outdiffusion from the AlGaAs, interfacing roughening, or dopant "surface riding" could be responsible for lowered mobilities in inverted HEMT. Recently it has been found that growth at lower temperatures than usual ($\sim$ 530°C) can improve inverted structures sufficiently for them to be useful in high performance devices [6].

The present control, reproducibility and uniformity of MBE material is such that fully functional 1 K HEMT SRAM's have been reported [7], 4 K HEMT SRAMs with > 90% bit yield have been operated [8], and 16 K HEMT SRAMs are being designed [9]. For the 4 K HEMT SRAM, the threshold voltage standard deviation across a 2" wafer was 20 mv for depletion mode devices and 12 mv for enhancement mode devices.

3 Areas Requiring Further Development

Despite the impressive highspeed device results which have been obtained using MBE, there are still several areas of this epitaxial technology requiring serious attention. This is particularly true as heterostructure devices move toward production which requires much higher throughput and better reproducibility of epitaxial layer characteristics than has been achieved so far in most research and development laboratories. Probably the four most important areas for improvement are the reduction of oval defects, improvement of AlGaAs quality, improvement of run-to-run reproducibility, and the design of apparatus capable of high throughput.

Oval defects are a surface defect characteristic of MBE material. They are not observed in LPE, VPE, or MOCVD material. Figure 4 shows an oval defect which had propagated through approximately 20 μm of GaAs, which had been doped and then stained to allow delineation of the growth surface micron-by-micron. Using techniques such as filling the Ga oven to the top, using solid, high purity Ga to load the Ga oven, growing slowly, and baking the growth chamber thoroughly, many laboratories have been able to reduce oval defect densities to below 500/cm^2 per micron of GaAs grown. The lowest values which have been achieved under practical growth conditions appear to be in the range of 200-300/cm^2-μm.

It is important to obtain lower oval defect densities because the threshold voltage and pinchoff of HEMT devices is affected when the gate of the device are placed over a defect [10]. It is strongly suspected that oval defects will affect the performance of HBTs also, although no data is available yet. Oval defect densities in the range of 10/cm^2 or below probably will be required for a reasonable yield of 16 K HEMT RAMs.

The difficulty in reducing oval defects stems from a lack of a microscopic model for their formation. This in turn is complicated by data which often appears conflicting. The measures which reduce oval defect densities in one laboratory may have little or no effect in another. This is probably because there are multiple origins for these defects, which vary from laboratory to laboratory, as well as several types of defects with similar appearance. The progress has been slow in this area.

The quality of the AlGaAs material used in heterostructure devices is also in need of improvement. Deep level densities can approach the free carrier density leading to short minority carrier lifetime, excess recom-

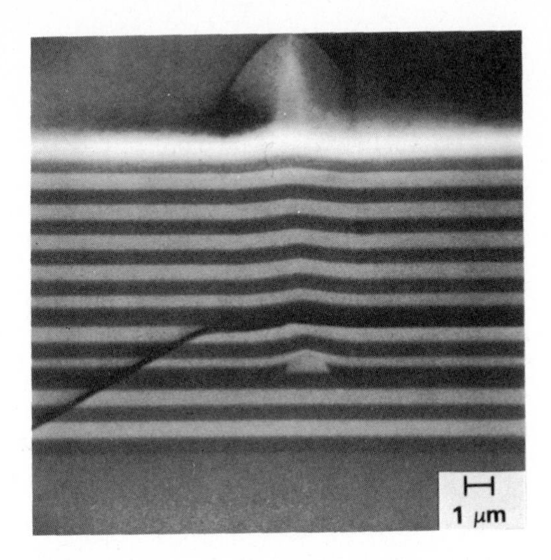

Fig. 4 Cleaved and stained multilayer structure comprising alternate 1 μm layers of n- and p-type GaAs. The nucleation and propagation of an oval defect can be seen in cross-section. The sample was tilted at 45°; the surface at the top of the photograph contains half of the cleaved-through oval defect.

bination at heterojunctions, and persistent photoconductivity effects. In HEMT devices, trapping of photogenerated or injected carriers in the AlGaAs layer can lead to threshold voltage shifts and catastrophic current collapse. This is especially true at 77 K, where effects can persist for a very long time.

It has been recognized for some time that AlGaAs MBE material may be improved by growing at relatively elevated temperatures (680°C-740°C) and by the use of As_2 beams instead of As_4. There are practical difficulties to this approach, such as increased dopant diffusion during growth, and Ga reevaporation from the epitaxial layer surface.

One possible solution is to construct an artificial alloy of AlGaAs by making a superlattice of very thin alternating layers of undoped AlAs and n-type GaAs [11]. If the layers are sufficiently thin, the wavefunctions of donors in the GaAs wells extend through the AlAs barriers and overlap, forming conduction bands. If the donor atoms are placed only in the GaAs layers, it appears that the donor-Al-Ga complexes which may form the deep levels in AlGaAs are largely avoided. Persistent photoconductivity effects are greatly reduced and doping efficiency increased. HEMT devices made using such a doped superlattice look promising [12].

In a production environment, reproducibility is essential, whereas in research it is an unexpected luxury. For device research, the usual day-to-day reproducibility of MBE (typically ± 5% to 10% for growth rate and doping density) allows one to investigate device physics or establish performance benchmarks with relative ease. However, when complex high performance circuits are being produced, tolerances tighten as it becomes necessary for devices to operate in circuits with small noise margins. Thus, tolerances of ± 2% may be necessary for circuit production.

Reproducibility of growth rates, doping densities, and alloy composition are primarily affected by the arrival rates at the substrate of Ga, Al, and the dopant atoms. Alloy composition can be affected at higher substrate temperatures by Ga loss, which requries good control of substrate temperature and As flux. Similarly, dopant incorporation can be affected to some extent by substrate temperature. With amphoteric dopants such as silicon, a change of As flux by a factor of two might lead in some cases to a 20% change in silicon donor incorporation. Therefore, for growth rates and doping densities to be reproducible to within ± 2%, the Ga, Al, and dopant fluxes must be reproducible to about ± 2%, while the arsenic flux must be reproducible within about ± 10% to ± 25%. Substrate temperature reproducibility of ± 5°C is usually sufficient. Substrate temperatures and arsenic fluxes can now be reproduced well enough to satisfy these conditions, with the possible exception of high-temperature AlGaAs growth. However, the Ga, Al, and dopant fluxes show day-to-day variations and long-term drift which takes them out of the desired range.

There are several possible approaches to improving reproducibility. Larger ovens with better thermal stability are being developed by MBE equipment manufacturers to reduce random temperature variations and long-term flux drift caused by charge depletion. Attempts may also be made to use dedicated ion guages to measure the fluxes from each oven directly, possibly leading to real-time feedback control of fluxes. Finally, calibration techniques are being refined to allow the relevant parameters to be determined rapidly and accurately.

One calibration technique of relatively recent origin is peculiar to MBE and bears special attention. This is the use of the oscillatory nature of RHEED pattern spot intensities to determine the absolute growth rate. It has been found that upon initiation of the growth of GaAs or AlGaAs, oscillations in the intensity of the diffracted electron beam occur. The period of the oscillations corresponds precisely to the time required to grow a single layer of the GaAs or AlGaAs crystal lattice. This method has been used in real time to grow precise bi-layer GaAs/ AlAs superlattices [13], and promises to become a rapid method for growth rate determination capable of the required ± 2% accuracy.

Finally, wafer throughput is important since this will directly affect the eventual cost of MBE wafers. Throughput may be increased by increasing growth rate, decreasing device thicknesses, improving wafer handling, and growing on several wafers simultaneously. Higher growth rates have been used successfully to grow GaAs by MBE up to 5 μm/h [14]. It remains to be seen how high the growth rate may be increased in practical situations; it seems clear that rates above the present values of ~ 1 μm/h may be considered.

Decreased device thickness may be achieved in some devices, such as the HEMT, where a large portion of the epilayer growth serves to create a buffer to separate device-active layers from effects of the substrate and substrate/epilayer interface. The thin double heterostructure HEMT shown in Fig. 3 achieves this by using low substrate temperatures and an AlGaAs layer for isolation from substrate effects. This structure takes only 14 minutes to grow at a GaAs rate of 1.5 μm/hr. It has also been reported from a number of laboratories that use of a superlattice buffer, consisting of many alternating thin undoped AlAs and GaAs layers can eliminate the need for thicker GaAs buffers. Other devices structures, such as the HBT shown in Fig. 1, do not contain layers which serve only as buffers and, therefore, cannot be thinned significantly without affecting device performance.

Improvement of wafer handling and design of MBE equipment to achieve uniform growth on several wafers simultaneously are engineering problems which are well within the scope of semiconductor equipment manufacturers' capabilities. This would require a major investment on their part, however, and probably will not occur until operation of existing apparatus in a pilot environment has demonstrated that reproducibility and oval defect considerations are understood sufficiently to allow the reliable production of MBE material for large circuits. One can then visualize, for example, an MBE growth system with a 3-wafer simultaneous growth capability and negligible wafer handing overhead time growing thin HEMT (3000 Å total thickness) structures at 3 μm/h. Such a situation would result in a structure growth time of 6 min, and a throughput of 30 wafers per hour.

4 References

1. P.M. Asbeck, D.L. Miller, R.J. Anderson, L.D. Hou, R. Demming, and F.H. Eisen, IEEE Electron Dev. Lett. EDL, 181 (1984).
2. P.M. Asbeck, D.L. Miller, R.J. Anderson, R. Demming, R.T. Chen, C.A. Leichti, and F.H. Eisen, Tech. Digest 1984 IEEE Gallium Arsenide Integrated Circuit Symposium, Boston, p. 133.
3. J.N. Miller, D.M. Collins, N.J. Moll, J.S. Kofol, presented at A.I.M.E Electronic Materials Conference, Burlington, VT, 1983 (unpublished).
4. D.L. Miller and P.M. Asbeck, to appear in J. Appl. Phys., March 1985.
5. K. Inoue and H. Sakaki, Japan. J. Appl. Phys. 23, L61 (1984).
6. K. Inoue, H. Sakaki, and J. Yoshino, Japan. J. Appl. Phys. 23, L767 (1984).
7. K. Nishiuchi, N. Kobayashi, S. Kuroda, S. Notomi, T. Mimura, M. Abe, and M. Kobayashi, IEEE Intl. Solid State Circuits Conf. Tech. Digest 1984, p. 48.
8. S. Kimoda, T. Mimura, M. Suzuki, N. Kobayashi, K. Nishiuchi, A. Shibatomi, and M. Abe, Tech. Digest 1984 IEEE Gallium Arsenide Integrated Circuit Symposium, Boston, p. 125.
9. M. Abe, private communication.
10. M. Shinohara, T. Ito, K. Wada, and Y. Imamura, Japan. J. Appl. Phys. 23, L371 (1984).
11. T. Baba, T. Mizutani, and M. Ogawa, Japan. J. Appl. Phys. 22, L627 (1983).
12. T. Baba, T. Mizutani, M. Ogawa and K. Ohata, presented at 42nd Annual Device Research Conference, June 18-20, 1984, Santa Barbara.
13. T. Sakamoto, H. Funabashi, J. Ohta, T. Nakagawa, N. Kawai, and T. Kojima, Japan. J. Appl. Phys. 23, L657 (1984).
14. Young G. Chai, Appl. Phys. Lett. 37, 379 (1980).

New High-Speed Quantum Well and Variable Grap Superlattice Devices

Federico Capasso
AT&T Bell Laboratories, Murray Hill, NJ 07974, USA

1. Band-gap engineering

Recent advances in high speed electronic and optoelectronic heterojunction devices are discussed. These include variable gap base transistors, new resonant tunneling transistors with quantum wells in the base regions and high-low avalanche detectors with superlattice grading regions.

These examples show that with appropriate combinations of quantum wells, graded regions and doping profiles the transport properties of a semiconductor structure can be modified and tailored to a specific device application (band-gap engineering, Capasso [1]-[3]).

Finally a recently developed technique to artificially tune band-edge discontinuities using doping interface dipoles is illustrated.

2. New Heterojunction Bipolar Transistors

The essential feature of the heterojunction bipolar transistor (HBT) relies upon a wide band gap emitter wherein part of the energy band gap difference between the emitter and base is used to suppress hole injection from the base into the emitter. This allows the base to be more heavily doped than the emitter leading to a low base resistance and emitter-base capacitance, both of which are necessary for high frequency operation, while still maintaining a high emitter injection efficiency [4].

Band-gap engineering can be used to design new HBTs. The band gap can be graded in the base to achieve a significant improvement in the speed. This device, along with electron velocity measurements in graded-gap p^+ AlGaAs, is discussed in this section.

A quantum well, or a superlattice in the base layer, on the other hand, make possible an entire new class of negative differential resistance devices based on resonant tunneling. These functional devices can have interesting applications in multiple-state logic and in other signal processing applications.

a. The Graded Gap Base Transistor

Kroemer [5] first proposed the use of a graded-gap p-type layer for the base of a bipolar transistor, to reduce the minority carrier (electron) transit time in the base (Fig. 1a). When the base-emitter and base-collector junctions are respectively forward- and reverse-biased, electrons are injected from the emitter into the base and move over to the collector layer.

With no grading in the heavily doped p-base minority carriers (electrons) are transported by diffusion, a relatively slow process. In addition, a fraction of the injected electrons recombine with holes in the base, thus reducing the base transport factor.

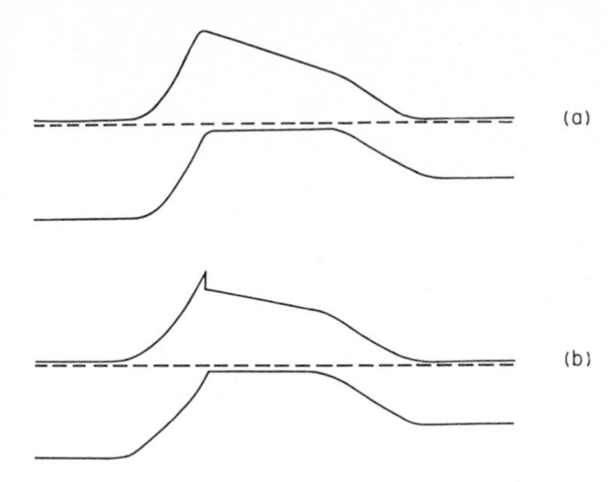

Fig. 1 Band-diagram of graded gap base transistor: (a) with graded emitter-base interface (b) with ballistic launching ramp for even higher velocity in the base.

The presence of band gap grading in the base creates a quasi-electric field acting on the electrons. These therefore move predominantly by drift (if the field is sufficiently strong) leading to a higher velocity in the base.

It can be easily shown that the ratio of the base transit times for an ungraded bipolar and a bipolar with a graded base is [6]

$$\frac{\tau_b}{\tau_b'} = \frac{E_{g2} - E_{g1}}{2kT} \tag{1}$$

where $E_{g2} - E_{g1}$ is the band gap difference across the base, T the lattice temperature, τ_b and τ_b' the base transit times for the transistor without and with grading in the base respectively. Thus the band-gap difference must be made as large as possible, without exceeding the intervalley energy separation ($\Delta E_{\Gamma L}$) of the material with gap E_{g1}, which would result in a strong reduction of the electron velocity and in the nonvalidity of Eq. (1). Using $E_{g2} - E_{g1} = 0.2$ eV, the transit time is reduced by a factor of $\simeq 4$ at 300 K over a bipolar with an ungraded base of the same thickness. This allows a precious tradeoff against the base resistance (R_b), making possible an increase of the base thickness and a consequent reduction of R_b, while still keeping a reasonable base transit time. This will increase the maximum oscillation frequency of the transistor, f_{max}.

The band-gap E_{g2}, on the emitter side of the base, should be smaller than the gap of the emitter to avoid back-injection of holes into the emitter and reduction in the current gain. Thus the structures of Fig. 1 combine the advantages of the wide-gap emitter bipolar, with that of a graded base. The emitter base junction in this case may be either graded (Fig. 1a) or abrupt (Fig. 1b). The latter structure in fact may allow even higher velocities in the base compared to that of Fig. 1a. With the abrupt emitter electrons can be launched ballistically in the base with initial velocities $\gtrsim 5 \times 10^7$ cm/sec; the quasi-field in the base maintains the velocity high, thus giving a shorter base transit time than the graded emitter bipolar [7].

Recently Capasso et al. [7] demonstrated that the quasi-field in the base strongly reduces slow diffusion effects. They measured the response time of a phototransistor with a 0.45μm wide graded gap base to a short picosecond laser pulse absorbed in the base layer. The symmetric ultrafast (FWHM $\approx$ 40 ps) scope limited pulse response

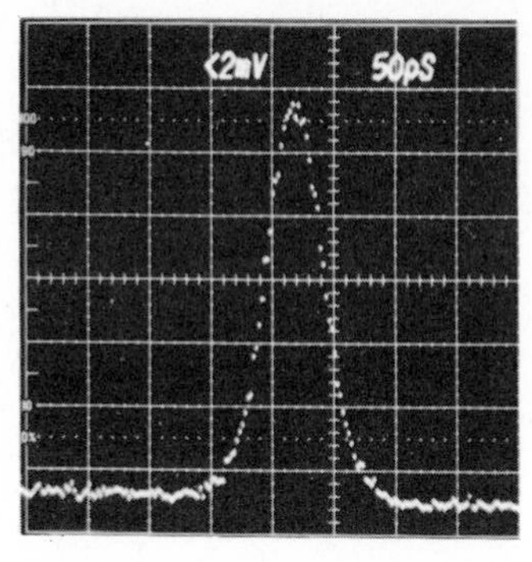

Fig. 2 Pulse response of a graded gap base phototransistor to a 5 ps laser pulse.

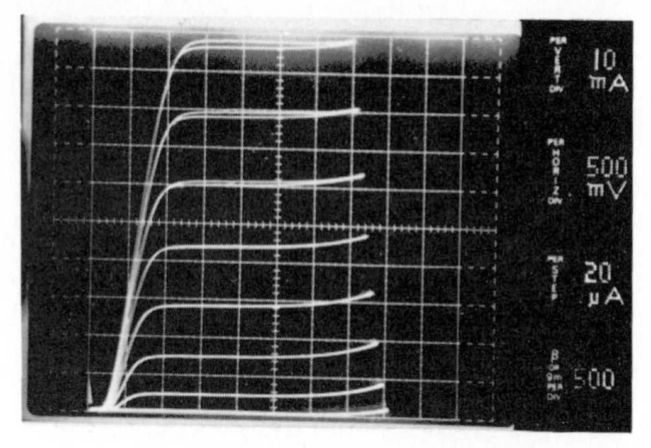

Fig. 3 Common-emitter characteristics of a graded base bipolar transistor.

(Fig. 2) gives evidence that transport in the base is drift limited and is not broadened by diffusion.

Operation of a graded base three terminal bipolar transistor was also demonstrated (Hayes et al., [6]; Miller et al., [8]). The latter authors reported a cut-off frequency $f_T = 16$ GHz in a 800Å base device.

More recently high current gain graded base bipolars, with good high frequency performance have been reported [9]. The base layer was linearly graded over 1800Å from x=0 to x=0.1 resulting in a quasi-electric field of 5.6 kV/cm and was doped with Be to $p=5\times10^{18}cm^{-3}$. The emitter-base junction was graded over 500 Å from X=0.1 to X=0.25 to enhance hole confinement in the base. The 0.2μm thick $Al_{0.25}Ga_{0.75}As$ emitter and the 0.5μm thick collector were doped n-type at $2\times10^{17}cm^{-3}$ and $2\times10^{16}cm^{-3}$, respectively. The $Al_xGa_{1-x}As$ layers were grown at a substrate temperature of 700 °C. It was found that this high growth temperature resulted in better $Al_xGa_{1-x}As$ quality as determined by photoluminescence. However, it is known that significant Be diffusion occurs during MBE growth at high substrate temperature and at high doping levels ($p>10^{18}cm^{-3}$). SIMS data also indicated a misplacement of the p-n junction into the wide band gap emitter at 700 °C substrate growth temperatures. Therefore, it was determined empirically that the insertion of an undoped setback layer of 200-500Å between the base and emitter to compensate for the Be diffusion resulted in significantly increased current gains. Zn diffusion was used to contact the base and provides a low base contact resistance.

The common emitter I-V characteristics of a test transistor with an emitter area of $7.5\times10^{-5}cm^2$ is shown in Fig. 3. It is seen that the current gain increases with higher current levels and that the collector current exhibits flat output characteristics. The maximum differential DC current gain is 1150, obtained at a collector current density of $J_c=1.1\times10^3Acm^{-2}$, which is the highest yet reported for graded band gap base HBT's. The slight negative resistance effects at high collector currents is due to thermal heating. A small offset voltage of about 0.2V is also evident. The maximum V_{ce} for these devices before collector breakdown was about 8V.

These gains were obtained with a dopant setback layer in the base of 300Å and can be compared with previous work which consistently resulted in current gains of <100 HBT's without the setback layer [6],[8]. Several transistor wafers were processed with undoped setback layers in the base of 200-500Å and all exhibited gain enhancement.

High frequency graded band gap base HBT's were fabricated using the Zn diffusion process. A single 5μm wide emitter stripe contact with dual adjacent base contacts were used. The areas of the emitter and collector junctions were approximately 2.3 $10^{-6}cm^2$ and 1.8 $10^{-5}cm^2$, respectively. The transistors were wire bonded in a microwave package and automated s-parameter measurements were made with an HP 8409 network analyzer. The frequency dependence of the small signal current gain and power gain (for a transistor biased at $I_c = 20$ mA, and $V_{ce} = 3V$) are shown in Fig. 4. The transistor has a current gain cutoff frequency $f_T \approx 5$ GHz and a maximum oscillation frequency of $f_{max} \approx 2.5$ GHz. Large signal pulse measurements resulted in rise times of $\tau_r \sim 150$ ps and pulsed collector currents of $I_c > 100$ mA which is useful for high current laser drivers.

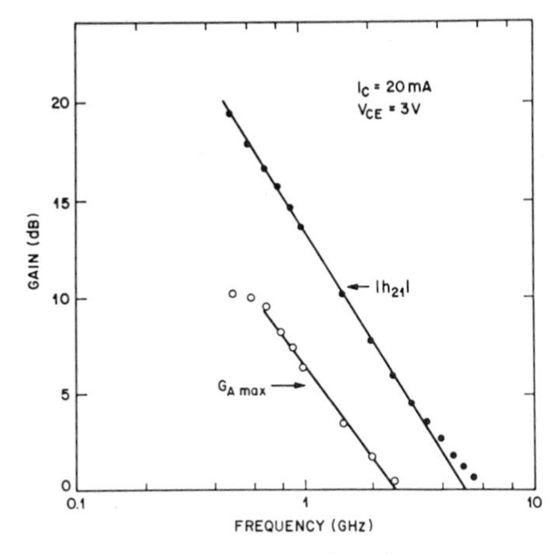

Fig. 4 Current gain $|h_{21}|$ and maximum available power gain vs. frequency of a graded base bipolar transistor.

b. Electron Velocity Measurements in Variable Gap AlGaAs

Recently Levine et al., [10],[11], using an all optical method, measured for the first time the electron velocity in a heavily p^+ doped compositionally graded $Al_xGa_{1-x}As$ layer, similar to the base of the bipolar transistor illustrated in Fig. 1.

The energy band diagram of the sample is sketched in Fig. 5 along with the principle of the experimental method. The measurement technique is a "pump and probe" scheme. The pump laser beam, transmitted through one of the AlGaAs window layers is absorbed in the first few thousand Å of the graded layer. Optically generated electrons, under the action of the quasi-electric field drift towards the right in Fig. 5 and accumulate at the end of the graded layer. This produces a refractive index change at the interface with the second window layer. This refractive index variation produces a reflectivity change that can be probed with the counter propagating probe laser beam. This reflectivity change is measured as a function of the delay between pump and probe beam using phase sensitive detection techniques. The reflectivity data are shown in Fig. 6 for a sample with a 1μm thick transport layer, graded from $Al_{0.1}Ga_{0.9}As$ to GaAs and doped to $p \cong 2 \times 10^{18}/cm^3$. This corresponds to a quasi-field of 1.2 kV/cm. The laser pulse width was 15 ps and the time 0 in Fig. 6 represents the center of the pump pulse as determined by two photon absorption in a GaP crystal cemented near the sample.

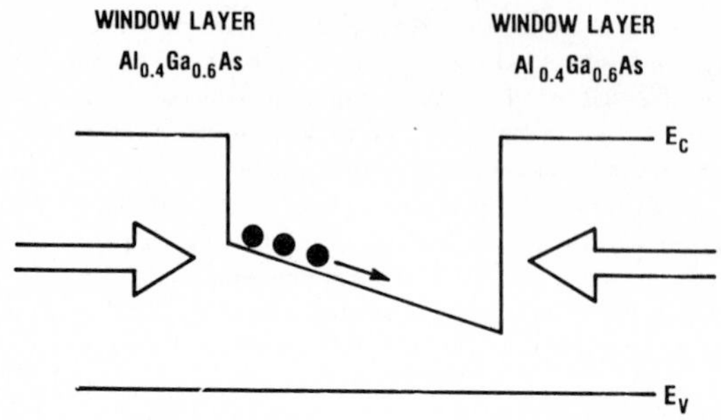

Fig. 5 Band diagram of sample used for electron velocity measurements and schematics of the experiment.

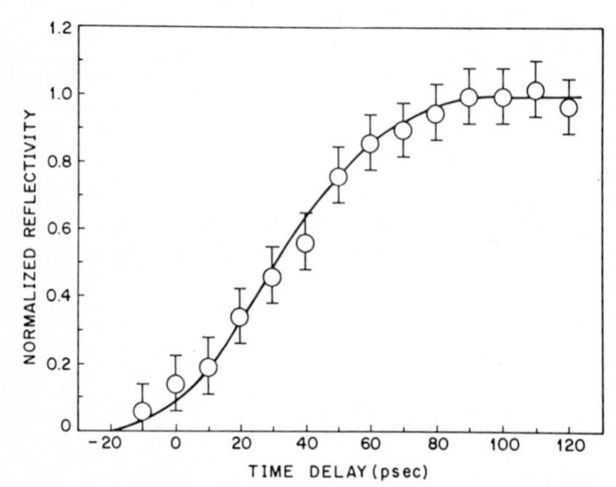

Fig. 6 Normalized experimental results for pump induced reflectivity change vs. time delay obtained in 1μm thick graded gap p+ AlGaAs, at a quasi-electric field F=1.2 kV/cm.

The transit time is approximately given by the shift of the half height point of the reflectivity curve from zero, which is $\tau=33$ ps. Taking as the drift length the graded layer thickness minus the absorption length of the pump beam ($1/\alpha \cong 2500\text{Å}$), one estimates a minority carrier velocity 2.3×10^6 cm/sec.

In these relatively thick samples diffusion effects are important and cause a spread in the arrival time of electrons at the end of the sample, given roughly by the 10-90% rise time of the reflectivity curve, i.e., 63 ps.

It is interesting to note that the drift mobility obtained from the measurement is $\mu_d = V/F = 1900$ cm^2/V/sec, which is comparable to the usual electron mobility of 2200 cm^{2}V/sec in GaAs at the doping level of the graded layer in GaAs.

Electron velocity measurements were also made in a 0.42μm thick strongly graded (F = 8.8 kV/cm) highly doped ($p = 4\times 10^{18}$ cm^{-3}) $Al_xGa_{1-x}As$ layer, graded from $Al_{0.3}Ga_{0.7}As$ to GaAs. A transit time of only 1.7 ps was measured, more than an order of magnitude shorter than that for F = 1.2 kV/cm, as shown in Fig. 7 corresponding roughly to a velocity $V\cong 2.5\times 10^7$ cm/sec. The velocity can be obtained rigorously and accurately (=10% error) from the reflectivity data, by solving the drift diffusion equation and taking into account the effects of the pump absorption length (especially

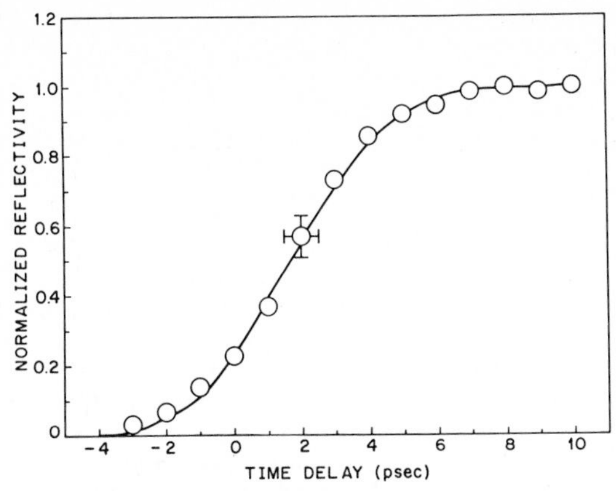

Fig. 7 Reflectivity change vs. time delay measured in a 0.4µm thick graded gap p+ AlGaAs layer at a quasi-electric field F=8.8 kV/cm.

important in the thin sample), the partial penetration of the probe beam in the graded material [11]. Including all these effects, one finds that the reflectivity data can be fitted using only one adjustable parameter, the electron drift velocity. This velocity is V $= 2.8\times10^6$ cm/sec for F=1.2 kV/cm and p $= 2\times10^{18}/cm^3$; and V $= 1.8\times10^7$ cm/sec for F=8.8 kV/cm and p $= 4\times10^{18}/cm^3$.

We see that when we increased the quasi-field from 1.2 to 8.8 kV/cm (a factor of 7.3) the velocity increased from 2.8×10^6 cm/s to 1.8×10^7 cm/s (a factor of 6.5). That is, we observed the approximate validity of the relation $V = \mu F$. In fact, using $\mu =$ 1700 cm^2/s (for p $= 4\times10^{18} cm^{-3}$) we calculate V $= 1.5\times10^7$ cm/s for F = 8.8 kV/cm in reasonable agreement with the experiment. It is worth noting that this measured velocity of 1.8×10^7 cm/s (in the quasi-field) is significantly larger than that for pure undoped GaAs where V $= 1.2\times10^7$ cm/s for an ordinary electric field of 8.8 kV/cm. In fact the measured high velocity is comparable to the peak velocity reached in GaAs for F = 3.5 kV/cm before the intervalley transfer occurs from the central to the L valley. It is noteworthy that the measured velocity is also comparable to the maximum possible phonon limited velocity in the central valley of GaAs. This is given by $v_{max} = [(E_p/m^*)\tanh(E_p/2kT)]^{1/2} = 2.3\times10^7$ cm/s, where $E_p = 35$ meV is the optical phonon energy and the effective mass $m^* = 0.067\, m_o$.

This high velocity can be understood without reference to transient effects since the transit time is much larger than the momentum relaxation time of 0.3 ps. The large velocity results from the fact that the electrons spend most of their time in the high velocity central valley rather than in the low velocity L valley. This may result from the injected electron density being so much less than the hole doping density that the strong hole scattering can rapidly cool the electrons without excessively heating the holes. Furthermore, the electrons remain in the central valley throughout their transit across the graded layer since the total conduction band edge drop ($\Delta E_c = 0.37$ eV) is comparable to the GaAs Γ-L separation ($\Delta E_{\Gamma L} = 0.33$ eV) and therefore they do not have sufficient excess energy for significant transfer to the L valley.

c. Resonant Tunneling Transistors with Quantum Well Base

In this section, we discuss a new class of resonant tunneling heterojunction devices which consist of heterojunction bipolar transistors, with a quantum well and a double barrier, or a superlattice in the base region (Capasso and Kiehl, [12]).

Pioneering work on resonant tunneling through a heterostructure quantum well was first done by Tsu and Esaki [13] and Chang et al. [14]. More recently Sollner et al. demonstrated resonant tunneling in AlGaAs/GaAs double barriers at tetrahertz frequencies [15] and a quantum well oscillator operating at 18 GHz [16]. In the above experiments, resonant tunneling was obtained by applying a voltage to the double barrier, to achieve matching between the Fermi level in the cathode and the resonant states of the well.

Recently however Ricco and Azbel [17] have pointed out that in order to achieve unity transmission at *all* the resonance peaks, the transmission of the left and right barrier must be equal at *all* the quasi-eigenstate energies. This is physically identical to what occurs in a Fabry-Perot resonator, with which resonant tunneling structures share profound analogies. If the transmissions of the two mirrors are made sufficiently different, the transmission at the resonant frequencies decreases significantly below unity. Application of an electric field to a symmetric double barrier introduces a difference between the transmission of the two barriers, thus decreasing below unity the overall transmission at the resonance peaks. Unity transmission can be achieved if the two barriers have different thicknesses; however using this procedure one can only optimize the transmission of one of the resonance peaks.

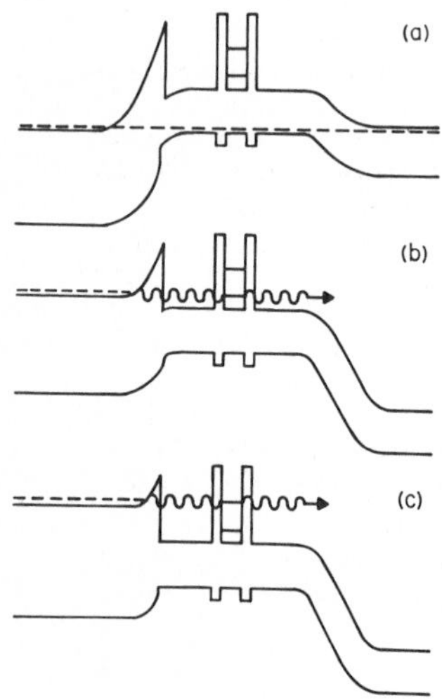

Fig. 8 Band diagram of RTT with tunneling emitter under different bias conditions: (a) in equilibrium; (b) resonant tunneling through the first level in the well; (c) resonant tunneling through the second level. (Not in scale.)

Fig. 8(a) shows the equilibrium band diagram of one of the devices. The structure is a heterojunction bipolar transistor with a degenerately doped abrupt emitter and a symmetric double barrier in the base. The base-emitter and base-collector junctions are then respectively forward and reverse-biased. As the base emitter voltage is increased the energy difference between the Fermi level in the emitter and the first resonant state of the quantum well decreases. When these two levels are matched electrons tunneling from the emitter region are injected in the first state of the well and undergo resonant tunneling through the double barrier with near unity transmission probability. Off resonance the transmission probability is typically $<<1$ and equal to the product of the transmission coefficients of the two barriers without the quantum well [17]. The collector current as a function of the base-emitter voltage V_{be} exhibits a series of peaks corresponding to the various quasi-stationary states of the well. Multiple negative conductance in the collector circuit can therefore be achieved. As a result of symmetry of the double barrier under operating conditions, current densities, negative conductance, and peak-to-valley ratios much larger than in conventional resonant tunneling structures should be possible. Thus, RTT's have potential for high performance oscillators.

The device of Fig. 8 can be implemented with AlGaAs/GaAs grown by molecular beam epitaxy. The wide gap degenerately doped emitter ($N_E \gtrsim 1 \times 10^{17}/cm^3$) would consist typically of $Al_xGa_{1-x}As$ ($x \gtrsim 0.3$) to ensure sufficient hole confinement in the base and high injection efficiency. The double barrier in the center of the base region consists of a GaAs well (typical thickness range 30-60Å) and of $Al_xGa_{1-x}As$ ($0.3 \lesssim X \lesssim 1$) rectangular barriers of equal thickness (typically 15 to 50Å). For example for two rectangular barriers, height 0.24 eV, (corresponding to $Al_{.30}Ga_{0.70}As$), width 20Å and separation 30Å , the first level has an energy $E_1 = 119.1$ meV. The double-barrier and the well should be undoped with ultra-low carrier concentration to minimize scattering and recombination. This is important since for high transmission, wavefunction coherence must be maintained. Alloy disorder in the barrier may contribute to scattering of the injected electrons. This can be minimized by the use of AlAs barriers.

The rest of the base layer outside the barrier region is instead heavily doped ($10^{18} \leq p \leq 5 \times 10^{18}/cm^3$) and it's total thickness should be typically 2000Å to provide the required low base resistance. The thickness of the base region between the double barrier and the emitter should be smaller than the scattering mean free path of the electron injected from the emitter but greater than the zero bias depletion width on the p side. A good choice for this thickness is $\approx$ 500-100Å , which also minimizes quantum size effects in this region.

To achieve high current at resonance, the width of the resonant peak should be of the order of the width of the energy distribution of the electrons in the emitter, which at 77 °K is comparable to the degeneracy ($E_F - E_c$). Assuming a GaAs well width $= 30$Å and $Al_{0.30}Ga_{0.70}As$ barriers 20Å thick, one obtains $\Delta E_1 = 50$ meV for the energy width of the first resonance. If the emitter doping level is $2 \times 10^{18}/cm^3$ the degeneracy $E_F - E_c$ is $\simeq 50$ meV $\cong \Delta E_1$ so that most electrons leaving the emitter will resonantly tunnel through the well.

With a tunneling emitter doped to $\gtrsim 10^{18}/cm^3$, collector current densities at resonance $\gtrsim 10^4 A/cm^2$ are estimated for the case of negligible electron recombination in the base.

To minimize thermionic emission of the electrons from the emitter and scattering effects in the base, these devices are preferably operating at 77 ° K.

The peak to valley ratio, i.e., the ratio of the currents at and off resonance, is given by $\approx 1/T_B^2$ where T_B is the barrier transmission coefficient [17]. For the previously considered example, this gives a peak-to-valley ratio of $\approx$ 36.

An alternative injection method is the nearly abrupt emitter which can be used to ballistically launch electrons into the resonant state with high momentum coherence. As the base-emitter voltage is increased the top of the launching ramp eventually reaches the same energy of the resonant state so that electrons can be ballistically launched into the resonant state (Fig. 9(a)).

If several peaks in the collector current vs. emitter-base voltage are desired, which is important for several device applications, the well should be relatively thick (100-200Å) and the barriers should have a high Al concentration. To achieve equally spaced resonances in the collector current, the rectangular quantum well in the base should be replaced by a parabolic one (Fig. 9(b)). Parabolic quantum wells have been recently realized in the AlGaAs system. Assuming the depth of the parabolic well to be 0.43 eV (corresponding to grading from AlAs to GaAs) and its width 400Å , one finds that the first state is at an energy of 11 meV from the bottom of the well and that the resonant states are separated by $\simeq 33.4$ meV. This gives a total of 12 states in the well.

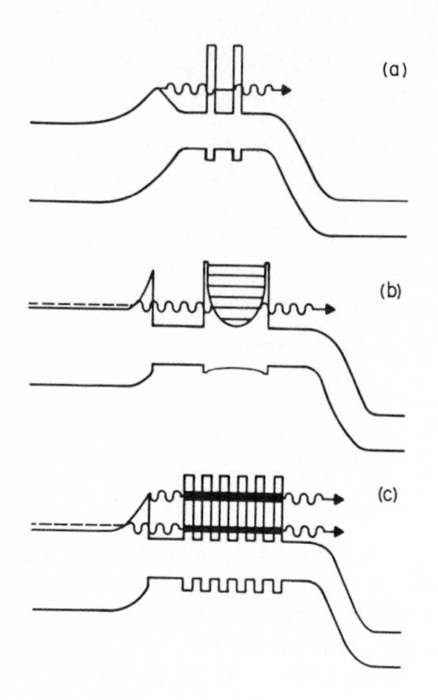

Fig. 9 (a) Band diagram of RTT with graded emitter (at resonance). Electrons are ballistically launched into the first quasi-eigenstate of the well. (b) RTT with parabolic quantum well in the base and tunneling emitter. (c) RTT with superlattice base.

Finally in Fig. 9(c) we illustrate another application, that of studying high energy injection and transport in the minibands of a superlattice, using ballistic launching. Minibands are formed when the barriers are sufficiently thin that the quasi-eigenstates of the wells are strongly coupled. For example in an AlAs/GaAs superlattice with 40Å barriers and wells the first miniband E_1 is at 0.1 eV from the bottom of the wells and the second one at 0.36 eV. The width of these two minibands is respectively $\simeq 10$ meV and $\simeq 50$ meV. To achieve ballistic launching into the first excited state of the superlattice it is necessary to choose an emitter composition such that $\Delta E_c \cong E_2 \cong 0.36$ eV. $Al_{0.5}Ga_{0.5}As$ would be the choice.

It is important to mention the time dependence of resonant tunneling. This feature has been stressed clearly only recently by Ricco and Azbel [17]. To achieve resonant tunneling the electron probability density $|\psi|^2$ must be peaked in the well. Therefore if initially there are no electrons in the double barrier and the carriers are made to tunnel by applying a positive base emitter-voltage, it takes a certain time constant to build up the probability density in the well, via multiple reflections, and to achieve high transmission at resonance. An identical situation is present in an optical Fabry-Perot. The above time constant τ_0 is of the order of $h/\Delta E$ where ΔE is the width of the resonant state. For $\Delta E \cong 50$ meV (as for one of the structures previously discussed) $\tau_0 \cong 1 \times 10^{-14}$ secs. Note however that τ_0 increases exponentially with barrier thickness. After a time of the order of a few τ_0 has elapsed, a quasi-steady state has been reached whereby electrons continuously enter in the well and exit from it to maintain a constant electron density in the well.

Estimates of the emitter-collector delay time τ_{ec} for the device of Fig. 9(a) indicate that cutoff frequencies $f_T = \dfrac{1}{2T_c\tau_{ec}}$ as high as 60 GHz should be achievable.

A particularly interesting heterojunction for RTT's is $Al_{0.48}In_{0.52}As/Ga_{0.47}In_{0.53}As$ because of the large conduction band discontinuity ($\Delta E_c = 0.55$ eV), and the low $Ga_{0.47}In_{0.53}As$ electron effective mass ($m^* = 0.041$).

The multiple resonant characteristic of the RTT is of interest for a variety of signal processing and logic applications. One class of applications takes advantage of the ability to achieve a multiple-valued voltage transfer characteristic with an RTT, as shown in

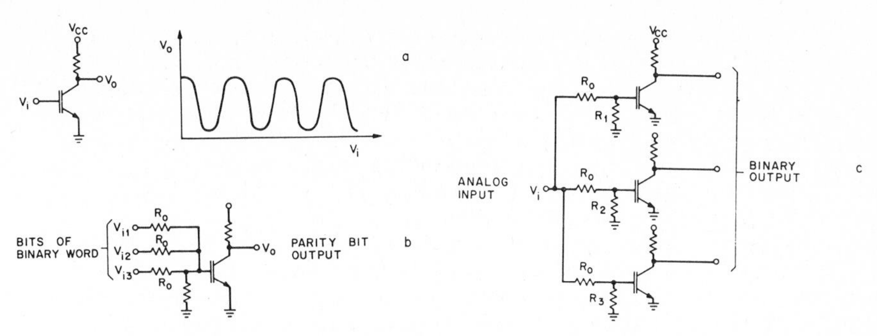

Fig. 10 (a) Schematic of multiple-valued voltage transfer characteristics of the RTT and corresponding circuit diagram; (b) Parity generator circuit; (c) Analog-to-digital converter circuit.

Fig. 10(a). In this configuration, the output voltage V_o takes on one of two values in accordance with the level of the input voltage V_I. Thus, the device provides a binary digital output for an analog input, or a multiple-valued digital input. This function, which is that of a threshold logic gate, is useful for a variety of signal processing applications. For example, a single device of this sort can be used to provide a parity generator as used in error detection circuits. In this application, the binary bits of a digital word are added in a resistive network at the input of the RTT as depicted in Fig. 10(b). This produces a binary output having a value that depends on whether the total number of 1's in the input word is odd or even. The advantage of this approach over conventional circuits is that the RTT implementation should be extremely fast since it uses a single high-speed switching device. Conventional transistor implementations require complex circuitry involving many logic gates with a consequent reduction in speed. By combining a number of RTT's in a parallel array, an A-to-D converter could also be realized. In this application, the analog input is simultaneously applied to an array of RTT's having different voltage scaling networks, as shown in Fig. 10(c) thereby producing an interlaced pattern of harmonically related transfer characteristics. The outputs of the RTT array constitute a binary code representing the quantized analog input level. Again, the circuitry involved in this approach is simple and should be very fast.

A second class of applications takes advantage of the ability to achieve a multiple-valued negative resistance characteristic. This type of characteristic is achieved at the emitter-collector terminals by holding the base-collector junction at a fixed bias V_{BC}, as shown in Fig. 11. With V_{BC} fixed, variations in V_{CE} produce variations in V_{EB} which cause the collector current to peak as V_{EB} crosses a tunneling resonance. When con-

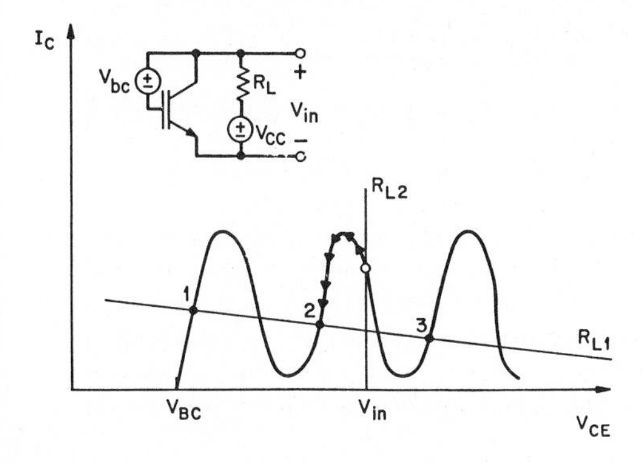

Fig. 11 Current voltage characterristic with multiple-valued negative resistance. R_{L1} and R_{L2} indicate load lines. Stable states are denoted by solid circles.

nected to a resistive load R_2 and voltage supply V_{CC}, as shown in the inset of Fig. 11, a device having N stable states is produced where N is the number of resonant peaks. The state of this latch can be set by momentarily applying a voltage V_{LS} to the circuit, forcing the operating point so that of the open circle in Fig. 11. When the input line is opened, it can easily be shown that the operating point moves along the indicated trajectory, finally latching at State 2. Thus, the RTT in this configuration can serve as a N-state memory element, providing the possibility of extremely high density data storage. Memories of this sort and other circuits based on a multiple-valued negative resistance, such as counters, multipliers, and dividers, have been of interest for some time [18]. However, since no physical device exhibiting multiple value negative resistance previously existed, such circuits were possible only with combinations of binary devices, such as conventional tunnel diodes. In order to achieve N states, N two-state devices were connected, resulting in a complex configuration with reduced density and speed. Because it allows multiple valued negative resistance to be achieved in a single physical element, the RTT could offer significant advantages.

3. Pseudoquaternary Semiconductors: Applications to High Speed Detectors

Recently Capasso et al. [19] have demonstrated a new superlattice (pseudoquaternary GaInAsP) capable of conveniently replacing conventional GaInAsP semiconductors in a variety of device applications. It is important to recall that these quaternary materials play a key role in optoelectronic devices for the 1.3-1.6 μm low-loss, low dispersion window of silica fibers.

The concept of a pseudo-quaternary GaInAsP semiconductor is easily explained. Consider a multilayer structure of alternated $Ga_{0.47}In_{0.53}As$ and InP. If the layer thicknesses are sufficiently thin (typically a few tens of angstroms) one is in the superlattice regime. One of the consequences is that this novel material has now its own band gap, intermediate between that of $Ga_{0.47}In_{0.53}As$ and InP. In the limit of layer thicknesses of the order of a few monolayers the energy bandgap can be approximated by the expression

$$\bar{E}_g = \frac{E_g(Ga_{0.47}In_{0.53}As)L(Ga_{0.47}In_{0.53}As)+E_g(InP)L(InP)}{L(Ga_{0.47}In_{0.53}As)+L(InP)} \tag{2}$$

where the L's are the layer thicknesses.

These superlattices can be regarded as novel pseudo-quaternary GaInAsP semiconductors. In fact, similarly to $Ga_{1-x}In_xAs_{1-y}P_y$ alloys, they are grown lattice matched to InP and their bandgap can be varied between that of InP and that of $Ga_{0.47}In_{0.53}As$. The latter is done by adjusting the ratio of the $Ga_{0.47}In_{0.53}As$ and InP layer thicknesses. Pseudo-quaternary GaInAsP is particularly suited to replace variable gap $Ga_{1-x}In_xAs_{1-y}P_y$. Such alloys are very difficult to grow since the mole fraction x (or y) must be continuously varied while maintaining lattice matching to InP.

Figure 12(a) shows a schematic of the energy-band diagram of undoped (nominally intrinsic) graded gap pseudo-quaternary GaInAsP. The structure consists of alternated ultrathin layers of InP and $Ga_{0.47}In_{0.53}As$ and was grown by a new vapor phase epitaxial growth technique (Levitation Epitaxy) [20]. Other techniques such as molecular beam epitaxy or metalorganic chemical vapor deposition may also be suitable to grow such superlattices. From Fig. 12(a) it is clear that the duty factor of the InP and $Ga_{0.47}In_{0.53}As$ layer is gradually varied, while keeping constant the period of the superlattice. As a result the average composition and bandgap [dashed lines in Fig. 12(a)] of the material are also spatially graded between the two extreme points (InP and $Ga_{0.47}In_{0.53}As$). In our structure both ten and twenty periods (1 period = 60Å) were

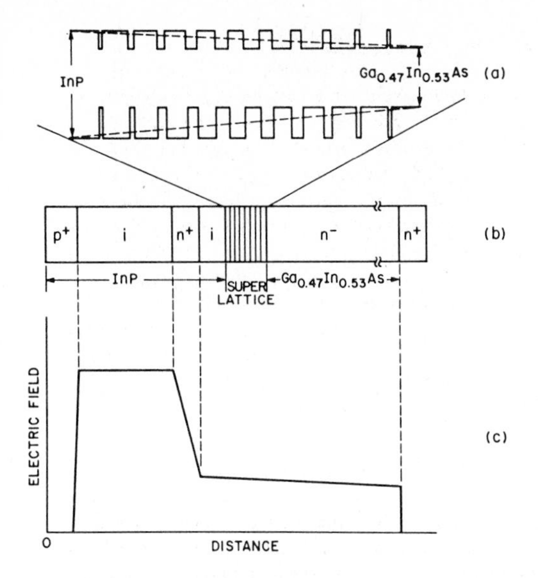

Fig. 12 (a) Band diagram of a pseudoquaternary graded gap semiconductor. The dashed lines represent the average bandgap seen by the carriers. (b) and (c) are a schematic and the electric field profile of a high-low APD using the pseudo-quaternary layer to achieve high speed.

used. The InP layer thickness was linearly decreased with distance from $\cong 50\text{Å}$ to $\cong 5\text{Å}$ while corresponding increasing the $Ga_{0.47}In_{0.53}As$ thickness to keep the superlattice period constant ($=60\text{Å}$).

The graded gap superlattice was incorporated in a long-wavelength $InP/Ga_{0.47}In_{0.53}As$ avalanche photodiode as shown in Fig. 12(b). This device is basically a photodetector with separate absorption ($Ga_{0.47}In_{0.53}As$) and multiplication (InP) layers and a high-low electric field profile (HI-LO SAM APD). This profile [Fig. 12(c)] is achieved by a thin doping spike in the ultralow doped InP layer and considerably improves the device performance compared to conventional SAM APD's [21]. The $Ga_{0.47}In_{0.53}As$ absorption layer is undoped ($n \approx 1\times 10^{15}/cm^3$) and 2.5 μm thick. The n^+ doping spike thickness and carrier concentration were varied in the 500-200Å and $5\times 10^{17}-10^{17}/cm^3$ ranges respectively (depending on the wafer), while maintaining the same carrier sheet density ($\cong 2.5\times 10^{12}/cm^2$). The N^+ spike was separated from the superlattice by an undoped 700-1000Å thick InP spacer layer. The p^+ region was defined by Zn diffusion in the 3-μm-thick low carrier density ($n^- \approx 10^{14}/cm^3$) InP layer. The junction depth was varied from 0.8 to 2.5 μm.

Similar devices, but without the superlattice region, were also grown.

Previous pulse response studies of conventional SAM APD's with abrupt $InP/Ga_{0.47}In_{0.53}As$ heterojunctions found a long ($>$ 10 ns) tail in the fall time of the detector due to the pile-up of holes at the heterointerface [22]. This is caused by the large valence-band discontinuity ($\simeq 0.45$ eV). It has been proposed that this problem can be eliminated by inserting between the InP and $Ga_{0.47}In_{0.53}As$ region a $Ga_{1-x}In_xAs_{1-y}P_y$ layer of intermediate band gap (Campbell et al., [23]). This quaternary layer is replaced in our structure, by the $InP/Ga_{0.47}In_{0.53}As$ variable gap superlat-

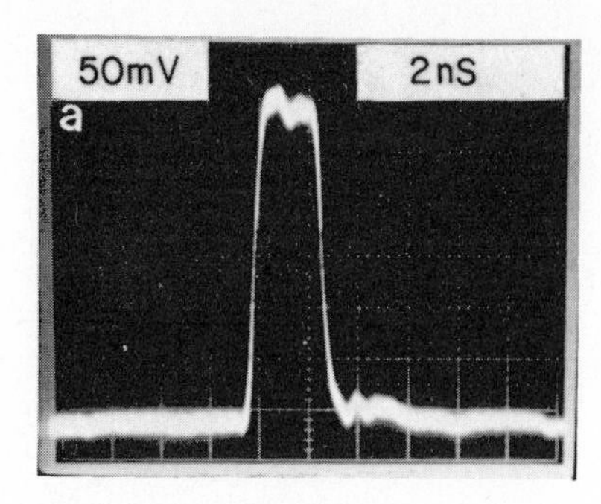

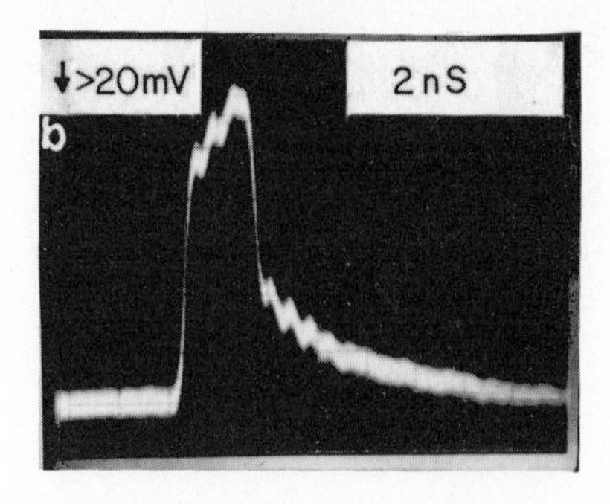

Fig. 13 Pulse response of a High-Low SAM avalanche detector with graded gap superlattice (a) and without (b) to a ≈2ns λ=1.55μm laser pulse. The bias voltage is -65.5 V for both devices. Time scale 2ns/div.

tice. This not only offers the advantage of avoiding the growth of the critical, independently lattice-matched GaInAsP quaternary layer, but also may lead to an optimum "smoothing out" of the valence-band barrier for reproducible high-speed operation. This feature is essential for Hi-Lo SAM APD's since the heterointerface electric field is lower than in conventional SAM devices.

For the pulse response measurement we used a 1.55-μm GaInAsP driven by a pulse pattern generator. Figure 13 shows the response to a 2-ns laser pulse of a NI-LO SAM APD with (a) and without (b) a 1300Å -thick superlattice. Both devices had similar doping profiles and breakdown voltages ($\simeq$80 V) and were biased at -65.5 V. At this voltage the ternary layer was completely depleted in both devices and the measured external quantum efficiency $\simeq$70%. The results ωo Fig. 13 were reproduced in many devices on several wafers. The long tail in Fig. 13(b) is due to the pile up effect of holes associated with the abruptness of the heterointerface.

In the devices with the graded gap superlattice [Fig. 13(a)] there is no long tail. In this case the height of the barrier seen by the holes is no more the valence-band discontinuity ΔE_v but

$$\Delta E = \Delta E_v - e\epsilon_i L \, , \tag{3}$$

where ϵ_i is value of the electric field at the InP/superlattice interface and L the thickness of the pseudo-quaternary layer.

The devices are biased at voltage such that $\epsilon_i > \dfrac{\Delta E_v}{eL}$ so that $\Delta E = 0$ and no trapping occurs.

In the devices with no superlattice instead $\Delta E = \Delta E_v$ for every ϵ_i so that long tails in the pulse response are observed at all voltages.

4. Doping Interface Dipoles: Tunable Heterojunction Barrier Heights and Band-Edge Discontinuities

It is clear from the material presented in the previous sections, that band discontinuities and, in general, barrier heights, play a central role in the design of novel heterojunction devices. For example their knowledge is absolutely essential for devices such as multilayer APD's and heterojunction bipolar transistors. If a technique were available to

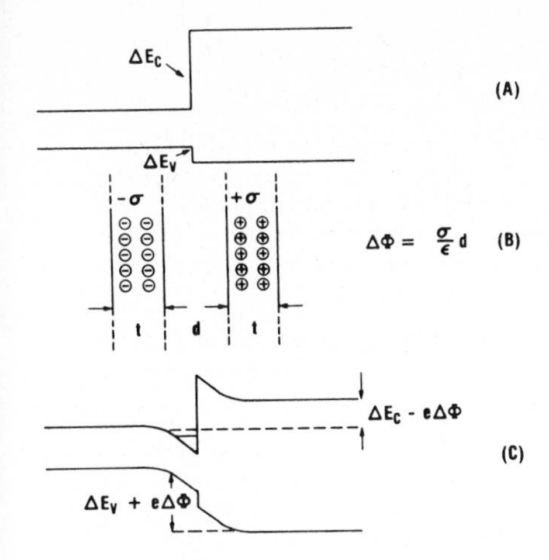

Fig. 14 (a) Band diagram of an intrinsic heterojunction. (b) Schematics of doping interface dipole. σ is the sheet charge density and $\Delta\Phi$ the dipole potential difference. (c) Band diagram of an intrinsic heterojunction with doping interface dipole.

artificially and controllably vary band offsets and barrier heights at abrupt heterojunctions, this would give the device physicist tremendous flexibility in device design, as well as many novel opportunities.

Recently Capasso et al. [24] demonstrated for the first time that barrier heights and band-discontinuities at an abrupt, intrinsic heterojunction can be tuned via the use of a doping interface dipole (DID) grown by MBE.

Compositional grading at the interface is an effective way to control barrier heights, but eliminates the abruptness of the heterojunction. In many cases one would like to preserve it while simultaneously being able to tune the barrier height. This concept is illustrated in Fig. 14. Figure 14(a) represents the band diagram of an abrupt heterojunction. The material is assumed to be undoped (ideally intrinsic) so that we can neglect band-bending effects over the short distance (a few hundred Å) shown here.

We next assume to introduce in situ, during the growth of a second identical heterojunction, one sheet of acceptors and one sheet of donors, of identical doping concentrations, at the same distance $d/2$ ($\lesssim 100$ Å) from the interface (Fig. 14). The doping density N is in the $1\times10^{17} - 1\times10^{19}/cm^3$ range, while the sheets' thickness t is kept small enough so that both are depleted of carriers ($t \lesssim 100$ Å). The DID is therefore a microscopic capacitor. The electric field between the plates is σ/ϵ, where $\sigma = eNt$. There is a potential difference $\Delta\Phi = (\sigma/\epsilon)d$ between the two plates of the capacitor. Thus the DID produces abrupt potential variations across a heterojunction interface by shifting the relative positions of the valence and conduction bands in the two semiconductors outside the dipole region (Fig. 14(c)). This is done without changing the electric field outside the DID.

The valence band barrier height at the heterojunction is increased by the DID to a value $\Delta E_v + e\Delta\Phi + \frac{\sigma}{\epsilon}t$. If $\Delta\Phi$ is dropped over a distance of a few atomic layers and the total potential drop across the charge sheets $\left[= \frac{\sigma}{\epsilon}t\right]$ is small compared to $\Delta\Phi$, the valence band discontinuity has effectively been increased by $e\Delta\Phi$.

The DID reduces the energy difference between the conduction band edges on both sides of the heterointerface to $\Delta E_c - e\Delta\Phi$. On the low gap side of the heterojunction a triangular quantum well is formed. Since typically the electric field in this region is

$\gtrsim 10^5$V/cm and $e\Delta\Phi \approx 0.1$-0.2 eV, the bottom of the first quantum subband E_1 lies near the top of the well. Therefore the thermal activation barrier seen by an electron on the low gap side of the heterojunction is $\cong \Delta E_c - e\Delta\Phi/_2$.

Electrons can also tunnel through the thin ($\lesssim 100$Å) triangular barrier, this further reduces the effective barrier height. In the limit of a DID a few atomic layers thick and of potential $\Delta\Phi$ the triangular barrier is totally transparent and the conduction band discontinuity is lowered to $\Delta E_c - e\Delta\Phi$. By inverting the position of the donor and acceptor sheets one can instead increase the conduction band discontinuity and decrease the valence band one.

Note that experimental evidence suggests that "natural" dipoles may occur at polar heterojunction interfaces causing the orientation dependence of band discontinuities (Grant et al. [25]).

To verify the barrier lowering due to the DID, heterojunction AlGaAs/GaAs pin diodes on p-type (100) GaAs substrates were grown by MBE. Two types of structures were grown: one with and the other without dipole. The one with dipole consists of four GaAs layers first, $p^+ > 10^{18}/cm^3$ (5000Å), undoped (5000Å), $p^+ = 5\times10^{17}/cm^3$ (100Å), forming the negatively charged sheet of the dipole, and undoped (100Å), followed by four $Al_{0.26}Ga_{0.74}As$ layers, undoped (100Å), $n^+ 5\times10^{17}/cm^3$ (100Å), forming the positively charged sheet of the dipole, undoped (5000Å), and $n^+ \gtrsim 10^{18}/cm^3$ (5000Å). The second type of structure is identical, with the exception that it doesn't have DID. They were grown consecutively in the MBE chamber without breaking the vacuum to ensure virtually identical growth conditions. Beryllium was used for the p-type dopant and silicon for the n-type. The substrate temperature was held at 590 °C during growth. The background doping of the undoped layers is $\lesssim 10^{14}cm^{-3}$. It is important to note that the charged sheets were introduced by controlling the aperture of the shutters of the ovens, without interrupting the growth of the GaAs and AlGaAs layers. This minimizes the formation of defects in the interface region.

The solid and dashed lines in Fig. 15(a) are respectively the band diagram of the diodes at zero applied bias, with and without dipole (not to scale). In the structure with the DID the electric field inside the dipole layer is strongly increased while it is slightly ($\cong 10\%$) decreased outside the dipole (compared to the structure without dipole) since the potential drop across the depleted i layer is identical to that of the diodes without dipole. Figure 15(b) gives the conduction band diagram (to scale) near the interfaces for the cases with and without dipole. The potential of the dipole is $\Delta\Phi = 0.14$ V; for ΔE_c we have used the value 0.2 eV, following the new band line-ups for AlGaAs/GaAs. The barrier height E_b is about a factor of 2 smaller than in the case without dipole ($=\Delta E_c$).

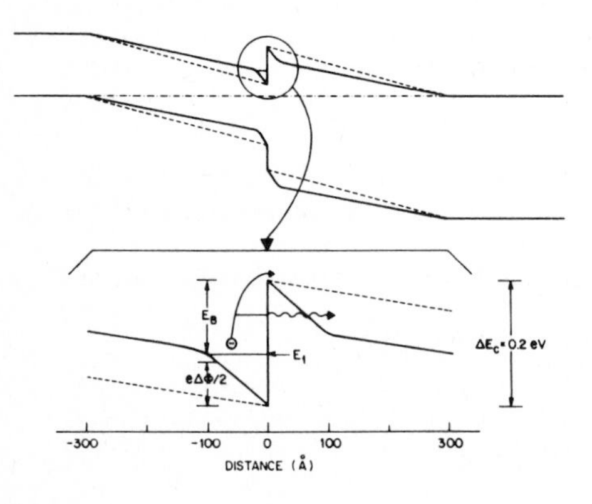

Fig. 15 Solid and dashed lines represent respectively the band diagram of the pin diodes with and without interface dipole (not in scale). Shown is also the band diagram of the conduction band near the heterointerface of the diodes with dipole and without (in scale).

We have measured the photocollection efficiency of the two structures; light chopped at 1 KHz and incident on the AlGaAs side of the diode was used and the short-circuit photocurrent was measured with a lock-in. Absolute efficiency data were obtained by comparing the photoresponse to that of a calibrated Si photodiode. In Fig. 16 we have plotted the external quantum efficiency η as a function of wavelength for devices with and without dipole. In the ones without dipole η is very small ($\lesssim 2\%$) for $\lambda \gtrsim 7100\mathring{A}$; this wavelength corresponds to the band gap of the $Al_{0.26}Ga_{0.74}As$ layer as determined by photoluminescence measurements. At wavelengths longer than this and shorter than $\approx 8500\mathring{A}$, photons are absorbed partly in the GaAs electric field region and partly in the p^{+}GaAs layer within a diffusion length from the depletion layer. Thus most of the photoinjected electrons reach the heterojunction interface and have to surmount the heterobarrier of height $\Delta E_c = 0.2$ eV to give rise to a photocurrent. Thermionic emission limits therefore the collection efficiency, which is proportional to $\exp(-\Delta E_c)/kT$ (Te. Velde, [26], Shik and Shmartsev, [29]). This explains the low efficiency for $\lambda > 7100\mathring{A}$ since ΔE_c is significantly greater than kT.

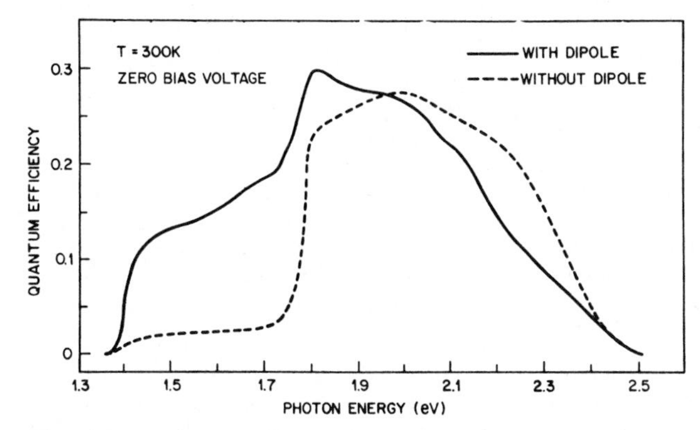

Fig. 16 External quantum efficiency of the heterojunctions with and without dipole at zero bias vs. photon energy.

For $\lambda < 7100\mathring{A}$ the light is increasingly absorbed in the AlGaAs as the photon energy increases and the quantum efficiency becomes much larger than for $\lambda > 7100\mathring{A}$ since most of the photocarriers don't have to surmount the heterojunction barrier to be collected. For $\lambda < 6250\mathring{A}$ the quantum efficiency decreases since losses due to recombination of photogenerated holes in the n^{+} AlGaAs layer and to surface recombination start to dominate (Womac and Rediker, [28]). The above behavior of the efficiency is consistent with predictions for abrupt AlGaAs/GaAs heterojunctions without interface charges.

The solid curve in Fig. 16 is the photoresponse in the presence of the DID. A striking difference is noted compared to the case with no dipole. While the quantum efficiencies for $\lambda \lesssim 7100\mathring{A}$ are comparable, at longer wavelengths it is enhanced by a factor as high as one order of magnitude in the structures with dipoles. This effect was reproduced in four sets of samples.

The physical interpretation is simple. The barrier height E_g has been lowered by $\cong 87$ meV (Fig. 15), which enhances thermionic emission across the barrier. Tunneling through the thin triangular barrier and hot electron effects due to the smaller reflection coefficient also contribute to the enhanced collection efficiency.

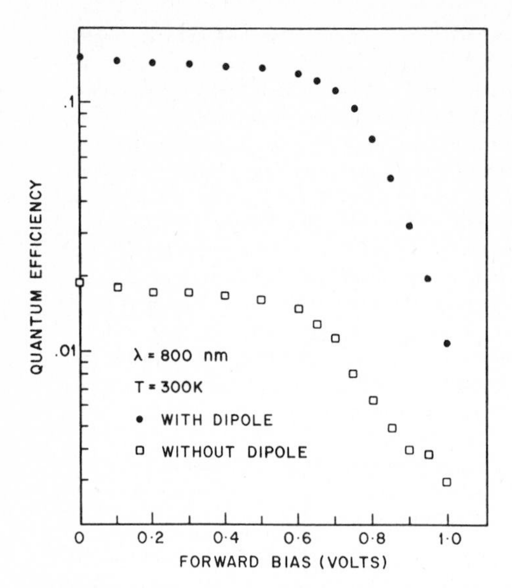

Fig. 17 External quantum efficiency vs. forward bias voltage.

Figure 17 shows the quantum efficiency vs. forward bias at $\lambda \cong 8000\text{Å}$ for the two structures. It decreases first gradually and then rapidly above a certain cutoff voltage. This behavior, due to band flattening, is well known and has been observed previously. The efficiency rapidly increased with reverse voltage in both structures and then saturated. Above 10V the quantum efficiency in the energy range 1.5-1.7 eV was identical in both structures and $\cong 60\%$. This is expected, since at fields $> 10^5$ V/cm the electrons acquire so much energy that the barrier height is no more a significant limiting factor to the efficiency.

The smallest size of the DID depends on the diffusion coefficient of the dopants, which depends on the doping density, the substrate temperature and the growth time. For Si and Be in the AlGaAs/GaAs system one should be able to place the doping sheets as close as 10Å, for substrate temperatures $\lesssim 600\,^\circ$C and growth times of $< 1/2$ hr. without significant interdiffusion.

An important application of DIDs is the enhancement of impact ionization of one type of carrier in superlattice and staircase avalanche photodiodes. Recently Capasso et al. [29] demonstrated that the difference between conduction and valence discontinuities in a heterojunction superlattice can lead to the enhancement of the ionization rates ratio. These experiments lead to the conception of the staircase solid state photomultiplier [30]. Introducing DID's at the band steps of these detectors can be used to further enhance the ionization probability of electrons at the steps, since carriers gain ballistically the dipole potential energy in addition to the energy ΔE_c obtained from the band step. In addition, the DID helps by promoting over the valence band barrier holes created near the step by electron impact ionization without sacrificing speed. Thus the dipole energy $e\Delta\Phi$ should equal or exceed the valence band barrier. This application is illustrated in Fig. 18. The top part of Fig. 18 gives the band diagram of the staircase solid state photomultiplier. Since the conduction band discontinuity exceeds the band gap after the step, while the valence band step is of the opposite sign to assist hole ionization and the electric field is too small to cause hole initiated ionization, only electrons

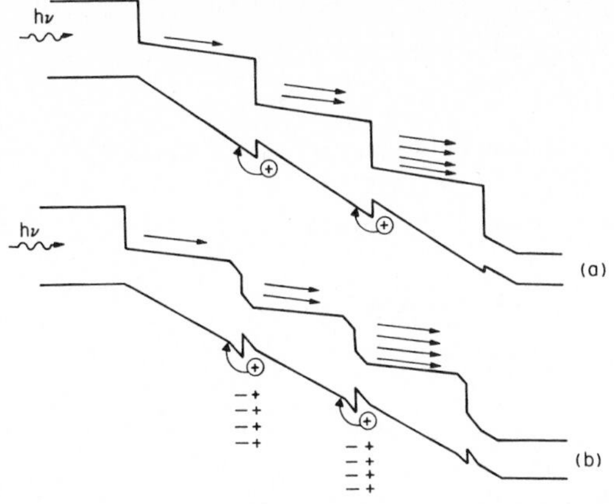

Fig. 18 Band-diagram of staircase avalanche photodiode (a) without DID; (b) with DID.

ionize and the steps are the analog of the dynodes in a photomultiplier. Adding dipoles at the step as shown in Fig. 18(b) will minimize hole trapping effects. In addition in the event that in the structure of interest the conduction band step is not quite equal to the required ionization energy in the material after the step, the dipole can help to compensate the small energy deficit (0.1-0.2 eV).

DID's can also be used at the interfaces of heterojunction and quantum well lasers to increase the confinement energies of carriers.

In conclusion DIDs represent a technique to effectively tune barrier heights and band discontinuities and as such have great potential for heterojunction devices.

1. Capasso, F. (1984a), Surf. Sci. *142*, 513.
2. Capasso, F. (1984b), Laser Focus, July issue.
3. Capasso, F. (1985), Physica *129*B, 92.
4. Kroemer, H. (1982), Proc. IEEE *70*, 13.
5. Kroemer, H. (1957) RCA Rev. *18*, 332.
6. Hayes, J. R., Capasso, F., Gossard, A. C., Malik, R. J., and Wiegmann, W. (1983a), Electron. Lett. *13*, 410.
7. Capasso, F., Tsang, W. T., Bethea, C. G., Hutchinson, A. L., and Levine, B. F., (1983b), Appl. Phys. Lett. *42*, 93.
8. Miller, D. L., Asbeck, P. M., Anderson, R. J., and Eisen, F. H., (1983), Electron. Lett. *19*, 367.
9. Malik, R. J., Capasso, F., Stall, R. A., Kiehl, R. A., Ryan, R. W., Wunder, R., and Bethea, C. G. (1985), Appl. Phys. Lett. *46*, 600.
10. Levine, B. F., Tsang, W. T., Bethea, C. G., and Capasso, F. (1982), Appl. Phys. Lett. *41*, 470.
11. Levine, B. F., Bethea, C. G., Tsang, W. T., Capasso, F., Thormber, K. K., Fulton, R. C., and Kleinman, D. A. (1983), Appl. Phys. Lett. *42*, 769.

12. Capasso, F. and Kiehl, R. A. (1985), J. Appl. Physics, August 1st issue.

13. Tsu, R. and Esaki, L. (1973), Appl. Phys. Lett. *22*, 562.

14. Chang, L. L., Esaki, L., and Tsu, R. (1974), Appl. Phys. Lett. *24*, 593.

15. Sollner, T. C. L. G., Goodhue, W. D., Tannenwald, P. E., Parker, C. D., and Peck, D. D., (1983), Appl. Phys. Lett. *43*, 588.

16. Sollner, T. C. L. G., Tannenwald, P. E., Peck, D. C., and Goodhue, W. D. (1984), Appl. Phys. Lett. *45*, 1319.

17. Ricco, B. and Azbel, M. Ya. (1984), Phys. Rev. *B29*, 1970.

18. Rine, C. Editor, Computer Science and Multiple-valued Logic (North Holland 1977), Amsterdam).

19. Capasso, F., Cox, H. M., Hutchinson, A. L., Olsson, N. A., and Hummel, S. G. (1984b), Appl. Phys. Lett. *45*, 1193.

20. Cox, H. (1985), J. Crystal Growth, *69*, 641.

21. Capasso, F., Cho, A. Y., and Foy, P. W. (1984a), Electron. Lett. *20*, 635.

22. Forrest, S. R., Kim, O. K. and Smith, R.G., (1982) Appl. Phys. Lett. *41*, 95.

23. Campbell, J., Dentai, A. G., Holden, W. S., and Kasper, B. L. (1983), Electron. Lett. *19*, 818.

24. Capasso, F., Cho, A. Y., Mohammed, K., and Foy, P. W. (1985a), Appl. Phys. Lett. *46*, 664.

25. Grant, R. W., Waldrop, J. R., and Krawt, E. A. (1983), Phys. Rev. Lett. *40*, 6.

26. Te. Velde, T. S. (1973), Solid-State Electron. *16*, 1305.

27. Shik, A. Ya. and Shmartsev, Y. V. (1984), Sov. Phys. Semicond. *15*, 799.

28. Womac, S. F. and Rediker, R. H. (1972), J. Appl. Phys. *43*, 4129.

29. Capasso, F., Tsang, W. T., Hutchinson, A. L., and Williams, . F. (1982d), Appl. Phys. Lett. *40*, 36.

30. Capasso, F., Tsang, W. T., and Williams, G. F. (1983), IEEE Trans. Electron Devices, *ED-30*, 381.

Electric Field-Induced Decrease of Excition Lifetimes in GaAs Quantum Wells

J.A. Kash and E.E. Mendez
IBM Thomas J. Watson Research Center, P.O. Box 218,
Yorktown Heights, NY 10598, USA
H. Morkoç
University of Illinois at Urbana-Champaign, Urbana, IL 61801, USA

Previously[1], the c.w. photoluminescence (PL) at helium temperatures for excitons confined in quantum wells has been seen to decrease sharply (i.e. is quenched) when an electric field is applied perpendicular to the layers. It was suggested there that the decrease might be due to spatial separation of the electrons and holes under the influence of the electric field. The electron-hole spatial overlap is reduced by the separation, thus increasing the exciton radiative lifetime. We have measured the exciton PL decay directly in order to examine the luminescence quenching in more detail.

We used MBE-grown quantum wells consisting of six $Ga_{0.65}Al_{0.35}As$ - GaAs quantum wells. Well thickness was ~30Å, with a barrier thickness of 100Å. The wells were clad between a 1μm GaAs buffer layer and a 0.1μm $Ga_{0.65}Al_{0.35}As$ layer on top. An electric field perpendicular to the layers was applied by means of a Schottky diode configuration, formed by evaporating a 60Å thick semitransparent Ni film on the heterostructures. The quantum wells were inside the space-charge region, estimated to be ~0.8μm, where the built-in field was ~10^4 V/cm. For the range of applied voltages V_{ext} used in this work, we estimate the total electric field to be $\lesssim 5\times10^4$V/cm.

The samples, held at T=8K, were excited with a synchronously-pumped cavity-dumped dye laser producing 10 psec pulses at a 700 kHz repetition rate. The laser wavelength was 5820Å, and the maximum pulse energy incident on the sample was 1 njoule. The PL was analyzed with a double-pass 0.75m monochromator and detected with a cooled photomultiplier tube. The time decay was measured by time-resolved photon counting described more fully elsewhere[2]. The system resolution was 200 psec (FWHM); decay times as short as 50 psec were inferred from deconvolution techniques.

Figure 1 shows the PL decay of the intrinsic n=1 exciton at several values of externally applied voltage V_{ext}. Note the substantial decrease in lifetime with applied field. The fastest decay is approaching the system resolution limit. In addition to the fast (τ < 400 psec.) initial decay, there is a slower component which probably involves the recombination of ground state electrons with carbon acting as an acceptor (e,A^0). This component, which is also quenched by the electric field, is very weak and will not be considered further here. The time-integrated PL spectra also decreases sharply with applied field[1]. Similar results are seen in other quantum well samples. In fig. 2, the luminescence decay time τ

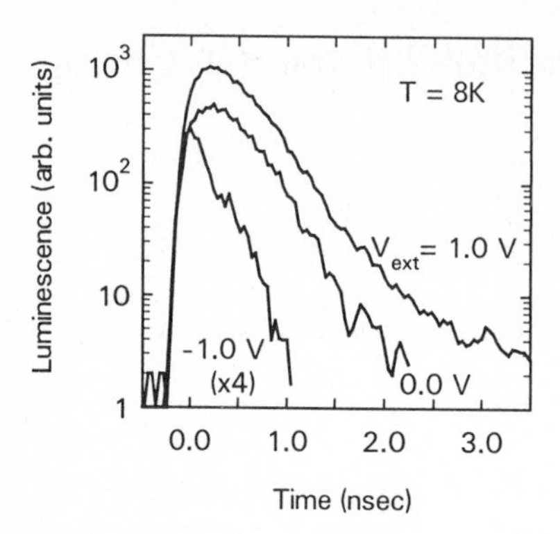

Fig.1. Photoluminescence decay of the intrinsic n=1 exciton at 7570A for several values of externally applied voltage V_{ext}. The average laser power here was 100 μwatts

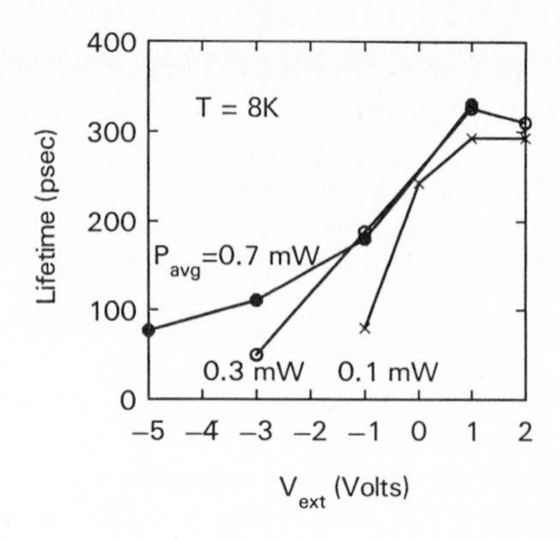

Fig.2. Luminescence decay time vs. V_{ext} for several average laser pump powers for the same sample as in Fig.1. The highest power corre-sponds to a peak exciton density of about 3×10^{18} cm^{-3}

vs. V_{ext} is shown for several average laser pump powers for the same sample. We see that the lifetime quenching is reduced at high laser intensities.

This time-resolved data is not consistent with the carrier separation separation model[1]. That model gives an increase in the exciton radiative lifetime τ_r with the field. As a result, the non-radiative decays (assumed to have a constant lifetime τ_{nr}) should dominate at high fields and the overall exciton lifetime $\tau = (\tau_r^{-1} + \tau_{nr}^{-1})^{-1}$ should increase. We observe a decrease in the overall lifetime τ, which suggests a shorter τ_{nr} with applied field. We propose two mechanisms which could be responsible for the observed decrease. First, the carriers, which are strongly polarized within the wells by the external field, can leave the wells by Fowler-Nordheim tunneling. The tunneling probability depends exponentially[3] on the combination $-(m^{1/2}\phi^{3/2}/\mathcal{E})$, where m is the effective mass, ϕ is the potential barrier with respect to the quantum level, and $\mathcal{E}$ is the electric field. Times of order 10^{-13} sec. can be obtained for $\mathcal{E}=5\times10^4$ V/cm and reasonable values of ϕ (e.g. ϕ_e = 0.06 eV, ϕ_h = 0.03 eV). Unfortunately, this "binding" time is very sensitive to ϕ and $\mathcal{E}$, so that a more quantitative analysis could not be made. The second explanation is field-induced leakage of the wave functions into the $Ga_{1-x}Al_xAs$ barriers. If τ_{nr} is proportional to the amount of the electron and hole wavefunctions in the barriers, then it will increase with field. A variational calculation, similar to the one described in [4], shows that this leakage will reduce the exciton lifetime. The reduction, however, is only a few percent, which is much less than the observed decrease. We conclude that Fowler-Nordheim tunneling is the principal cause of the lifetime quenching.

The reduced lifetime quenching seen at high laser powers can be explained by exciton screening. The photo-created excitons are polarized by the electric field, and the net exciton polarization (which depends upon the exciton density) can then screen out the electric field. If n is the exciton concentration and α the polarizability, then screening of the external field will be important for $\alpha n \sim 1$.

From the variational calculation[4] of the wavefunctions, we estimate the polarizability of the excitons perpendicular to the layers to be $\alpha \approx 10^{-19}$ cm^3. Thus the screening should become important at exciton densities comparable to the highest densities ($\sim 3 \times 10^{18}$ cm^{-3}) used in these experiments.

We thank L.F. Alexander for preparing the Schottky diode structures. The work at IBM has been sponsored in part by the U.S. Army Resarch Office. The work at the University of Illinois has been sponsored by the Joint Services Electronics Program.

References

1. E.E. Mendez, G. Bastard, L.L. Chang, L. Esaki, H. Morkoc, and R. Fischer, Phys. Rev. B26, 7101 (1982).
2. J.A. Kash, J.H. Collet, D.J. Wolford, and J. Thompson, Phys. Rev. B27, 2294 (1983).
3. R.A. Smith, *Wave Mechanics of Crystalline Solids, Second Edition,* (Chapman and Hall, London, 1969), p. 60.
4. G. Bastard, E.E. Mendez, L.L. Chang, and L. Esaki, Phys. Rev. B28, 3241 (1983).

Reduction of Electon-Phonon Scattering Rates by Total Spatial Quantization

M.A. Reed, R.T. Bate, W.M. Duncan, W.R. Frensley, and H.D. Shih
Central Research Laboratories, Texas Instruments Incorporated, P.O. Box 225936, M/S 154, Dallas, TX 75265, USA

The physics of spatially quantized systems has been the subject of intense investigation since the introduction of MBE and MOCVD. Quantum wells have been exhaustively examined since the seminal works of Dingle et al[1]. More recently, studies on quantum wires[2] have yielded interesting new properties. Here, we present data on a completely spatial quantized system that shows unique photoluminescence structure resulting from an electron or hole phonon scattering bottleneck. This paper discloses evidence of the modification of carrier-phonon scattering rates by the imposition of complete spatial quantization on GaAs-AlGaAs quantum wells. The photoluminescence spectra exhibit striking structure in the normal intrinsic exciton luminescence of the confined quantum well states which is best explained by a bottleneck for electron/hole energy loss. We believe that this bottleneck is a direct consequence of the quantization of the electronic and/or phonon dispersion relations. This photoluminescence structure does not occur for samples of higher dimensionality.

The samples used in these experiments were grown by MBE and are shown schematically in Fig. 1(a). The samples were grown on (100) Cr-doped GaAs substrates and consisted of a 1 micron GaAs MBE buffer layer followed by a 1

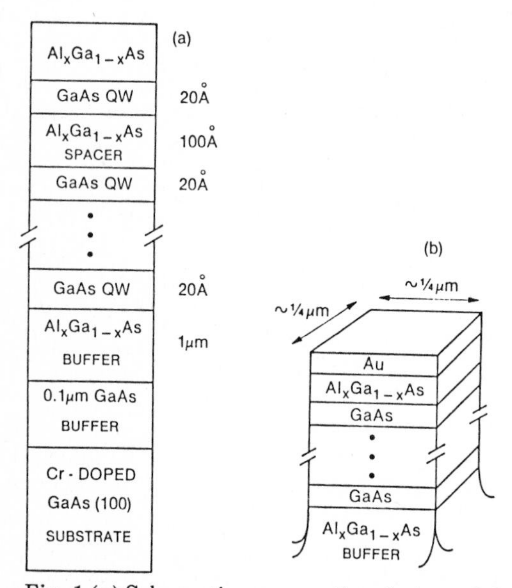

Fig. 1 (a) Schematic cross-sectional view of the GaAs quantum well MBE samples (b) Schematic cross-sectional view of a quantum dot (0 DOF) structure

micron $Al_xGa_{1-x}As$ buffer, 20 Å GaAs wells, and 100 Å $Al_xGa_{1-x}As$ (x=0.3) barriers. All samples were nominally undoped. Patterning of the bulk multiple quantum well samples was done by direct electron-beam writing in a film of polymethylmethacrylate (PMMA) on the sample surface. Conventional lift-off techniques were used to define 1000 Å thick Au metal patterns on the sample surface. The metal patterns were transferred to the underlying sample by a $BC\ell_3$ reactive ion etch with the metal serving as an etch mask. Figure 1(b) shows a schematic of a single quantum dot structure. Quantum wire (1 degree of freedom, or 1 DOF) structures were fabricated simultaneously on the same sample as the quantum dot (0 DOF) structures. Arrays of these structures were fabricated to achieve sufficient photoluminescence intensity.

Photoluminescence measurements were performed in a helium flow Janis optical cryostat. The sample was excited by the 2.54 eV line of a focused Ar+ laser and the photoluminescence radiation was collected by a 1.0 m Chromatix spectrometer of 7 Å/mm focal plane dispersion. Conventional detection techniques were used.

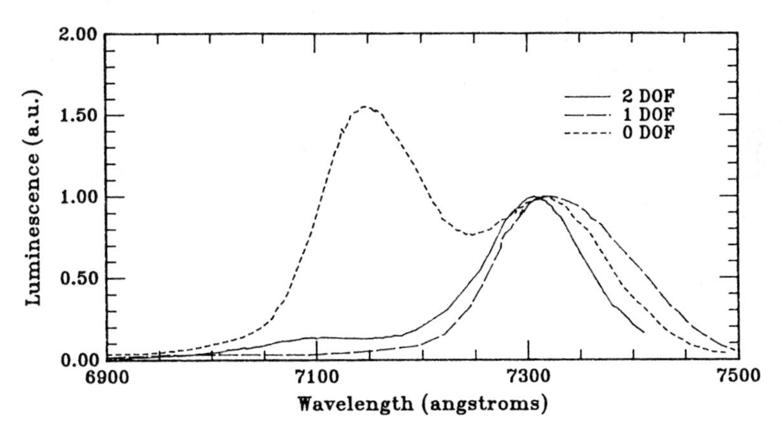

Fig. 2 Photoluminescence spectra of GaAs - AlGaAs quantum wells (2 DOF), quantum wires (1 DOF), and quantum dots (0 DOF) at T=4.2°K

Figure 2 shows the photoluminescence spectra of sample A for the three distinct DOF cases previously discussed. The 7300 Å luminescence peak is the (n=1) electron - (n=1) heavy hole recombination radiation, prominent in all three DOF cases. We have normalized the three different DOF spectra to the same intensity at this wavelength for comparison purposes. The small shifts in the 7300 Å peak positions are attributable to residual surface damage or strain resulting from the plasma etch. The loss in photoluminescence signal was readily accounted for by loss of sample volume, indicating that non-radiative loss mechanisms due to sidewall damage are not large.

The outstanding structure at ~7100 Å in the 0 DOF spectra, also seen at smaller relative intensity in the 2 DOF case, is the (n=1) electron - (n=1) light hole recombination radiation. The absence of this structure in the 1 DOF case is attributable to a decrease in carrier density by the attenuation of the excitation radiation in the top Au metal mask. The attenuation is the same in the 0 DOF (quantum dot) case, yet the (n=1) electron - (n=1) light hole recombination has increased x10 over the 2 DOF case. The appearance of the light hole peak in the quantum dot case was not a local effect; all sections of the sample investigated exhibited the same spectrum. The 1 DOF and 0 DOF structures were fabricated adjacent to each other on the sample to eliminate any systematic errors due to the plasma etch.

Figure 3 schematically diagrams the relaxation mechanisms in the luminescence process of the ground state of a quantum well. $\tau_{hh,intra}$, $\tau_{lh,intra}$, and $\tau_{e,intra}$ are the intrabranch phonon relaxation times of heavy holes, light holes, and electrons, respectively, down their respective branches. τ_{inter} is the light-to-heavy hole scattering time. τ_{recomb} is the electron - hole radiative recombination time. We shall assume that the heavy and light hole - electron recombination times are approximately equal and energy independent. Relaxation mechanisms involving any existing excited states, the split-off band, or interface defects, have been ignored. Carrier-carrier scattering has also been neglected in this treatment, but will be approximately constant in the three different dimension cases.

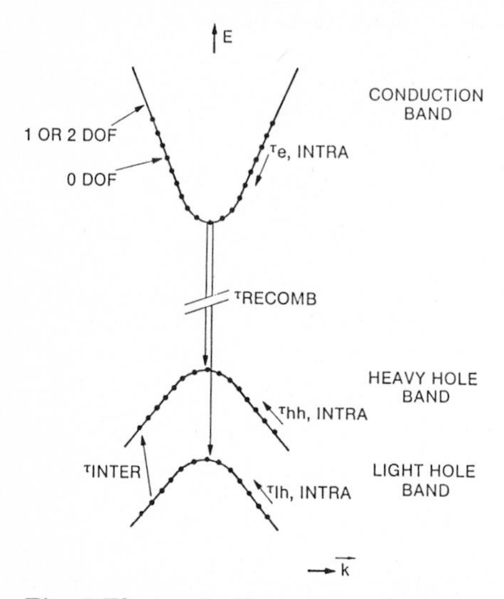

Fig. 3 Electronic dispersion relation of the n=1 ground state for quantum wells and wires (solid lines) and for quantum dots (points)

Upon initial excitation, the three DOF cases have similar thermalization dynamics. The crossover from essentially 3-dimensional behavior to the specific DOF case is complex and will not be treated here. When the carriers reach the region $E \sim \hbar\omega_{LO}$, the branch crossing from the light hole to heavy hole branch can occur[3]. The interbranch scattering time, τ_{inter}, will be limited by the phase space available at the crossover. If τ_{inter} is sufficiently large, a "bottleneck" results and an appreciable population of light holes can be formed. Additionally, the elimination of selected small **k**-modes in the phonon spectrum will enhance the light-hole transient population.

Observation of this bottleneck in light-to-heavy hole scattering by photoluminescence in bulk GaAs is impractical since $\tau_{recomb}/\tau_{inter} \sim 10^3$. We have reduced this rate not only by the fabrication-imposed increase of τ_{inter} but also by the decrease of τ_{recomb} due to the enhancement of the radiative transition probability in a quantum well with sufficiently narrow well thickness[4]. Band filling can be eliminated as an alternative explanation since the linewidths are approximately the same, and excitation power dependence exhibited no change in the heavy hole peak/light hole peak ratio.

An intriguing possibility is the direct observation of spatial quantization from the discrete (multiply degenerate) electronic levels in the 0 DOF quantum dots. The spectrum was investigated in sufficiently high resolution to observe fine structure in the quantum well luminescence with the sample immersed in superfluid helium. The negative results for this search are best explained by an effect analogous to inhomogeneous line broadening; the variation of the dot size across the array will produce fluctuations in the discrete line positions. For the experimental resolution of 0.1 Å and the above dot parameters, a fluctuation of $\sim 1\%$ is sufficient to mask the discrete lines.

In summary, we have observed the photoluminescence from a series of spatially quantized quantum structures and believe we have observed the first evidence for the reduction of carrier-phonon scattering via total spatial quantization by a fabrication-imposed potential. The effect of the discrete dispersion relations of the electrons and/or phonons is to modify the relaxation kinetics of carriers in the quantum structures, and this is observable in the photoluminescence spectrum.

We wish to thank K. Bradshaw for electron beam fabrication and T. Kaluza, P. Tackett, A. Wetsel, and J. Williams for technical assistance. We are grateful for illuminating discussions with L. Cooper, J. Erskine, G. Iafrate, L. Kleinman and P. Stiles. This work was supported in part by the Office of Naval Research and in part by the U.S. Army Research Office.

References

1. R.Dingle, A.C.Gossard, and W.Wiegmann, *Phys. Rev. Lett.* **34**, 1327(1975).
2. W.J.Skocpol, L.D.Jackel, E.L.Hu, R.E.Howard, and L.A.Fetter, *Phys. Rev. Lett.* **49**, 951(1982).
3. E.M. Conwell, *High Field Transport in Semiconductors*, Solid State Physics Supplement **9**, F. Seitz, D. Turnbull, H. Ehrenreich, eds. (Academic Press Inc., New York, 1967).
4. J. Christen, D. Bimberg, A. Steckenborn, and G. Weimann, Appl. Phys. Lett. **44**, 84 (1984).

Hot Electron Diffusion in Superlattices

J. Ho and R.O. Grondin
Center for Solid State Electronics Research, Arizona State University,
Tempe, AZ 85287, USA

Abstract

Hot electron diffusion in superlattices is studied by the Monte Carlo technique. As the applied electric field increases, the diffusion coefficient decreases. It has values comparable to those of the constituent materials.

Hot electron diffusion in a lateral surface superlattice (LSSL) is studied by the Monte Carlo technique. The LSSL is a surface oriented two-dimensional system with superstructure in both dimensions. Reich et al [1] found that the LSSL should exhibit a negative differential mobility due to the onset of Bloch oscillations. This negative differential mobility may make this structure useful as a negative resistance oscillator and amplifier at millimeter-wave frequencies. In order to properly model its application as such an oscillator, one additionally needs a relation between the electron diffusion coefficient and the field. Carrier diffusion in superlattices has not yet been studied. Here we use the Reich model,and estimate values of the diffusion coefficient by Monte Carlo techniques.

The diffusion coefficient is estimated in two ways. First, the diffusion coefficient was estimated by observing the spread of an ensemble of electrons whose velocity and position are allowed to evolve in time. The ensemble is initially located at an arbitrary location in real space. The initial K-vectors are chosen in accordance with an equilibrium Maxwellian distribution. Then the ensemble is allowed to evolve through an initial transient into a steady-state condition. The steady-state condition is identified by waiting for the drift velocity to settle. Then, a steady-state ensemble corresponding to a delta function in real space is created, by assigning each electron a zero real-space position without altering its steady-state k-vector value. We then observed the speed of this initial delta function into an approximately gaussian shape, and estimated the diffusion coefficient by using the standard time-of-flight estimator

$$D = \frac{\langle \Delta X^2 \rangle}{2t} \qquad (1)$$

where D = diffusion coefficient (cm^2/sec)
ΔX = deviation of an electron's real space position from the mean value (cm)
t = time of flight (sec)

and $\langle \ \rangle$ denotes an ensemble average.

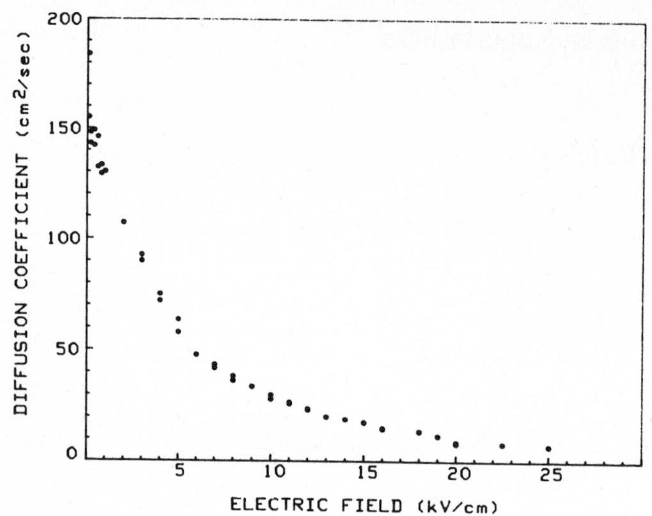

Figure 1. Electron Diffusion Coefficient versus Electric Field.

The result is shown in Figure 1. We can see that diffusion coefficient decreases as the field increases. Statistical variance on the order of 10% is associated with these values. As is shown in Figure 2, there indeed is a linear relationship between $\langle \Delta X^2 \rangle$ and time. The slope of the least mean square error fit shown in Figure 2 yields a diffusion coefficient lying within the 10% spread seen in our estimates.

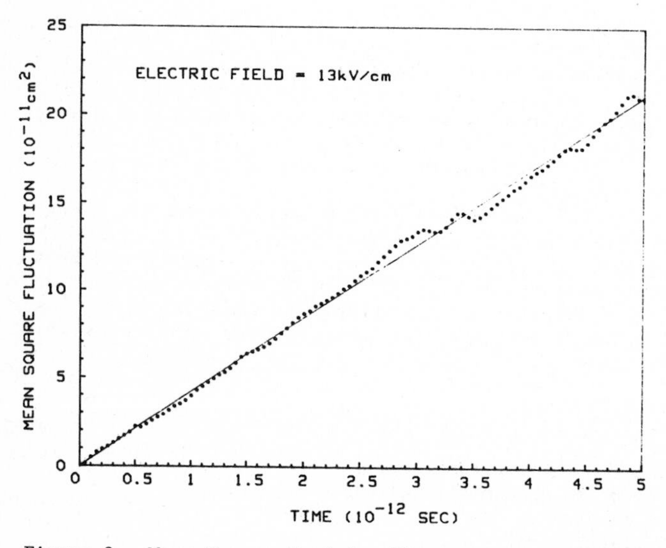

Figure 2. Mean Square Position Fluctuations versus Time.

The second method used to estimate the diffusion coefficient is an integration of the velocity autocorrelation function [2]. Diffusion is related to an integral of this function by

$$D = \int_0^{\infty} \phi_{\Delta V}(\theta)d\theta \qquad (2)$$

where $\phi_{\Delta V}$ is the velocity fluctuation autocorrelation function. This function is defined by

$$\phi_{\Delta V}(\theta) = \langle \Delta V(t)\, \Delta V(t+\theta) \rangle \qquad (3)$$

where ΔV is the fluctuation of an electronic velocity away from the mean velocity. The result of this integration differed from values of figure 1 by approximately 5%. We attribute this difference to both numerical error and to the use of finite integration time instead of the infinite integration time found in equation (2). Reich et al.[1] have demonstrated that the velocity fluctuation autocorrelation function exhibited significant peaks at periods corresponding to that of the Bloch oscillation. We show some sample autocorrelation functions in figure 3.

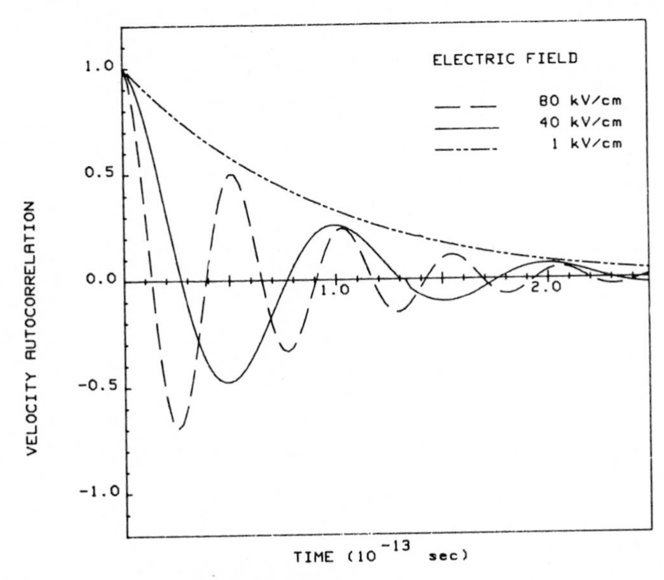

Figure 3. Velocity Fluctuation Autocorrelation Function for several fields.

We believe that the low values of diffusion coefficient at high electric fields are a consequence of Bloch oscillations. This can be visualized in several fashions. First, as an electron undergoes a periodic movement in real space during a Bloch oscillation, it in this sense is spatially localized. Alternatively, it can be viewed as the result of the damped, oscillatory shape of $\phi_{\Delta V}$. This shape is the direct result of Bloch oscillations [1].

References

1. R.K. Reich, R.O. Grondin and D.K. Ferry: Physical Rev. B 27, 3483 (1983)

2. R.O. Grondin, P.A. Blakey, J.R. East and E.D. Rothman: IEEE Trans. Electron Devices ED-28, 914 (1981)

Time-Resolved Photoluminescence of $GaAs/Al_xGa_{1-x}As$ Quantum Well Structures Grown by Metal-Organic Chemical Vapor Deposition

J.E. Fouquet and A.E. Siegman
Stanford University, Stanford, CA 94305, USA
R.D. Burnham and T.L. Paoli
Xerox Palo Alto Research Center, Palo Alto, CA 94305, USA

I. Introduction

Until this work [1], time-resolved photoluminescence from metal-organic chemical vapor deposition (MOCVD)-grown $GaAs/Al_xGa_{1-x}As$ quantum wells has not been extensively studied. We have observed time-resolved and time-integrated photoluminescence of MOCVD quantum well structures from 17K to room temperature over a wide range of excitation densities. These structures were grown in a reactor which routinely produces high quality laser material.

II. Experiment

The multiple quantum well (MQW) structure consisted of 40 100Å GaAs wells between 200Å $Al_{.24}Ga_{.76}As$ barriers, surrounded by $Al_{.5}Ga_{.5}As$ cladding regions. The single quantum well (SQW) structure consisted of one GaAs well between two 600Å $Al_{.24}Ga_{.76}As$ barriers, surrounded by $Al_{.5}Ga_{.5}As$ cladding regions. The active and cladding regions were not intentionally doped.

The samples were excited by an actively mode-locked krypton-ion laser system at 647 nm. Time-resolved single photon counting detection was used to determine the photoluminescence decay rate. The ultimate resolution of the detection system was about 100 psec.

III. Results

Time-resolved photoluminescence at room temperature depends strongly on excitation density in both the MQW [1] and SQW structures. The observed decrease in decay rate with increasing excitation is characteristic of carriers saturating a limited number of trap sites. At low excitations, most of the carriers fall into traps and thus a fast decay rate is observed. The MQW results are quite unusual, however, in that tens of nanoseconds after the establishment of an initial slow exponential decay rate, the decay rate increases. This behavior has not been previously observed in bulk GaAs or $GaAs/Al_xGa_{1-x}As$ quantum wells, to the best of our knowledge. This increase in decay rate may be due to carriers emptying out of traps with relatively short holding times. The SQW structure also exhibits trap saturation. However, traps are relatively easier to saturate in the SQW structure than in the MQW structure, since a slower decay is obtained in the SQW structure at low excitation levels.

Time-integrated photoluminescence (area under spectral peak) was examined as a function of excitation density at a number of different temperatures for the MQW sample. At room temperature the photoluminescence is relatively low at low excitations because many of the carriers fall into the trap states and become unavailable for radiative recombination at the $n = 1$ transition. At high excitation the trap states can be saturated and photoluminescence efficiency is much higher. At low temperatures the photoluminescence efficiency at low excitations is much higher than at room temperature. Photoluminescence does not decrease significantly from the lowest temperatures to above 150K. Thus the trap does not become active until relatively high temperatures.

The photoluminescence spectrum for low excitation at 17K of the same multiple quantum well sample has a large secondary peak at 1.548 eV and a low energy tail, illustrating the relatively large role of impurity species. Time-resolved photoluminescence at 17K at the wavelength of the main peak is shown in Fig. 1 (note time scale). The decays are faster than at room temperature and highly nonexponential, indicating that not just simple excitonic recombination is being observed. The spectral peak at 1.548 eV exhibits a slower decay than the main peak.

Interfaces between different materials play a major role in MQW structures. Thus, the unusual increase in photoluminecence decay rate observed in the MQW at room temperature could be due to traps located at interfaces and may be specific to MOCVD growth. Bulk $Al_xGa_{1-x}As$

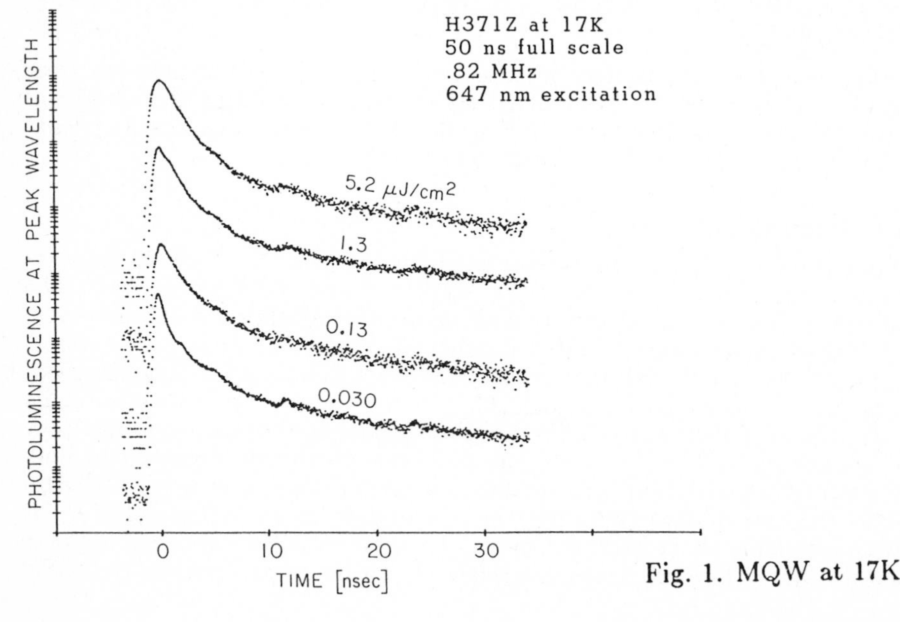

Fig. 1. MQW at 17K

often suffers from impurity-related problems, so it could also be the source of the traps.

A sample consisting of 1200Å of bulk $Al_xGa_{1-x}As$ of the same Al mole fraction x as the barriers in the MQW structure between two $Al_{.5}Ga_{.5}As$ cladding layers was investigated. If the traps are in the bulk material, one would expect the trapping to be worse in this bulk sample than in the MQW structure. If the traps are at the interfaces, the trapping problem should be milder in the bulk sample, which has just two interfaces at the cladding layers.

Time-resolved photoluminescence showed the faster initial decays at low excitation characteristic of trap saturation and showed increases in decay rate at long delays after excitation qualitatively similar to the MQW results. However, the onset of a serious increase in photoluminescence decay rate occurred at lower excitations than in the MQW sample. Thus the traps are more likely to be interface traps because they are harder to saturate in the MQW sample than in the bulk sample.

Longitudinal optical (LO)-phonon replicas of various emission lines are often observed in bulk III-V compounds, but not in quantum wells. Holonyak *et al.* [2] have reported lasing in quantum wells on a transition roughly 36 meV below the n=1 heavy hole transition, and attributed it to stimulated phonon emission. (36 meV is the energy of an LO-phonon in bulk GaAs.) Other researchers have not been able to reproduce this result. The low excitation MQW spectrum at 34K, Fig. 2, shows a prominent

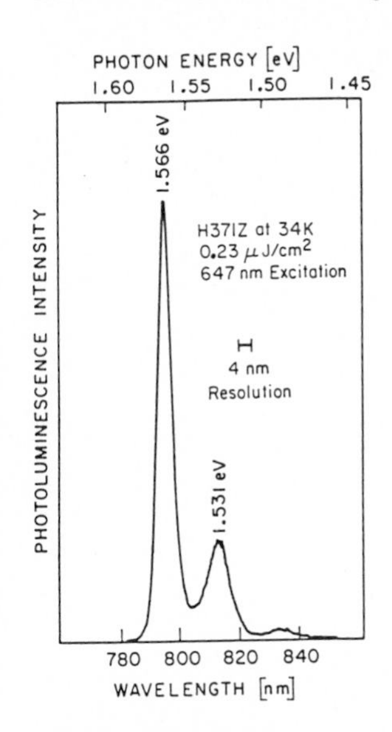

Fig. 2. Upper-left: low excitation

Fig. 3. Lower-left: high excitation ➤

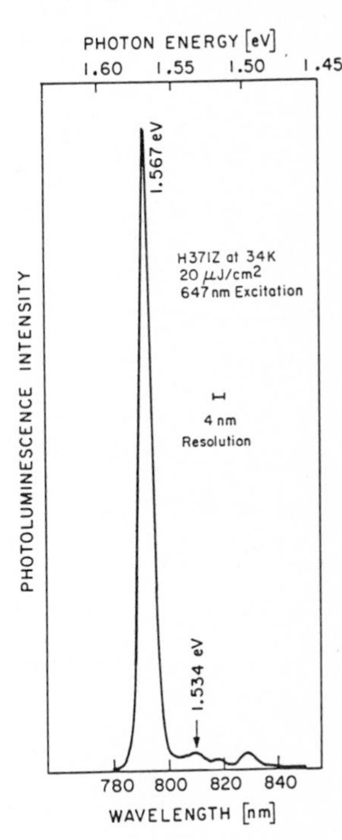

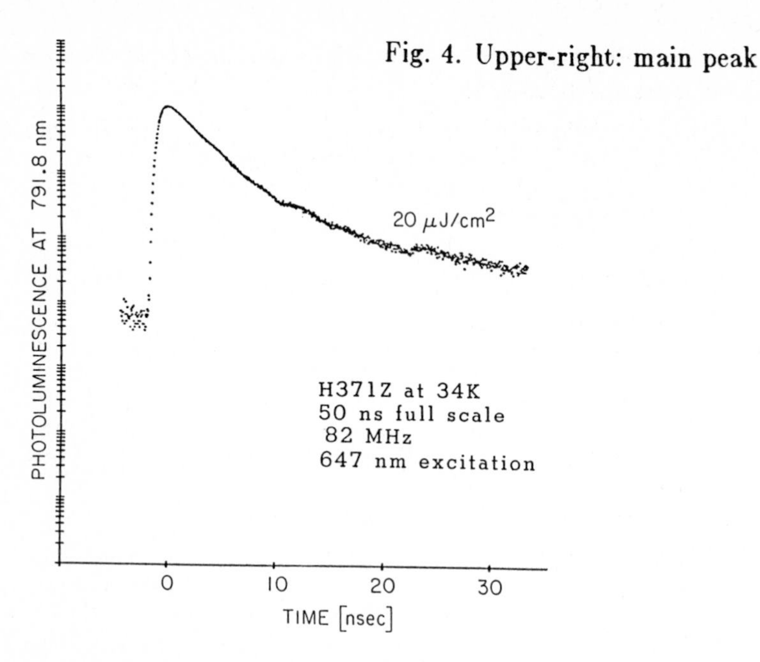

Fig. 4. Upper-right: main peak

peak at an energy (1.531 eV) appropriate for an LO-phonon sideband of the main peak (1.566 eV). However, time-resolved photoluminescence shows that this peak can not be a sideband. The 1.566 eV feature decays rapidly (Fig. 4) while the 1.531 eV feature decays much more slowly (Fig. 5). An LO-phonon sideband should decay at least as rapidly as the 1.566 eV feature.

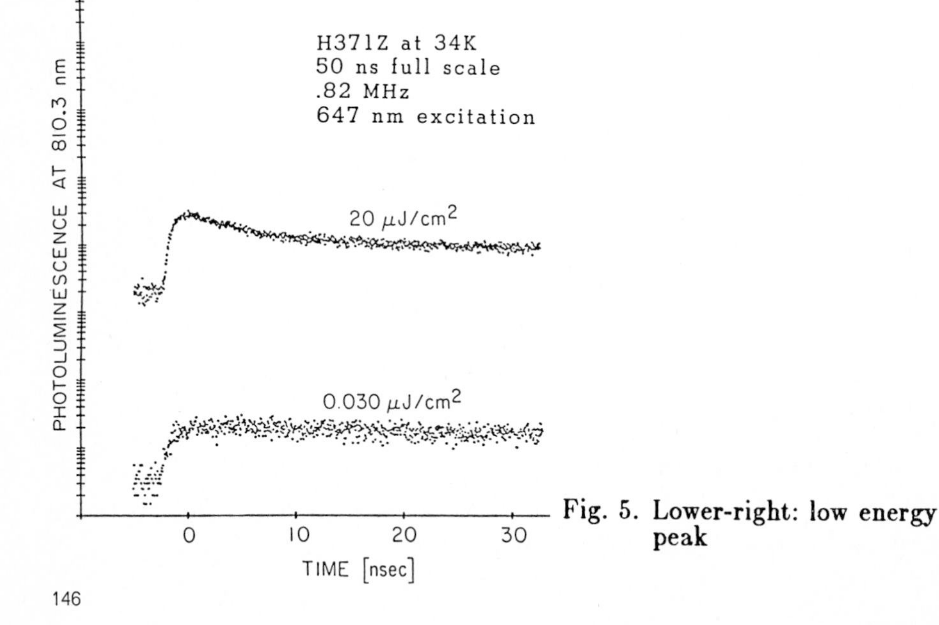

Fig. 5. Lower-right: low energy peak

Time-integrated photoluminescence also indicates this feature is not due to LO-phonon emission. Comparison of Fig. 3 at $20\mu J/cm^2$ excitation and Fig. 2 at $0.23\mu J/cm^2$ excitation shows that the 1.531 eV peak plays a proportionately larger role at low excitation, suggesting this peak is due to an impurity, whereas a stimulated process should be stronger at high excitation.

IV. Conclusion

Room temperature photoluminescence in these samples is characteristic of trap saturation. These traps are probably interface traps rather than bulk traps, are not active at low temperatures and probably have limited holding times.

References

1. See also J. E. Fouquet, A. E. Siegman, R. D. Burnham, and T. L. Paoli, *Appl. Phys. Lett.* **46**, 374 (1985).
2. Nick Holonyak, Jr., Robert M. Kolbas, Russell D. Dupuis, and P. Daniel Dapkus, *IEEE J. Quant. Electron.* **QE–16**, 170 (1980).

A Study of Exciton and Carrier Dynamics and a Demonstration of One-Picosecond Optical NOR Gate Operation of a GaAs-AlGaAs Device

N. Peyghambarian, H.M. Gibbs, and J.L. Jewell*
Optical Sciences Center, University of Arizona, Tucson, AZ 85721, USA
A. Migus and A. Antonetti
Laboratoire d'Optique Appliquée, Ecole Polytechnique-ENSTA, F-91120 Palaiseau, France
D. Hulin and A. Mysyrowicz
Groupe de Physique des solides, Ecole Normale supérieure, 2 Place Jussieu, F-75005 Paris, France

Abstract

Dynamics of exciton screening by either free carriers or excitons in a GaAs-AlGaAs superlattice is investigated. The speed of a GaAs-AlGaAs optical-logic gate is time resolved.

We report the first time-resolved observation of a GaAs-AlGaAs multiple-quantum-well (MQW) optical NOR gate. The ≃ 1-ps speed of this room-temperature nonlinear etalon is the fastest for such a low-power optical-logic device. About 100-fJ/μm^2 energy per unit area was used for the logic operation.

The device consists of alternating layers of 152-Å GaAs and 104-Å AlGaAs (≃ 1.5-μm total GaAs thickness) grown by molecular-beam epitaxy on a GaAs substrate. The substrate was etched away and the remaining flake sandwiched between two dielectric coated glass slides with ≃ 90% reflectivities. The amplified output of a cw mode-locked ring dye laser [1] (pulses of ≃ 100-fs duration at 620 nm) was divided into two parts, each focused on a 2-cm-long water cell to generate white-light-continuum pulses (also of ≃ 100-fs duration). One of the continuum pulses was used as a broadband probe; and a narrow-band interference filter was inserted in the second continuum beam to generate a pump beam with frequency above the bandgap of GaAs (wavelength ≃ 760 nm). The pump frequency was below the bandgap of AlGaAs, so only GaAs layers were excited. The probe beam was focused to a 50-to-100-μm spot. The pump beam overlapped the probe and was focused to a much larger spot. The focal spot was magnified by a factor of 40, and a 200-μm pinhole was scanned across this image to probe small areas of the device (effective sample size of ≃ 5 μm). The probe light was detected either by a combination of optical multichannel analyzer, spectrometer, and photodiode array, or, by a single photodiode and spectrometer.

The origin of the optical nonlinearity in our device is the presence of an exciton resonance below the bandgap of GaAs at room temperature. The device is adjusted in the absence of the pump beam, so that one of the Fabry-Perot transmission peaks is on the low-frequency side of the exciton resonance. The pump beam creates carriers, screening the Coulomb

---

Presently at AT&T Bell Laboratories, Holmdel, New Jersey 07733

interaction between electron and hole in the exciton orbit, and causing saturation of the exciton absorption. If the device is originally transmitting, the application of the pump shifts the Fabry-Perot peak away from the laser wavelength, causing the transmission to decrease. This is the physical mechanism for the operation of our optical NOR gate.

Dynamic measurements of exciton screening by injecting either carriers or additional excitons give information about the saturation time of the exciton resonance, its recovery time, dissociation time of excitons into free carriers, and the formation time of excitons from free carriers [2]. The exciton screening in MQW samples is instantaneous within the resolution of the measuring system of ≃ 150 fs [2,3]. Since the index change occurs as fast as the exciton saturates, the speed of the optical gate should be limited by the cavity buildup time of $(2n\ell/(1-R)c) \simeq 300$ fs (assuming an effective mirror reflectivity of 0.6 from the measured finesse of ≃ 6). However, the data (Figs. 1 and 2) suggest [4] that the shift of the Fabry-Perot peak lasts for $\lesssim 3$ ps. This sluggishness in the response might be attributed to the induced absorption occurring on the tail of the exciton resonance. The measurements of the exciton dynamics suggest that this absorption lasts for a few picoseconds. The index change caused by the induced absorption might partially cancel that caused by the exciton saturation until the induced absorption decreases. In fact, the next Fabry-Perot order at about a 200 Å longer wavelength, where the exciton absorption has dropped considerably, shows a faster response of ≃ 1 ps. (The NOR-gate contrast for this transmission peak was poor because of its greater detuning from the exciton resonance.)

In conclusion, we have performed measurements of exciton and carrier dynamics and demonstrated for the first time ≃ 1-ps operation of an

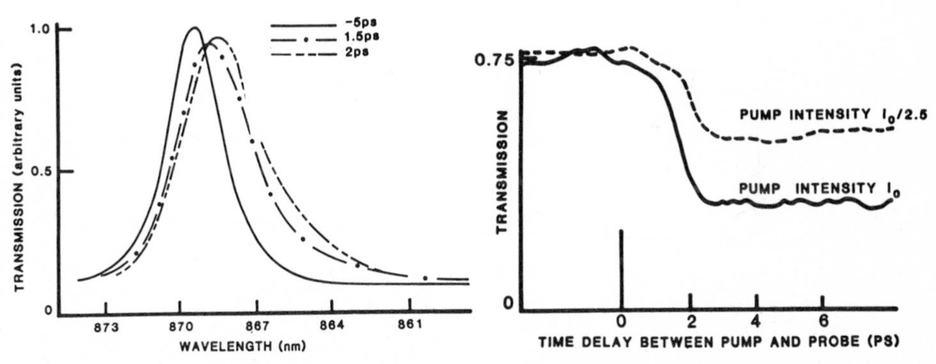

Fig. 1. Position of the Fabry-Perot transmission peak at different times before and after the arrival of the pump beam (the exciton absorption peak wavelength is about 860 nm)

Fig. 2. Normalized transmission of the device as a function of time at two pump intensities, I_0 and $I_0/2.5$. The spectrometer was tuned to 70% to 80% of the transmission point in Fig. 1. The better contrast observed here (compared to that of Fig. 1) occurs because here an electronic discriminator only allows detection of laser pulses with intensities above a certain value. This reduces the apparent broadening of the average shifted Fabry-Perot peak caused by variations in the pump pulse intensity.

optical NOR gate using a GaAs-AlGaAs nonlinear etalon at room temperature and a femtosecond laser system. These measurements show that an etalon peak can be shifted in ≃ 1 ps. This corresponds to the switch-on time of an optically bistable device (the fastest measured switch-on time previously reported for a bistable device is detector-limited 200 ps) [5]. However, because of the long carrier lifetime of a few nanoseconds, the repetition rate of the NOR gate is presently limited by a few hundred megahertz. That is, even though the gate responds in 1 ps, the next gating operation cannot be performed for a few nanoseconds. Several techniques may be employed to reduce the carrier lifetime and to effectively increase the repetition rate of the device. For example, proton bombardment of the material allowed shortening of the recovery time of the GaAs-AlGaAs MQW structure used as a saturable absorber [6]. This technique may be used to reduce the switch-off time of a GaAs bistable switch and increase the repetition rate of a GaAs NOR gate if thermal problems associated with the increase of the unsaturable background absorption caused by the broadening of the bandtail (which results because of the proton damage) can be avoided. The NOR gate contrast is only ≃ 2.5 because our probe pulses are shorter than the cavity response time of the etalon, but we have observed recently a contrast of more than 4 to 1 in a similar device with less than 3 pJ of energy incident. An array of such devices might allow parallel logic operations,and may also be used as an all-optical addressable spatial light modulator.

The Arizona portion of this research has been supported by grants from NSF, AFOSR, ARO, and NATO 882/83. The authors would like to thank A. C. Gossard and W. Wiegmann for the MQW crystals.

References

1. A. Migus, J. L. Martin, R. Astier, A. Antonetti, and A. Orszag, in Picosecond Phenomena III (Springer, New York 1982) p. 6

2. N. Peyghambarian, H. M. Gibbs, J. L. Jewell, A. Antonetti, A. Migus, D. Hulin, and A. Mysyrowicz, Phys. Rev. Lett. 53, 2433 (1984)

3. C. V. Shank, R. L. Fork, R. F. Leheny, and J. Shah, Phys. Rev. Lett. 42, 112 (1979)

4. A. Migus, A. Antonetti, D. Hulin, A. Mysyrowicz, H. M. Gibbs, N. Peyghambarian, and J. L. Jewell, Appl. Phys. Lett. 46, 701 (1985)

5. S. S. Tarng, K. Tai, J. L. Jewell, H. M. Gibbs, A. C. Gossard, S. L. McCall, A. Passner, T. N. C. Venkatesan, and W. Wiegmann, Appl. Phys. Lett. 40, 205 (1982)

6. Y. Silverberg, P. W. Smith, D. A. B. Miller, B. Tell, A. C. Gossard, and W. Wiegmann, included in this publication

Exciton-Exciton Interaction in GaAs-GaAlAs Superlattices

D. Hulin, A. Migus, A. Antonetti, A. Mysyrowicz
Laboratoire d'Optique Appliquée, Ecole Polytechnique-ENSTA,
F-91120 Palaiseau, France
H.M. Gibbs and N. Peyghambarian
Optical Sciences Center, University of Arizona, Tucson, AZ 85721, USA
H. Morkoç and W.T. Masselink
University of Illinois, Urbana, IL 61801, USa

Multi-quantum well structures (MQWS) have been shown to be a very attractive material for optical devices. One of the most interesting features is their large non-linear optical response in the frequency region of the excitons . This bound state of an electron in the conduction band and a hole in the valence band leads to strong absorption lines in the optical spectra of MQWS ; these lines are very sensitive to any perturbation introduced by optical means in the system, giving rise to associated large changes in the optical response (1). Further, excitons are observed even at room temperature, in contrast with the case of almost all bulk semiconductors. It is very important to characterize the exact nature of the exciton modification and its time evolution, both from a fundamental point of vue and for any application to an optical device.

In presence of free carriers, the exciton binding is reduced, due to the screening of the Coulomb interaction between the bound electron and hole. Figure 1 shows the transmission spectra of a MQWS at low temperature in presence of increasing densities of free carriers. The sample consists of 100

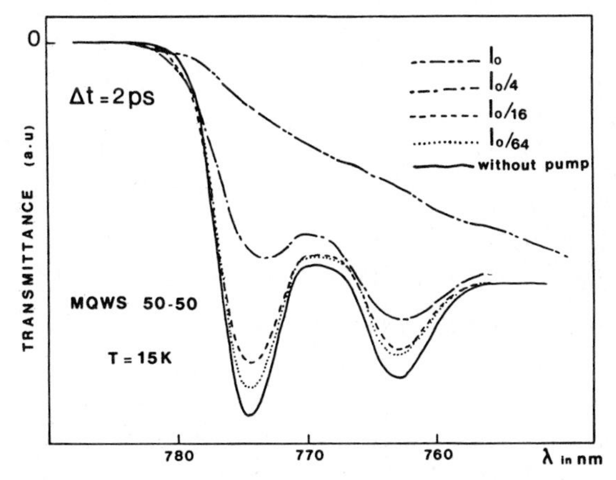

Figure 1 - Time-resolved transmission spectra of the (50 Å - 50 Å) MQWS at low temperature for different input intensities. The spectra have been recorded 2 ps after the pump pulse to have only free carriers present in the sample. I_0 is the maximum pump intensity.

layers of 50 Å thick GaAs and 50 Å thick $Ga_{1-x}Al_xAs$ (x=0.3). Free carriers are optically created by band-to-band excitation (non-resonant pumping). At low temperature, these electrons and holes will tend to decrease their energy by forming bound pairs, i.e. excitons ; therefore, the non-equilibrium situation corresponding to free carriers in the system can be examined only a short time after excitation. In this experiment, we use laser pulses of 150 fs duration to perform pump and probe experiments : the pump pulse at a selected wavelength is obtained through an interferential filter placed on the wavelength continuum generated by focusing the original 620 nm laser pulse in a water cell (2) As can be seen from the figure, the oscillator strength of the excitonic lines is reduced in presence of free carriers, but their energy positions remain unchanged. The two lines correspond to the heavy and light holes and behave similarly.

In this sample (t_{well} = 50 Å, $t_{barrier}$ = 50 Å), the situation is very different if the excited particles are no more free carriers but excitons. It is possible to create directly a large number of excitons by resonant pumping, but they are also very easy to obtain at large density just by waiting a time long enough after a non-resonant excitation to assure that the created hot carriers have relaxed to a thermal equilibrium where they have formed bound pairs. Figure2 shows optical spectra recorded 100 ps after excitation at low temperature for different initial carrier densities. The oscillator strength of the excitonic lines has almost recovered, the remaining screening being possibly due to some carriers still present in the medium. The major feature is now that the absorption lines appear at energies shifted towards higher values (blue shift). It has already be

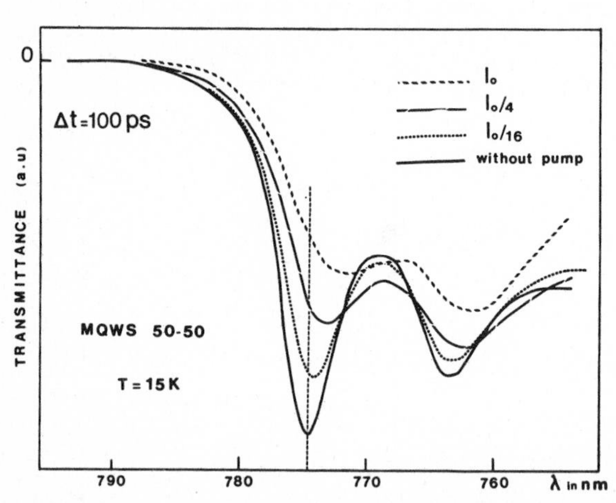

Figure 2 - Time-resolved transmission spectra of the (50 Å - 50 Å) MQWS at low temperature for different input intensities. The spectra have been recorded 100 ps after the pump pulse, when most of the carriers have paired into excitons.

shown (3) that this displacement increases with the number of excitons, as can be also inferred from the figure. This blue shift has not be reported in other semiconductors like bulk GaAs for example. This new effect is related to the structure of multi-quantum wells. In such a structure, the small thickness of the wells in which the excitons are confined allows one to describe their behavior as a two-dimensional one. But there is also a noticeable characteristic in this sample : the small thickness of the barriers does not prevent interactions between the wells which cannot be any more considered as isolated. It is then important to know whether this blue shift originates only from the two-dimensional nature of the MQWS or if the superlattice nature of the sample might play a role.

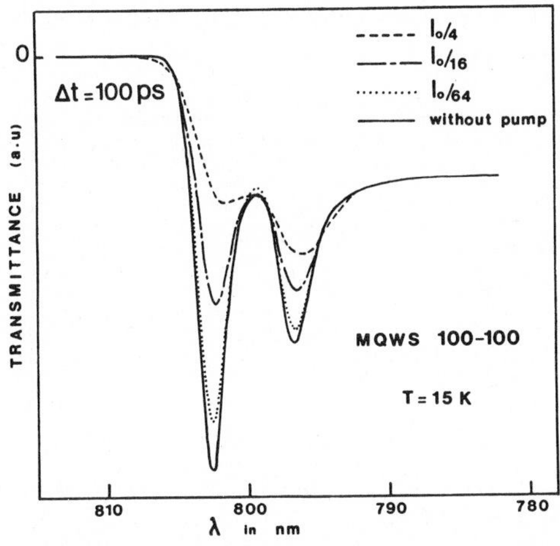

Figure 3 - Time-resolved transmission spectra of the (100 Å - 100 Å) MQWS at low temperature for different input intensities. The spectra have been recorded 100 ps after the pump pulse, when most of the carriers have paired into excitons.

We have performed the same type of experiments in samples of different well and barrier thicknesses. Figure 3 shows the results obtained under similar experimental conditions for a MQWS with barriers and wells 100 Å thick. The GaAs layer is still sufficiently narrow to be considered as a quantum well, but the interactions from one well to another are now reduced. The results indicate clearly that the shift has strongly decreased, if not disappeared. This blue shift effect is then only observable in superlattices. In such a case, the mean distance between two wells ($t_{well} + t_{barrier}$) is significantly smaller than the Bohr radius of an exciton in GaAs. The same experiments performed in a (t_{well} = 100 Å, $t_{barrier}$ = 25 Å) structure has not given conclusive results due to the fact that in this sample the excitonic lines are very easily screened by the small number of remaining carriers, preventing a reliable observation of the shift. However, the experimental results obtained in a (t_{well} = 40 Å, $t_{barrier}$ = 100 Å) structure indicate that the blue shift is still present (even if slightly smaller than in the case 50 Å - 50 Å) although the sample has the same barrier thickness as in the case 100 Å - 100 Å where no shift is observed. This last point can be understood if one consider the exciton interactions between neighboring wells.

The exciton wave function is built with the electron and hole wave functions which have sinusoïdal dependences in the well and exponential tails in the barrier region. Whatever the exact nature of the exciton interaction between two neighbouring wells, the resulting effect will depend upon the overlap of the wavefunctions from one well to the next one. This overlap is linked to the thickness d of the barrier and to δ, the evanescence length of the wave function in the barrier. δ is increasing when the width of the well is reduced :

$1/\delta = k \; tg \; k \; L/2$ where L is the well thickness and k is a parameter related to the energy of the excitonic level in the well ; k is also decreasing when L decrease due to the dependence of the exciton energy with L (4). A more detailed treatment will be presented elsewhere.

In conclusion, we have studied the displacement of the excitonic lines in MQWS in presence of a high density of excitons. The influence of well and barrier thicknesses have been explored and related to exciton-exciton interaction between neighbouring wells.

(1) D.S. Chemla, Helv. Phys. Acta, 56, p.607, 1983.
(2) A. Migus, A. Antonetti, J. Etchepare, D. Hulin, A. Orszag, JOSA -B, 2 to appear April 1985.
(3) N. Peyghambarian, H.M. Gibbs, J.L. Jewell, A. Antonetti, A. Migus, A. Mysyrowicz, Phys. Rev. Lett. 53. p.2433 (1984).
(4) R. Dingle, W. Wiegmann, C. Henry, Phys. Rev. Lett. 33, p.827 (1974).

Part IV

Picosecond Diode Lasers

An InGaAsP 1.55 μm Mode-Locked Laser with a Single-Mode Fiber Output

G. Eisenstein, S.K. Korotky, R.S. Tucker, U. Koren, R.M. Jopson, LW. Stulz, J.J. Veselka, and K.L. Hall
AT&T Bell Laboratories, Crawford Hill Laboratory, Holmdel, NJ 07733, USA

Mode-locked InGaAsP lasers emitting in the 1.3-1.6 μm wavelength region are attractive as potential sources for very-high bit-rate fiber communication applications. Previous mode-locking of InGaAsP lasers produced pulse widths of roughly 20 ps [1,2,3], at repetition rates as high as 10 GHz [1]. Shorter pulses [4], and higher repetition rates [5] have been demonstrated only in AlGaAs lasers. In most reported mode-locked semiconductor laser structures, the output pulses are taken from a cleaved laser chip facet. In order to couple the pulses to a fiber, it is necessary to use some form of a laser-fiber mode-matching technique. This will, in general, add coupling losses (typically 3-5 dB) as well as coupling instabilities.

We describe a new 1.55 μm InGaAsP-optical fiber composite-cavity laser. This structure combines the compact construction of a fiber resonator with the efficiency and flexibility of an integrated single-mode fiber pigtail with a biconic connector at the output port. We have actively mode-locked this laser to produce 4.8 ps pulse widths, at repetition rates ranging from 2 GHz to 20 GHz. The pulse peak powers delivered into the output fiber were measured to be 56 mW at 2 GHz and 18 mW at 20 GHz.

The composite-cavity consists of a 1.55 μm InGaAsP gain element and a single-mode fiber resonator. It is illustrated in Fig.1.

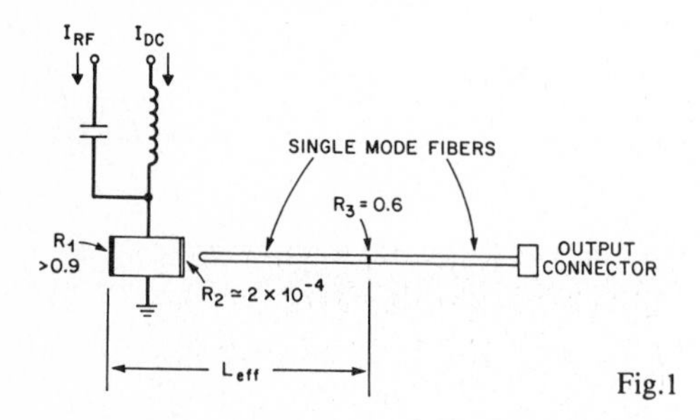

Fig.1

The gain element is a three channel buried crescent (TCBC) [6] laser chip with both facet reflectivities modified by dielectric coatings. One chip facet is coated with a multilayer dielectric mirror of reflectivity $R_1 > 0.9$. The second facet is coated with a single layer antireflection (AR) film [7]. The residual reflectivity is $R_2 = 2\times10^{-4}$. The fiber resonator is a short (5 cm) single-mode fiber with a microlens on one end and a partially reflecting dielectric mirror ($R_3 = 0.6$) on the other. The AR coated facet of the gain medium is optically coupled to the microlensed fiber resonator to form a single cavity of length L_{eff} between mirrors R_1 and R_3. The output pulses are coupled out through the partially reflecting mirror R_3, where the fiber resonator is spliced to a similar fiber using a V-grooved silicon chip. The output fiber is terminated by a standard biconic connector.

The round trip time in the cavity is $\tau_r = 500$ ps resulting in a cavity resonance frequency $f_r = 2$ GHz. The laser can be mode-locked by driving the gain element with a sinusoidal RF current at a

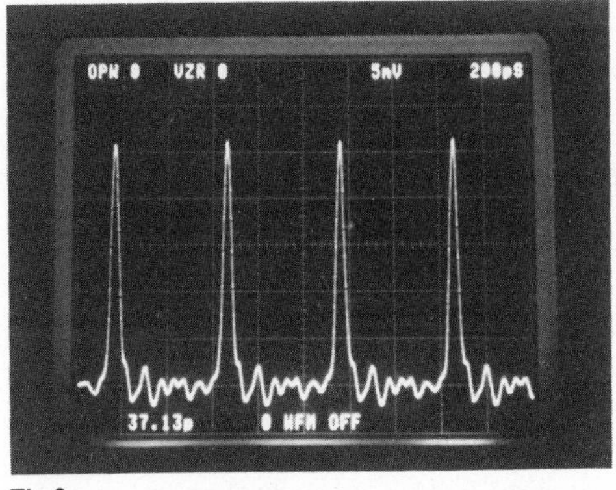

Fig.2

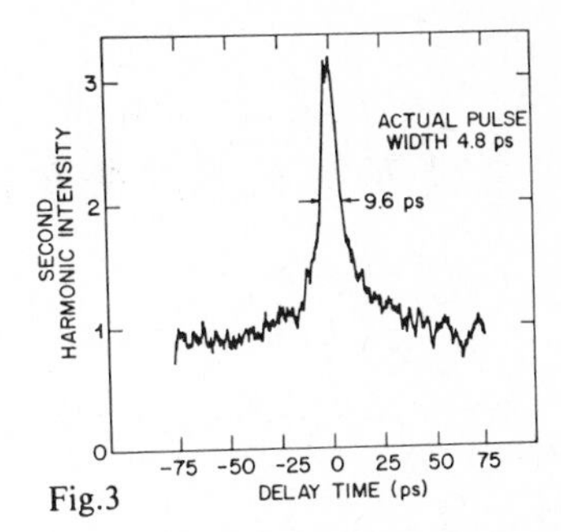

Fig.3

frequency f_r or the nth harmonic of this frequency. Alternatively, the sinusoidal current can be replaced by a pulse train from a comb generator at the same repetition frequency. The length of the resonator fiber has been carefully selected such that f_r falls within the useful operating range of an available comb generator.

Fig.2 shows a mode-locked pulse train obtained when the laser is driven with 45 ps wide electrical pulses from a comb generator at a repetition rate equal to the fundamental cavity resonance frequency f_r = 2 GHz. The pulses were detected with a 22-GHz bandwidth InGaAs pin detector [8] and a sampling oscilloscope. The observed modulation depth is close to 100%. The autocorrelation function of a single pulse obtained by second harmonic generation (SHG) in a standard, two beam interferometer with collinear beams incident on a phase matched $LiIO_3$ crystal is shown in Fig.3. The autocorrelation width is 9.6 ps FWHM. We have found that the trace can be fitted reasonably well by the autocorrelation of a single-sided exponential pulse. This is consistent with previous results in AlGaAs lasers [4]. Assuming this pulse shape, the actual pulse width is 4.8 ps FWHM. The corresponding pulse peak power into the output fiber is 56 mW.

Mode-locking at high repetition rates is obtained by driving the laser with a sinusoidal RF current at a harmonic of the fundamental cavity resonance frequency $n \times f_r$. In this case there are n pulses circulating simultaneousely in the cavity. Each pulse has a random phase relationship with the $n-1$ following pulses. It is phase-locked to the nth following pulse in the train.

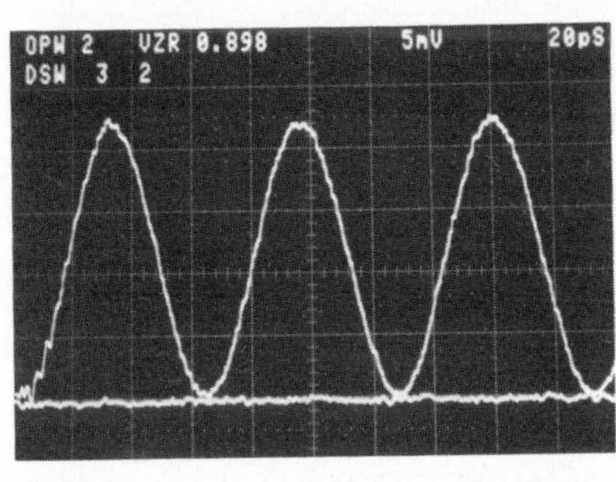

Fig.4

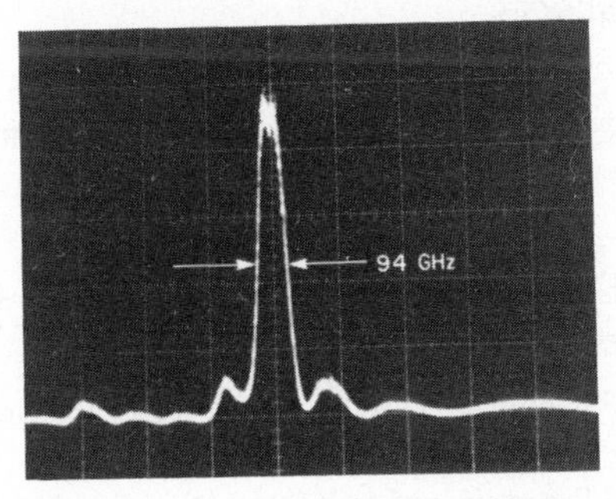

Fig.5

A mode-locked pulse train obtained at the 8th harmonic (16 GHz) is shown in Fig.4. The observed pulse train is detection system limited, but it is clear that the pulses have a high on/off ratio. The optical spectrum of this pulse train is shown in Fig.5. There is a central cluster of locked modes surrounded by smaller clusters of modes. These are separated from the central cluster by a

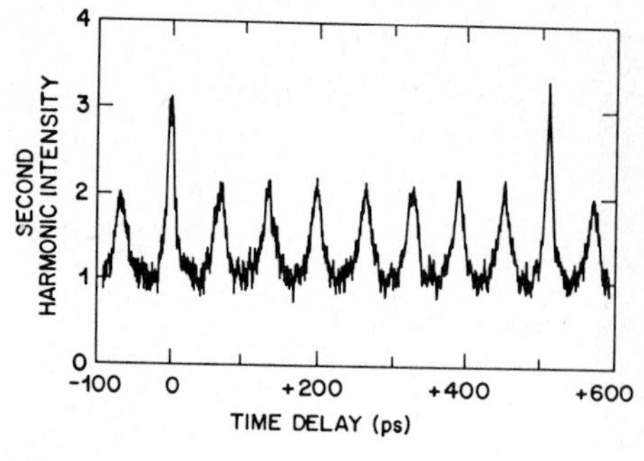

Fig.6

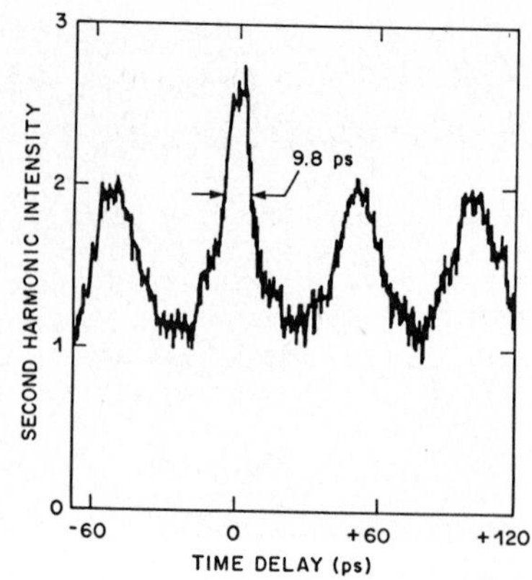

Fig.7

wavelength difference characteristic of the gain element length $\Delta\lambda = 13$ Å. The width of the central cluster of modes is 94 GHz.

An SHG correlation trace of the 16 GHz pulse train is shown in Fig.6.

The autocorrelation of a single pulse is centered at zero time delay. Its width is 9.4 ps which translates to a pulse width of 4.7 ps assuming a single-sided exponential pulse shape. Following are seven cross-correlation functions separated by $\tau_r / n = 62.5$ ps. These represent the cross correlation functions of a pulse with each of the next seven pulses in the pulse train with which it has a random phase relationship. The 9th correlation function centered at a time delay $\tau_r = 500$ ps is the cross-correlation of the two phase locked pulses. The 1st and 9th correlation functions have the expected 3:1 peak to valley intensity ratios. The ratio is only 2:1 for the seven central correlation functions, possibly due to amplitude fluctuations combined with phase incoherence between the pulses. The time-bandwidth product of a single pulse is 0.44, compared to 0.11 for a transform limited single-sided exponential pulse.

Mode locking has also been achieved at the 10th harmonic (20 GHz). A correlation trace of part of the pulse train, including the autocorrelation and three cross-correlation functions, is shown in Fig.7. The pulse width is 4.9 ps and the correlation functions are separated by 50 ps. The peak pulse power at 20 GHz was measured to be 18 mW in the output fiber.

REFERENCES

[1] R.S.Tucker, G.Eisenstein, and I.P.Kaminow., Electron. Lett., Vol. 19, pp.552-553, 1983.

[2] R.C.Alferness, G.Eisenstein, S.K.Korotky, R.S.Tucker, L.L.Buhl, I.P.Kaminow, C.A.Burrus, and J.J.Veselka., App. Phys. Lett., Vol. 45, pp.944-946, 1984.

[3] L.A.Glasser., Electron. Lett. Vol. 14, pp.725-726, 1978.

[4] J.P.Van der Ziel., J. App. Phys., Vol. 52, pp.4435-4446, 1981.

[5] K.Lau, and A.Yariv., Eighth Conf. on O.F.C., paper No. ME4, Feb. 1985.

[6] G.Eisenstein, U.Koren, R.S.Tucker, B.L.Kasper, A.H.Gnauck, and P.K.Tien., App. Phys. Lett., Vol. 45, pp.311-313, 1984.

[7] G.Eisenstein, and L.W.Stulz., App. Opt. Vol. 23, pp.161-164, 1984.

[8] C.A.Burrus, J.E.Bowers, and R.S.Tucker. submitted to Electron. Lett. 1985.

Fast Multiple Quantum Well Absorber for Mode Locking of Semiconductor Lasers

Y. Silberberg and P.W. Smith
Bell Communications Research, Inc., Holmdel, NJ 07733, USA
D.A.B. Miller and B. Tell
AT&T Bell Laboratories, Holmdel, NJ 07733, USA
A.C. Gossard and W. Wiegmann
AT&T Bell Laboratories, Murray Hill, NJ 07974, USA

1. Introduction

Passive mode locking of a semiconductor diode laser using GaAs/GaAlAs multiple quantum well (MQW) material as a saturable absorber has been reported recently [1]. The strong room-temperature excitonic absorption line in this material [2] saturates at relatively low powers, and it can be matched to the GaAs laser wavelength. Such mode-locked lasers are potentially important sources for future ultra-fast photonic signal processing.

In order to obtain mode locking, the relaxation time of the saturable absorber must be faster than that of the gain [3]. The relaxation time of the gain in a GaAs diode was recently measured to be 375 ps [4]. The absorption recovery time in a MQW absorber is governed by the carrier recombination time, which was measured to be 30 ns in our MQW samples [2]. Some shortening of the relaxation time can be achieved be tight focusing of the light on the MQW sample, so that carrier diffusion out of the excited region is faster than their recombination rate. This method, however, requires operating the laser just above threshold, since at high light powers the absorber becomes totally bleached. It would be much better to find a method to speed up the intrinsic carrier recombination time. Ion bombardment is well known to reduce carrier lifetimes in many semiconductors [5], but it can also affect the absorption near the bandedge, and reduce or eliminate the exciton absorption feature.

In this paper we report our studies of proton bombardment of MQW samples, which proves to be a practical method of obtaining fast-responding saturable absorbers. We also report the achievment of a stable, continuous subpicosecond pulse train by external compression of a mode-locked laser pulse. Pulses as short as 0.83 ps have been measured.

2. Proton Bombardment of MQW Structures

The main objective of this study is to determine if a significant shortening of carrier recombination time by ion bombardment is possible before the excitonic absorption line is affected. Our MQW samples consisted of 80 periods of 102 Å GaAs layers, alternated with 101 Å $Ga_{.71}Al_{.29}As$ layers, grown by molecular beam epitaxy on a GaAs substrate. We have chosen protons as the bombarding ions, in order to assure good penetration through the total MQW thickness of about 2 microns. The samples were bombarded with different doses of 200 kev protons. Some of the samples were later annealed for 10 minutes at 300°C, in order to eliminate very shallow traps which may anneal spontaneously. Such annealed samples should have better long-term stability.

The recovery time of the excitonic absorption was measured using a conventional pump-probe scheme. We have used a tunable synchronously mode-locked dye laser, which produced 5-7 ps pulses spaced by 12 ns. Table 1 summarizes the lifetime measurements for different proton doses. As expected, annealed samples have longer lifetimes than their unannealed counterparts, due to the removal of some damage centers. Values in parentheses refer to times measured in samples for which the excitonic nonlinearity is significantly reduced.

Table 1. Absorption recovery time of proton bombarded MQWs.

Dosage [cm^{-2}]	Recombination time [ps]	
	Annealed	Unannealed
10^{10}	...	530
10^{11}	>3000	450
10^{12}	560	200
10^{13}	150	(27)
10^{14}	(33)	(<10)

We have also measured linear and nonlinear absorption spectra of the samples. The nonlinear absorption is defined as the change in transmission of a weak probe beam due to the presence of a pump beam. These linear and nonlinear spectra are shown in Fig. 1. It is observed that up to a dosage of about 10^{12} protons/cm^2 no significant changes occur in either spectra. Higher bombardment doses cause a decrease of the peak absorption and a broadening of the resonance, as well as a reduced nonlinear response.

From the results presented in Table 1 and Fig. 1 it can be concluded that a substantial reduction of the absorption recovery time can be induced by proton bombardment before a significant

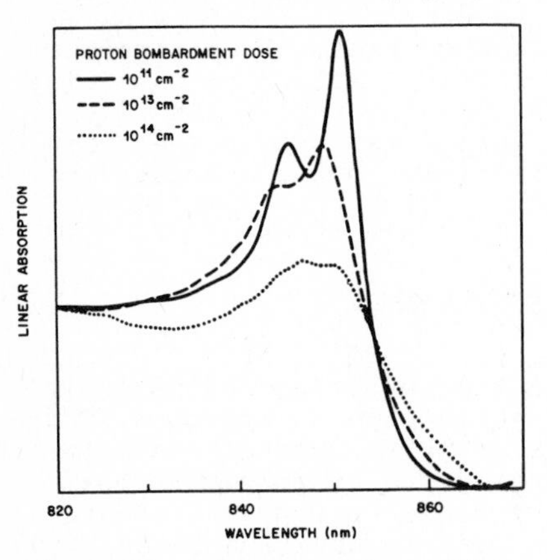

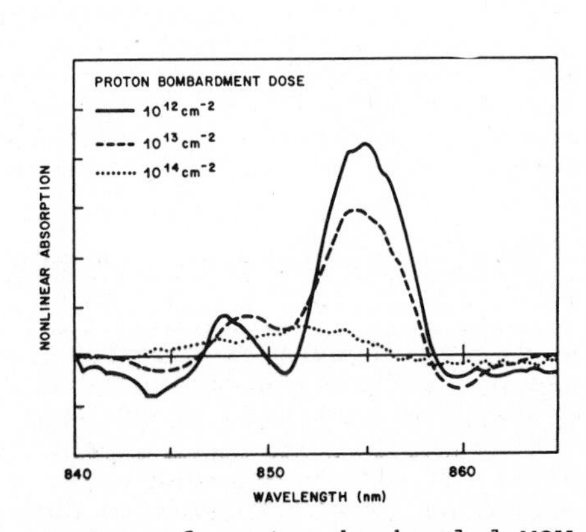

Figure 1. Linear and nonlinear spectra of proton bombarded MQWs.

change in the linear or nonlinear absorption characteristics occurs. A dose of 10^{13} protons/cm^2 seems to give a sufficiently short lifetime for mode-locking puposes, i.e. shorter than the gain recovery time (150 ps in annealed samples versus 375 ps gain recovery time), without causing a major reduction in the nonlinearity.

3. Mode Locking with Proton Bombarded Samples

Our experiments were conducted using a linear cavity laser, similar to that reported in Ref. [1]. One facet of a GaAs diode laser was anti-reflection coated, and an external cavity was constructed with the MQW absorber epoxied on the external mirror. Mode locking was obtained when the laser was tuned to the vicinity of the excitonic absorption line. Unlike our previous experiments with tight focusing as a response time reduction mechanism [1], mode locking was easy to achieve. The repetition rate of the laser was typically 1 GHz, corresponding to one pulse in the 15cm laser cavity. The output intensity was about 1 mW.

Second harmonic autocorrelation measurements were used to evaluate the pulse duration. It was found that the shortest pulses obtained were approximately 2 ps long. The spectral width of the lasing was 35 Å. This means that the pulses are not transform limited. Careful analysis of the interferometric autocorrelation signal suggested that the signal was broadened by phase ultrastructure and not by amplitude noise. We therefore attempted to remove the quadratic phase term, i.e. the linear chirp, by using the grating pair technique. We actually used a single grating, retro-reflecting the beam onto it, as shown in Fig. 2. We found that the laser pulses could be compressed by about a factor of 3.

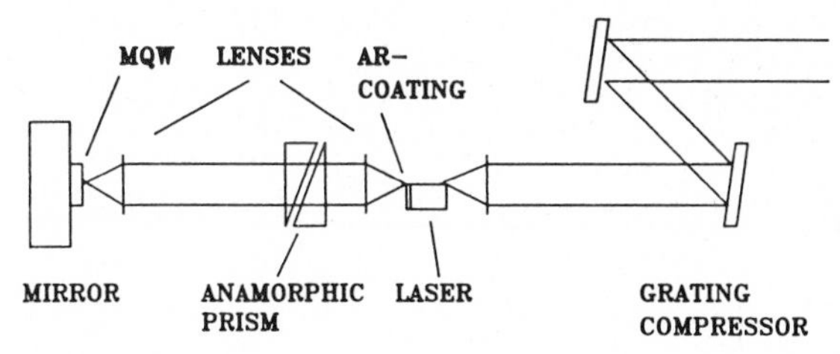

Figure 2. The laser cavity and the external pulse compression setup

Figure 3 shows an autocorrelation trace of a compressed pulse. The triple peaked autocorrelation is probably caused by a satellite pulse which trails the main pulse. The FWHM of this main pulse, assuming a Gaussian pulse shape, is 0.95 ps. Our best result so far is 0.83 ps. Even these compressed pulses are 1.5-2x the transform limit. Additional compression was not possible, however: the remaining broadening is probably due to higher order phase terms. The phase structure of the laser pulse could be caused by group velocity dispersion or by nonlinear self phase modulation in the laser cavity.

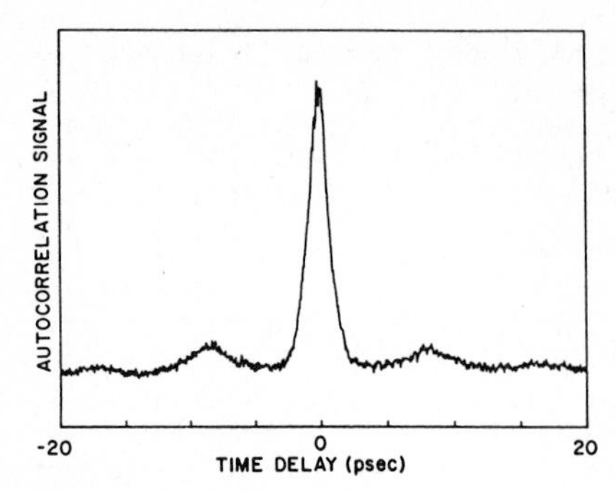

Figure 3. Autocorrelation trace for an externally compressed laser pulse.

4. Conclusion

We have demonstrated that proton bombardment is an efficient way to decrease the recovery time of MQW saturable absorbers, and that the recovery time could be reduced to 150 ps without substantially affecting the saturation energy. This finding could be important for a number of possible applications of this material, since most require fast response in order to compete with existing technologies. We have also demonstrated the usefulness of the fast responding absorber for mode locking of semiconductor lasers, and reported the shortest pulses yet obtained from these lasers by compressing 2 ps output pulses down to 0.83 ps.

1. Y. Silberberg, P. W. Smith, D. J. Eilenberger, D. A. B. Miller A. C. Gossard and W. Wiegmann: Opt. Lett. 9 , 507 (1984).
2. D. S. Chemla, D. A. B. Miller, P. W. Smith, A. C. Gossard and W. Wiegmann: IEEE J. Quantum Electron. QE-20 , 265 (1984).
3. H. A. Haus: IEEE J. Quantum Electron. QE-11 , 736 (1975).
4. W. Lenth: Opt. Lett. 9 , 396 (1984).
5. P. R. Smith, D. H. Auston, A. M. Johnson and W. M. Augustyniak Appl. Phys. Lett. 38 , 47 (1981).
6. E.B. Treacy: IEEE J. Quantum Electron. QE-5 , 454 (1969).

Suppression of Timing and Energy Fluctuations in a Modelocked Semiconductor Laser by cw-Injection

Finn Mengel
Telecommunications Research Laboratory, Borups Alle 43,
DK-2200 Copenhagen, Denmark
Chinlon Lin* and Niels Gade
Electromagnetics Institute, Technical University of Denmark,
DK-2200 Lyngby, Denmark

1. Introduction

Active modelocking of semiconductor lasers in an external cavity is a simple and useful technique for generation of short optical pulses in the wavelength region of interest for optical communications. Up to now, most attention has been paid to the time-averaged behaviour of pulse width, optical spectrum and time-bandwidth product, although noise, both at base-band and optical frequencies, is of major importance in communication and logic applications. In this paper, we investigate the noise behaviour of a modelocked GaAlAs laser and demonstrate a partial suppression by cw injection locking. A similar technique has previously been demonstrated with CO_2 lasers [1].

2. Experiment

A Hitachi 850 nm CSP laser with AR-coating of estimated reflectivity around 1% was placed in a 15 cm cavity with a 30X microscope objective and a 1180 ℓ/mm grating and synchronously modulated at 1 GHz. A similar, but uncoated single mode laser was used as master oscillator with injected power in the -10 - -20 dBm range. With master and slave frequencies properly aligned, the output pulses advance by roughly 30 ps relative to the unlocked state, corresponding to a phase-shift of 10 degrees at 1 GHz. As shown in fig. 1, this effect could be used to lock the master laser to the slave cavity.

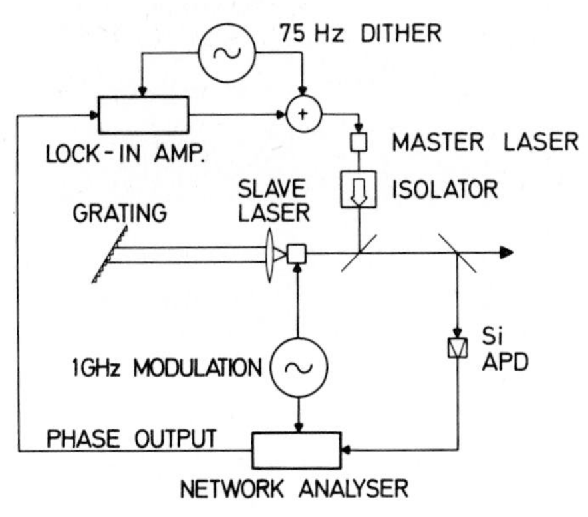

Fig. 1 Block diagram of injection-locked external cavity modelocked laser with stabilisation loop minimizing the delay between electrical pump and optical output.

* Permanent address: AT&T Bell Laboratories, Holmdel, NJ 07733, USA

The pulse train is analyzed by an 8 GHz Si pin diode and a spectrum analyzer, a Fabry-Perot interferometer or an intensity correlation interferometer. The baseband spectrum consists of a series of delta-like peaks at harmonics of the modulation frequency, surrounded by noise bands [2]. In fig. 2, we show the spectra around the first and seventh harmonic without (a) and with (b) injection. We observe a symmetric noise band of bandwidth around 20 MHz increasing with harmonic number n as n^2, indicating pure timing noise [2]. The broadband structure with a bandwidth of ~100 MHz becomes progressively more asymmetric with increasing harmonic number in contrast to the results of [2] for Argon and dye lasers. This can be explained by considering a model pulse train at repetition frequency f_p modulated in amplitude and arrival time at some frequency f_m. We can derive an approximate expression for the first sidebands in the power spectrum [3]

$$S(nf_p \pm f_m)/S(nf_p) = \frac{1}{4} n^2\beta^2 + \frac{1}{4} m^2 \pm \frac{1}{2} nm\beta\sin\theta \qquad (1)$$

where m and β are the AM and FM modulation indices and θ the phase delay between the FM and AM modulation. In addition to the constant AM-term m^2 and quadratic timing $n^2\beta^2$ discussed in [2], we get a correlation term which for $\theta \neq 0$ causes asymmetry between the two sidebands. Comparing with the experimental spectra, this suggests that the timing and amplitude fluctuations are at least partially correlated, although pulse width fluctuations may also be present.

With injection, both types of noise decrease by roughly 10 dB while the pulse width increases from 30 to 34 ps (assuming Gaussian shape).

The optical noise can be seen in the Fabry-Perot spectrum in fig. 3, displayed on a 250 MHz oscilloscope. The blurring is due to the optical carrier frequency fluctuating from pulse to pulse. Injection locking stabilizes the spectrum partiality, although complete noise suppression only occurs for high injection levels, causing significant pulse broadening.

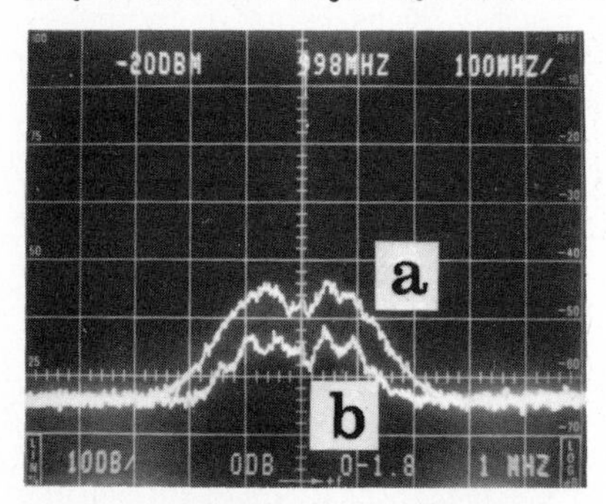

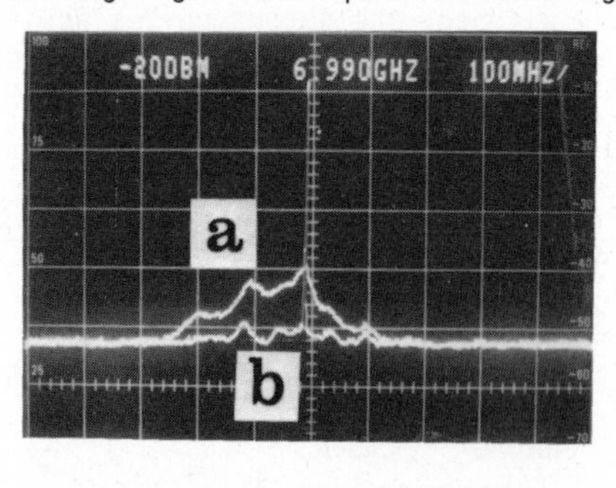

Fig. 2 Baseband noise spectra at the first and seventh harmonic of the modulation frequency (1 GHz) without (a) and with (b) injection

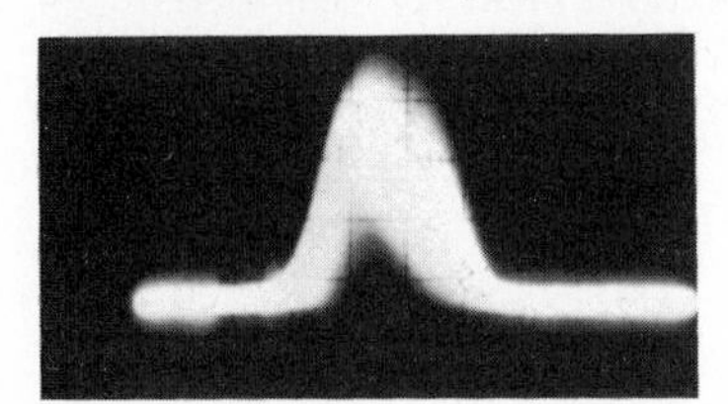

Fig. 3 Fabry-Perot spectrum of noisy pulse train with 250 MHz detection

3. Theory

Numerical simulations have been carried out based on the following equation for the complex electric field $E(t) = A(t)e^{j\omega_s t}$ [4]:

$$A(t+\tau_{in}) = r_1 e^{-j2k(\omega_s,N)l} \cdot \left\{ r_2 A(t) + A_{inj} e^{j(\omega_i - \omega_s)t} + (1-r_2^2) r_3 \cdot h(t) * A(t-\tau) e^{-j\omega_s t} \right\} + F_N(t) \qquad (2)$$

where $k(\omega_s, N)$ is the complex propagation constant in the diode at the stationary frequency ω_s and carrier density N, h(t) is the impulse response of the intracavity filter and A_{inj} and ω_i are the amplitude and frequency of the injection signal. F_N is a Gaussian noise term, describing spontaneous emission. Other quantities are defined in fig. 4. The carrier density is governed by a standard rate equation, see e.g. [5].

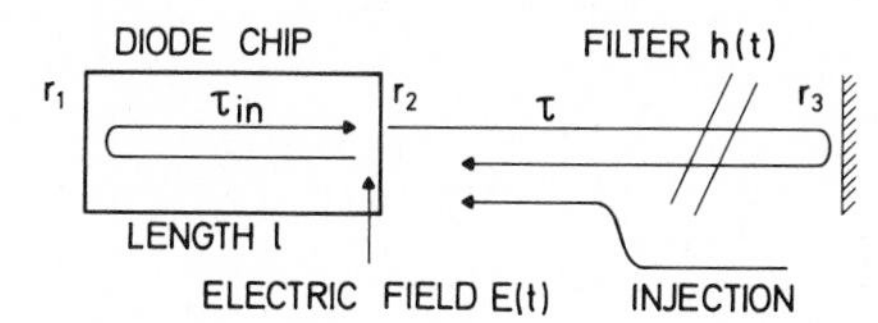

Fig. 4 Model for external cavity laser with Fabry-Perot filter

With dc-excitation, this set of equations can be analyzed by methods similar to those reported in [5] and can be shown to exhibit both stable and unstable behaviour. With sinusoidal current modulation, the numerical simulations result in a pulse train that can be analyzed with respect to pulse width, chirp, optical and baseband spectra etc. In fig. 5, we show a typical result for a laser with an imperfectly coated facet without and with injection. With injection, the noise decreases by more than 10 dB at the cost of a doubling of the pulse width and a deterioration of the extinction ratio. For other values of cavity detuning, injection frequency and level, one may get an *increase* in noise in agreement with experimental findings. The simulations clearly demonstrate the tradeoff between noise and pulse width and the sensitivity to injection frequency mismatch.

For a perfectly coated laser (r_2=0) the model predicts pulse widths limited by the external filter, since we neglect changes in gain during one diode transit time τ_{in}. For 50 GHz filter width we get a pulse width of 30 ps. For an uncoated diode (r_2=0.56) we may reach the same *average* pulse width but with a very high noise level. Injection at a relative power level of 1% decreases the noise by 20 dB, giving stable pulses of 80 ps width. Thus, injection locking may allow the use of uncoated lasers for some applications, where minimum pulse width is not imperative.

5. Conclusion

We have investigated the noise behaviour of modelocked semiconductor lasers and demonstrated a partial noise suppression by injection locking. Experiment and theory demonstrate a tradeoff between noise suppression, pulse width and extinction ratio.

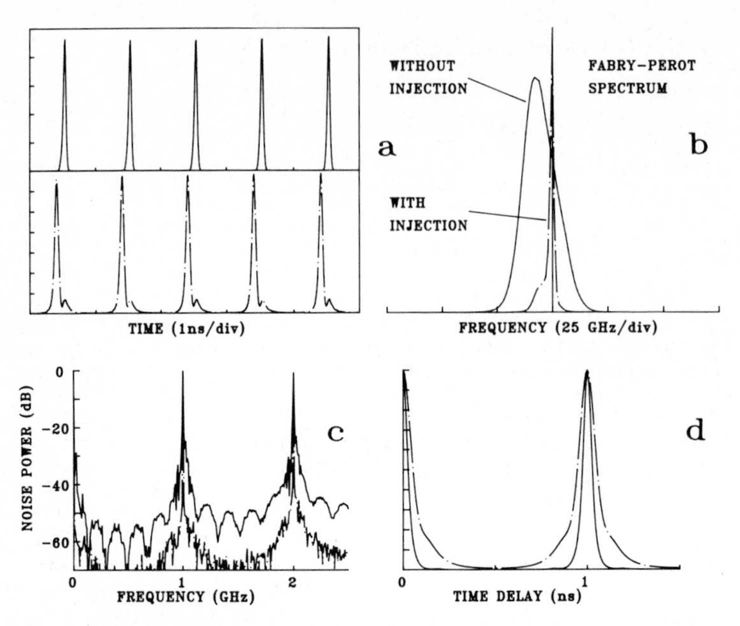

Fig. 5 Simulation of modelocking at 1 GHz with 1% residual reflection without (full lines) and with 5% injected power (dash-dotted lines). Filter bandwidth 50 GHz, a) Intensity waveforms, b) Average optical spectra, c) Baseband noise spectra in 7.8 MHz bandwidth, d) Autocorrelation function.

Acknowledgements

The numerical simulations were performed in corperation with B. Tromborg. This work was supported by the Danish Technical Research Council.

References

1 F.A. Van Goor: Opt. Commun. 45, 404 (1983).
2 J. Kluge, D. Weichert and D. von der Linde: Opt.Commun. 51, 271 (1984).
3 H. Olesen and G. Jakobsen: IEEE J.Quantum Electron. QE-18, 2069 (1984).
4 B. Tromborg, H. Olesen and J.H. Osmundsen. To be published.
5 B. Tromborg, J.H. Osmundsen and H. Olesen: IEEE J.Quantum Electron. QE-20, 1023 (1984).

Parametric Oscillations in Semiconductor Lasers

D. Haas, J. McLean, J. Wurl, and T.K. Gustafson
Department of Electrical Engineering and Computer Sciences and the Electronics Research Laboratory, University of California, Berkeley, CA 94720, USA
C.L. Tang
Department of Electrical Engineering, Cornell University, Ithica, NY 14853, USA

Externally induced optical parametric interactions within a laser cavity can induce diverse behavior depending upon the nature of the interaction and the oscillation conditions of the laser cavity. We have been particularly interested in the possibility of utilizing such interactions for the generation of short optical pulses from semiconductor lasers. Parametric interactions become simpler to implement as the round trip cavity time is decreased, thus increasing the modulation frequency required. This is in contrast to gain switching and other conventional microwave injection techniques, which become more difficult as the frequency is increased [1]. Consequently, parametric interactions should be advantageous when the oscillation frequency is greater than a few tens of gigahertz.

Among the possible interactions, that utilizing the $\chi^{(3)}$ nonlinearity appears to be most advantageous. This can be employed to mix two externally injected laser lines, resulting in a travelling refractive index modulation wave. In conjunction with the stimulated emission of the laser itself, a parametric source polarization is developed, which results in cavity oscillations if the frequency of the refractive index wave is resonant with the cavity.

The evolution of the pulse through the non-linear medium, and in particular a semiconductor-guided mode structure, can be described in terms of a basic pulse propagation equation for the slowly varying complex electric field mode amplitude E [2]. This, for z the propagation direction, is given by

$$\left(\frac{\partial}{\partial z} + k_1 \frac{\partial}{\partial t}\right) E(z,t) = \alpha_g T_2^2 \frac{\alpha^2 E(z,t)}{\partial t^2} - i\alpha_m[1-\cos(\omega_m t - \Delta\beta z)]E(z,t) + (\alpha_g - \alpha_\ell)E(z,t), \tag{1}$$

Here it has been assumed that the dispersion due to the gain profile dominates the modal dispersion [3]. The second term on the right-hand side of Eq. (6) represents parametric modulation obtained through non-linear mixing of the two injected fields, ω_m being their difference frequency and $\Delta\beta$ the difference in the z-directed propagation coefficients. A Lorentzien gain profile of width $1/T_2$ and peak center gain α_g has been assumed. α_ℓ is the effective distributed loss and k_1 the inverse of the pulse group velocity.

The steady-state driven field-envelope solution is obtained by assuming a profile traveling with speed $\nu = \omega_m/\Delta\beta$, that is,

$$E(z,t) = E(t - \frac{z}{\nu}) = E(\eta).$$

Substituting into Eq. (1), one obtains for this steady-state solution [4,5]

$$E(\eta) = \exp(-1/2\ \omega_p^{\ 2}\eta^2)\exp\left[-1/2\left(\frac{1}{\nu} - k_1\right)\frac{1}{\alpha_g T_2^2}\eta\right], \tag{2a}$$

with

$$\omega_p = \left(\frac{i\alpha_m}{2\alpha_g}\right)^{1/4}\left(\frac{\omega_m}{T_2}\right)^{1/2} \tag{2b}$$

and

$$1 - \frac{\alpha_\ell}{\alpha_g} - \frac{1}{4(\alpha_g T_2)^2}\left(k_1 - \frac{1}{\nu}\right)^2 = (\omega_p T_2)^2. \tag{2c}$$

ω_p determines the pulse width and chirp, and Eq. (2c) the complex gain parameter α_g. It is seen from this that the threshold gain and the pulse width are increased if velocity match is not achieved. It should also be emphasized that this solution is an accurate description as long as the pulse delay in Eq. (2a) is much less than the microwave period, so that a quadratic approximation to the modulation term, is valid.

The usual solutions are obtained by requiring that the velocity of the index modulation wave be equal to the pulse group velocity (resonant condition),

$$\frac{1}{\nu} = \mathrm{Re}(k_1) = \frac{n}{c} + \mathrm{Re}(\alpha_g T_2) = \frac{1}{\nu_g},$$

thereby ensuring that the driven pulse solution has a velocity equal to the linear modal group velocity. This can in principle be established by proper design of the waveguide for the injected modulating beams. Since the two modulating beams share a common waveguide, ν is approximately equal to the group velocity of the mode.

To obtain a numerical estimate of the possibility of mode locking, we use Eq. (2b) to solve for α_m, the peak of the modulation in terms of pulse width, gain and modulation frequency assuming a perfect velocity match. For a GaAs laser 1000 μm in length, the round-trip cavity time is 24 psec if $\alpha_g T_2$, which is << (n/c), is neglected. Thus ω_m, the modulation frequency required, is 2.6×10^{11} rad/sec, $2\alpha_g \approx 60\ \mathrm{cm}^{-1}$, and $\omega_g = 1/T_2 \approx 2 \times 10^{13}$ rad/sec [6]. Thus if we take a pulse width τ_p of the order of 4.2 psec where the pulse width (FWHM) $\tau_p = 2\{(\ln 2)^{1/2}/[\mathrm{Re}(\omega_p^2)]^{1/2}\}$, Eq. (2b) gives $\alpha_m \approx 0.11\ \mathrm{cm}^{-1}$ for α_m assumed real ($\chi^{(3)}$ real), and $\mathrm{Re}(\omega_p^2) = \mathrm{Im}(\omega_p^2)$, consistent with Eq. (2b). Equation (2c) shows that, since $|\omega_p| \ll \omega_g$, taking a real α_g is a good approximation for a velocity match.

The electric-field amplitude necessary for locking can be obtained from α_m since Eq. (1) using $P_{NL}(z,t) = \varepsilon_0 \chi^{(3)} E^3$ gives [7]

$$6 \times 1/2\ \frac{\omega_0}{cn}\chi^{(3)} E_m^2 = \alpha_m,$$

where the two modulating fields are assumed to be of equal amplitude, E_m. Assuming a mode overlap of unity for the modulating and lasing transverse modes, and a value of $\chi^{(3)} = 1.4 \times 10^{-19}(m/V)^2(10^{-11}$ e.s.u.) [8], one obtains

$$E_m^2 = \frac{n\alpha_m\lambda}{\pi\chi^{(3)}} = 1.3 \times 10^{13}\left(\frac{V}{m}\right)^2 .$$

This implies an intensity in the GaAs of 6.7 MW/cm^2, which is a reasonable requirement. For a 5-µm-radius guided beam, one obtains a total of 5.3 W with an index change $n\alpha_m\lambda$ of the order of 3.4×10^{-6}. Since α_m scales as τ_p^{-4}, 10-psec pulses would require ≈ 0.3 W of power with an index change of $\approx 2 \times 10^{-7}$.

It is anticipated, for various reasons, that the power can be lowered. With modulating fields having frequencies close or above the band gap $\chi^{(3)}$ should be larger [9] than the above value due to resonant enhancement or parametric amplification. Multi-quantum-well interactions [10], the free-electron contribution [8] and nonparabolic band bending [11] provide additional means of increasing $\chi^{(3)}$. By modulating at extremely high harmonic frequencies one could expect also to decrease the power considerably. Ultimately picosecond pulses with a watt or less of external power should be obtainable by utilizing a combination of these various techniques. Such power levels are available from stabilized quarternary lasers, which would result in an all-solid-state pulse generator.

We are experimentally attempting the locking while doing further numerical calculations. In particular, both the modal dispersion and the material refractive-index dispersion are expected to be influential since the cavity consists solely of GaAs. The gain profile of the GaAs is also being portrayed more accurately than a Lorentzian in Eq. (1). Other nonlinear optical interactions, such as two-photon absorption in the laser cavity, are also being considered.

Acknowledgement

Research sponsored by National Science Foundation Grant ECS-8318682.

References

1. P. T. Ho, L. A. Glasser, E. P. Ippen, and H. A. Haus, Appl. Phys. Lett. 33, 241 (1978); H. Yukoyama, H. Ito and H. Inaba, Appl. Phys. Lett. 40, 105 (1982); C. Harder, J. S. Smith, K. Y. Lau, and A. Yariv, Appl. Phys. Lett. 42, 733 (1983), and the references therein; R. A. Elliot, Huang DeXiu, R. K. DeFreez, J. M. Hunt, and P. G. Rickman, Appl. Phys. Lett. 42, 1012 (1983); K. Y. Lau, C. Harder, and A. Yariv, Appl. Phys. Lett. 44, 273 (1984); K. Y. Lau, A. Yariv, OFC/OFS 85, paper ME4, San Diego, CA, Feb. 11-13, 1985.
2. T. K. Gustafson, J. P. Taran, H. A. Haus, J. R. Lifsitz, and P. L. Kelly, Phys. Rev. 177, 306 (1969); H. Haus, Fields and Waves in Optical Electronics (Prentice-Hall, Englewood Cliffs, NJ, 1984).
3. Background material dispersion in GaAs is estimated to be $\approx 10^4$ psec nm^{-1} km^{-1} [estimated from data of D. Sell et al., J. Appl. Phys. 45, 2650 (1974)] in contrast to $\alpha_g T_2^2$, which gives $\approx 10^5$ psec nm^{-1} km^{-1}. Thus this approximation is expected to be valid.

4. D. J. Kuizenga and A. E. Siegman, IEEE J. Quantum Electron. QE-6, 697 (1970).
5. E. T. Whittaker and G. N. Watson, A Course in Modern Analysis (Cambridge U. Press, Cambridge, 1950), Chap. X.
6. H. C. Casey, Jr., and M. B. Panish, Heterostructure Lasers Part B (Academic, New York, 1978), Fig. 7.9-2.
7. M. D. Levenson, Introduction to Nonlinear Laser Spectroscopy (Academic, New York, 1982).
8. C. K. N. Patel, R. E. Slusher, and P. A. Fleurey, Phys. Rev. Lett. 17, 1011 (1966).
9. Y. J. Chen and G. M. Carter, Appl. Phys. Lett. 41, 307 (1982).
10. D. S. Chemla, T. C. Daman, D. A. B. Miller, A. C. Gossard, and W. Wiegmann, Appl. Phys. Lett. 42, 864 (1983); D. A. B. Miller, D. S. Chemla, D. J. Eilenberger, P. W. Smith, A. C. Gossard, and W. T. Tsang, Appl. Phys. Lett. 41, 679 (1982).
11. P. A. Wolff and G. A. Pearson, Phys. Rev. Lett. 17, 1015 (1966).

Part V

Optoelectronics and Photoconductive Switching

Ultrafast Traveling-Wave Light Modulators with Reduced Velocity Mismatch

Masayuki Izutsu, Hiroshi Haga, and Tadasi Sueta
Faculty of Engineering Science, Osaka University, Toyonaka,
Osaka 560, Japan

Since the concept of optical guided-wave devices was introduced, remarkable progress has been made in the characteristics of opto-electronic devices. One of the most important features of waveguide-type components is the possibility of realizing higher performances compared with the bulk type counterpart. It is of great interest in applications like ultrafast signal processing, communication, and sensors.

Optical guided-wave technology offers the compactness of devices, which yields potentially high efficiency and fast response time. While, at the same time, it is especially important to develop suitable design methods for both optical and electrical circuits to realize the device of excellent performance.

Waveguide modulators have mainly been constructed as a lumped type by using the modulating electrodes like a lumped capacitance, so that the bandwidth is limited by the electrode capacitance and the transit time of the light beam through the device. For high-speed and efficient operation of the guided-wave modulator, we introduced the traveling-wave structure[1], wherein the modulator electrodes are used as a transmission line for the modulating wave. If the velocity matching condition is fulfilled between light and modulating waves, bandwidth limitations do not exist. And, even if the velocity matching is not achieved, wider bandwidth is expected with a traveling-wave structure.

One of the serious problems for high-speed components is the performance of electric circuit to which the time and frequency responses are strongly related. The electrode of coplanar stripline structure,which is widely used for the guided-wave modulators has a broad bandwidth, and also its characteristic impedance can easily be matched to the input and output cables. However, it is difficult to suppress discontinuities at the transitions between cables and electrodes, which cause unwanted reflection and radiation of the modulating wave and result in the unsatisfactory modulator response.

To reduce the discontinuity effect, we proposed the use of an asymmetric strip line[2] which consists of a strip electrode placed at the edge of the ground plane electrode. The large ground plane provides an easier transition between the two different kinds of transmission line.

The use of traveling-wave structure with an asymmetric figure of electrodes has resulted in a dramatic improvement in the modulation performance. Applying this scheme, the frequency response of the $LiNbO_3$ waveguide modulators has been smoothed over. The operation bandwidth has been expanded up to 18 GHz and a temporal pulse waveform less than 50 ps created by the modulator has already been observed by the streak camera[3].

The response time, or the width of an impulse response, t_r of the traveling-wave modulator is defined by the difference of transit time for lightwave t_o (= $n_o L/c = L/v_o$) and modulating wave t_m (= $n_m L/c = L/v_m$) through the device, where n_o and n_m are the effective indices for light and modulating waves, v_o and v_m are their velocities, and L is the electrode length. For a $LiNbO_3$ waveguide modulator with the coplanar parallel strip electrode, n_o and n_m are nearly equal to 2.2 and 4.2, respectively, so that the guided lightwave travels about two times faster than the modulating microwave. The response time is calculated as 66 ps and, in terms of the frequency response, the 3dB bandwidth becomes 6.6 GHz (for L = 1 cm).

With the reduced interaction length, faster operation will be achieved, although higher drive power is needed to decrease the modulation efficiency since required drive power is proportional to $1/L^2$.

To realize efficient and fast or wide band operation, two approaches are considered[4-6]. In the first type, discontinuities along with the electrode are introduced periodically as a slow-wave structure for the modulating wave. The modulating wavefront outdistanced by that of the lightwave is further retarded to readjust it to the next equivalent optical wavefront to cancel out effectively an accumulated distance between two wavefronts during their propagation. In this case, the discontinuities of the electrode yields an unevenness in the frequency response which results in the complicated time response of the modulator.

The second approach is the reduction of velocity mismatch between light and modulating waves. If the velocity of the modulating microwave can be increased about 15 %, nearly 50 % broadening of the bandwidth is expected for the $LiNbO_3$ traveling-wave modulator.

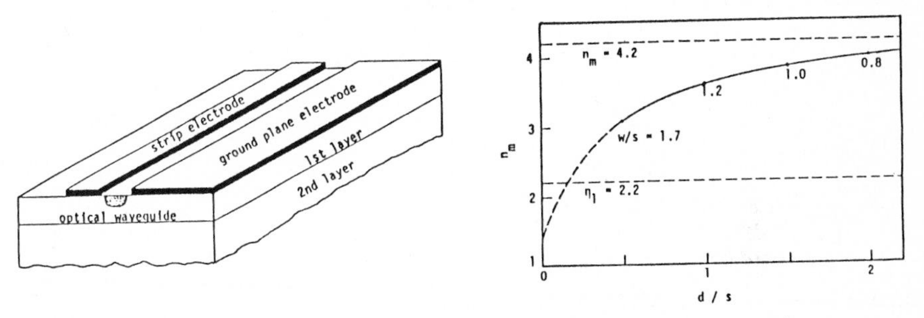

Fig. 1 Traveling-wave modulator with double-layer substrate and calculated effective index for the modulating wave vs. the thickness of the first layer.

For the reduction of velocity mismatch, we have considered using the double-layer substrate as shown in Fig. 1. The waveguide is in a thin layer of the high index crystal of which thickness d is in the same order to the electrode separation s. A portion of the modulating field energy spreads to the low index bottom layer to decrease the effective index. The calculated effective index for the modulating wave versus the thickness of the high index film is also shown in the figure. In Fig. 2, calculated field distribution in the double layer substrate is compared with the usual uniform substrate one.

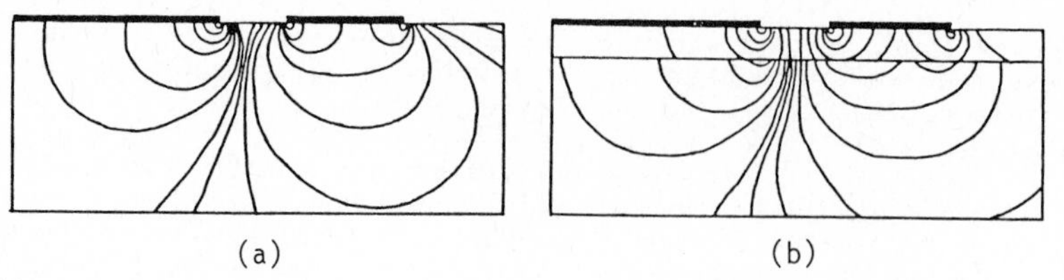

Fig. 2 Calculated examples of field distributions for single (a) and double (b) layer substrates. Electric field components normal to the surface are indicated.

The change of the effective index was measured experimentally. The $LiNbO_3$ film was supported by the glass block which had a long hole just below the optical and microwave waveguides on the crystal film, so that the underlayer was air. For a 28 µm wide electrode of 13 µm separation on a 20 µm thick crystal film, the measured characteristic impedance of the line was between 45 and 55 ohms. The effective index for the microwave was measured as 3.2 which implies the bandwidth is doubled. Modulation experiment remain for future work.

Figure 3 shows another model to reduce the velocity mismatch. A hollow channel is made on the crystal surface between two electrodes, so that the effective index for the microwave is decreased. At the same time, unwanted coupling of the light beams in parallel waveguides is suppressed by the groove to achieve high extinction ratio with shortened waveguide separation. The device employs the Mach-Zehender interferometer structure with the asymmetric X junction as the output coupler to have two complementary outputs.

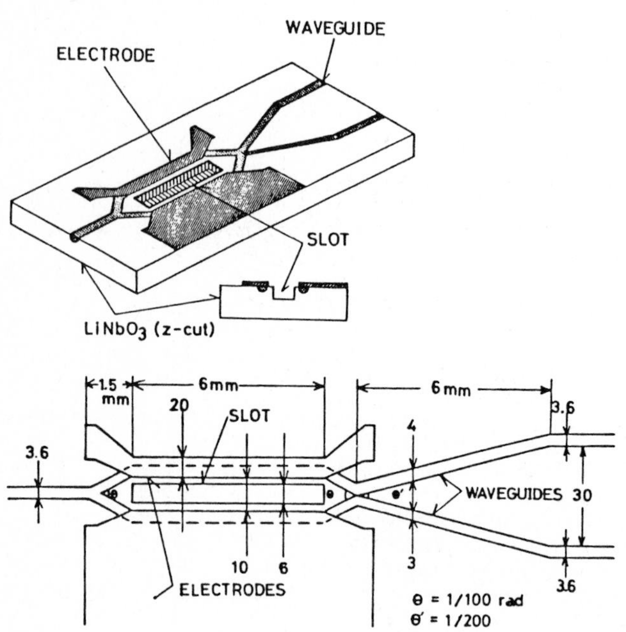

Fig. 3 Traveling-wave modulator with grooved substrate

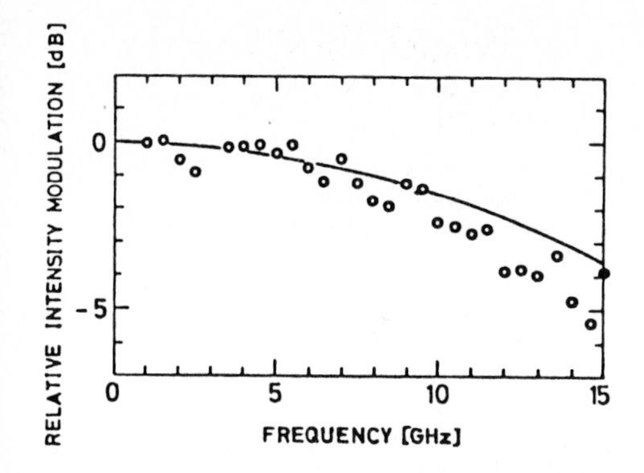

Fig. 4
Measured frequency response

The 6 μm wide groove was made by the reactive ion etching (by CF_4 gas) to 4.3 μm deep. Aluminium electrode 3 μm thick was fabricated with 80 nm SiO_2 buffer layer. The 6 mm long strip electrode was 20 μm wide and 10 μm apart from the ground plane. The characteristic impedance was about 40 Ω. The electrode with the 10 μm wide strip showed 50 ohm impedance while the insertion loss was higher than 10 dB at 10 GHz. For 20 μm wide electrode, it was 2 dB at 2 GHz and 5 dB at 10 GHz. The transit time measurement by the TDR indicated that the modulating wave traveled through the electrode in 75 and 85 ps for the device with and without the groove on the surface. From these values, it was estimated that the effective index for the microwave decreased to 3.8 from the conventional value 4.2, so that the 20 % broadening of the bandwidth is expected.

The measured half-wave voltage was 3 V for 630 nm TM-like mode and the extinction ratio was -18 dB. The modulation frequency response at the microwave region is shown in Fig. 4. Deviation of the measured values from the calculated curve in the frequency range above 8 GHz is attributable to the effect of insufficient impedance matching between electrode and cables. The measured 3 dB bandwidth is 12 GHz and, consequently, the $P/\Delta f$ figure is 4.2 mW/GHz for 88 % intensity modulation.

In conclusion, the velocity mismatch in the traveling-wave modulator was reduced by the groove fabricated on the surface of the waveguide substrate. The performance of the modulator was improved remarkably compared with that of our previous version of which $P/\Delta f$ figure was 11.3 mW/GHz. It is also notable that the separation of parallel waveguides, or electrodes, can be reduced without the unwanted optical coupling by the hollow channel.

REFERENCES

[1]M.Izutsu,T.Sueta:Trans.IECE Japan,J64-c,264/271,1981.
[2]M.Izutsu,T.Sueta:IEEE J.Q.E.,19,668/674,1983.
[3]M.Izutsu,T.Sueta:Appl.Phys.,5,307/315,1975.
[4]E.A.J.Marcatili:Appl.Opt.,19,1468/1476,1980.
[5]P.L.Liu:J.Opt.Commun,2,2/6,1981.
[6]R.C.Alferness,S.K.Korotky,E.A.J.Marcatili:IEEE J.Q.E.,20,301/309,1984.

Modulation of an Optical Beam by a Second Optical Beam in Biased Semi-Insulating GaAs

L.M. Walpita, W.S.C. Chang, H.H. Wieder, and T.E. Van Eck
Department of Electrical Engineering and Computer Sciences, C-014,
University of California, San Diego, La Jolla, CA 92093, USA

The potential of high speed and large information capacity of optical systems has initiated much interest in optical device research[1,2,3]. A variety of bulk optical semiconductor bistable devices have been demonstrated[2]. Further, multiple-quantum-well (MQW) structures have been shown to have nonlinear properties which will be useful in device applications[3,4,5]. We have observed modulation of radiation at wavelength λ_1 (below and near the bandgap), induced by a second radiation at wavelength λ_2 (above the bandgap) in Semi-Insulating GaAs (SI GaAs) in the presence of electric fields. This effect could potentially be used to obtain AND gate optical logic function. This property is expected to exist in other direct bandgap III-V compound semiconductors such as InP and to much larger extent in MQW structures. Here we discuss some of the results obtained during the experimental investigation on SI GaAs.

In the initial experiments, we have used an undoped Semi-Insulating GaAs sample of which both sides are polished. The sample has InSnO and Al electrodes deposited on both surfaces and a window is left in the Al electrode for light to pass through. A beam of light from a LED source (10mwatts) centered at λ_1 = 0.88 μm is transmitted through a 400 micron thick sample and is focused onto a detector (figure 1). Some part of the LED light fell below the absorption edge (figure 2). We applied a voltage

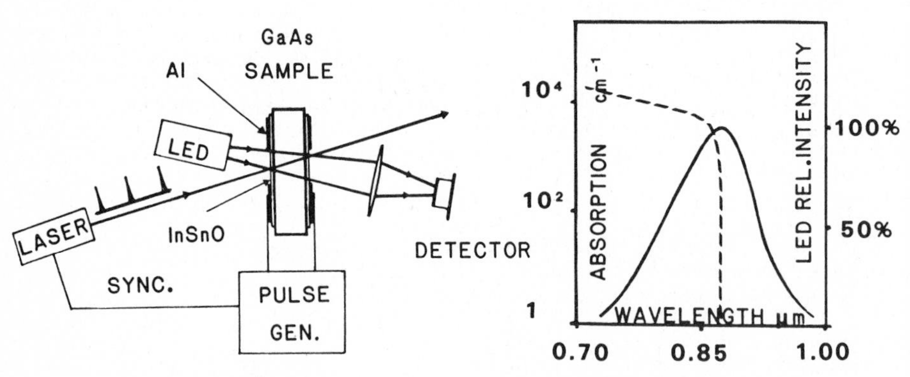

Fig. 1 Schematic of the experimental setup.

Fig. 2 The wavelength spectrum of the LED in relation to the absorption edge of GaAs in the ideal situation[6] is shown. The LED spectrum was obtained from the RCA data sheet. GaAs sample behaves as a short wavelength filter.

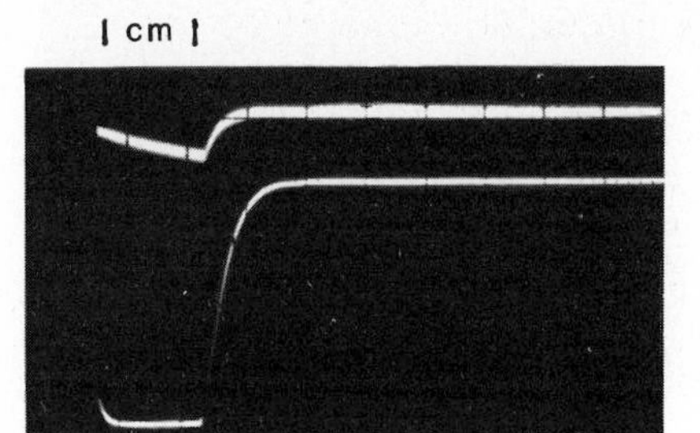

Fig. 3 The absorption of LED caused by a voltage pulse is shown. The bottom trace is the voltage pulse with horizontal scale 15μsec./cm and vertical scale 150v/cm. The upper trace represents the absorption with vertical scale 10%/cm.

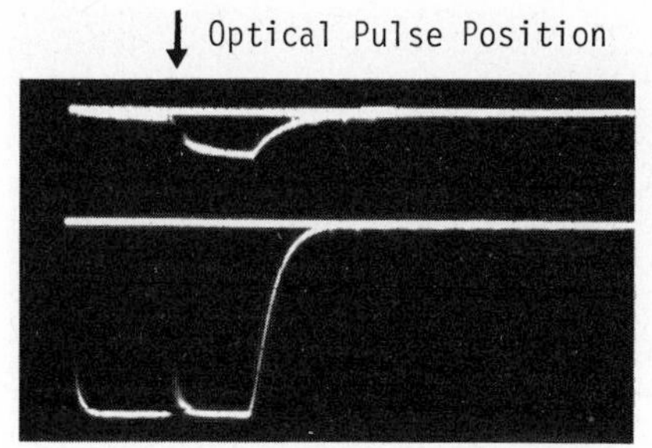

Fig. 4 The effect of the optical pulse of 1 kwatt/cm^2 on the LED absorption is shown. The bottom trace is the voltage pulse with horizontal scale 15μsec./cm and vertical scale 150v/cm. The upper trace represents the absorption caused by both optical and voltage pulse, with vertical scale 75%/cm.

pulse of a few tens of μ sec. duration at 400 volts (corresponding to an electric field of 10 kV/cm) on to the InSnO/Al electrodes; the LED light is absorbed as shown in figure 3. We measured an increase in attenuation of about 4 to 5% at 400 volts, and it agreed only roughly with other available published electroabsorption data[6], because the LED has an extended spectral width and because the absorption edge of SI GaAs is not very abrupt.

When another optical pulse (λ_2 = 0.885 ± .01 μm) of maxiumum intensity 1kwatt/cm^2 is transmitted through the sample, the voltage measured at the electrode and the absorption of the LED light is instantaneously reduced (see figure 4). Within the limits of the switching speed established by the RC time constant of the pulsed voltage circuit, the modulation of the LED light is instantaneous. This effect performs essentially the function of an optical 'AND' gate. This modulation effect is presumed to be due to photo current generated by the λ_2 pulse which has caused a reduction in voltage as shown in figure 4. However, this modulation may be influenced by other non-linear phenomena associated with the trapping processes in the presence of applied field.

A number of side effects have also been observed which could again be related to the above mentioned processes: (a) We note in figure 3 that the LED light attentuation is much lower at the onset of the voltage pulse compared to that after 25 μsec. of pulse duration. (b) After the optical pulse the electroabsorption continues to increase beyond what is possible with just the voltage pulse alone (figure 4). (c) A non-linear dependence of this increase in absorption is also observed; at optical pulse intensity 8 watts/cm^2 and voltage 200v, a 7% absorption is seen, while at 600 watts/cm^2 and for the same voltage only 20% absorption is seen.

We have performed additional experiments in order to clarify some of these side effects. (1) In the first experiment we used two voltage pulses, the second pulse following the first one after a time delay of a few micro seconds. We then observed the LED light absorption caused by the

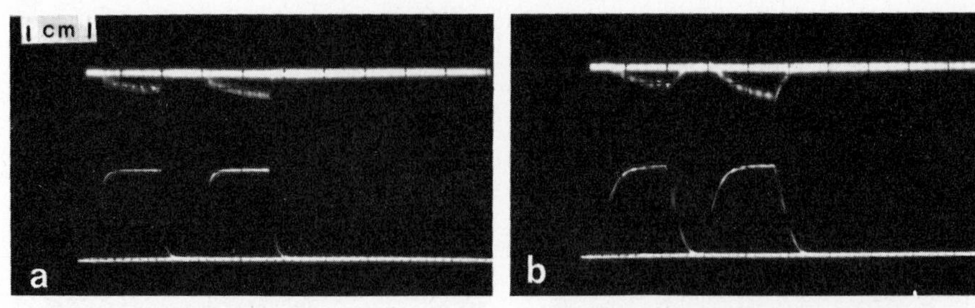

The effect of an external RC time constant on the second voltage pulse is shown. The bottom trace is the voltage pulse with horizontal scale 30μsec/cm and vertical scale 150v/cm. The upper trace represents absorption with vertical scale 8%/cm.

(a) No external capacitor across the sample.
(b) With external capacitor (0.005 μF) across the sample.

two pulses separated by 25 micro. secs; the absorption during the second pulse was not significantly affected by the presence of the first pulse (figure 5a). However, when a capacitor of 5×10^{-3} μfarad is introduced across the sample to change the effective time constant, the absorption in the second pulse was clearly affected by the first one (Figure 5b). The introduction of the capacitor increased the RC time constant and caused a residue electric field, due to the first pulse to be present at the start of the second pulse. This resulted in additional absorption in the second pulse. (2) In a second experiment we observed the photo current, the attentuation of LED light (λ_1) and the reduction of applied voltage caused by the pulse light (λ_2) as a function of λ_2 wavelength (figure 6). Clearly, the optical pulse-induced modulation of the LED light is strongest at the band edge of the semiconductor sample. The reduction of the cross modulation effect for λ_2 longer than 8900 Å can be attributed to the decreased absorption in λ_2 below the bandgap. In general, one

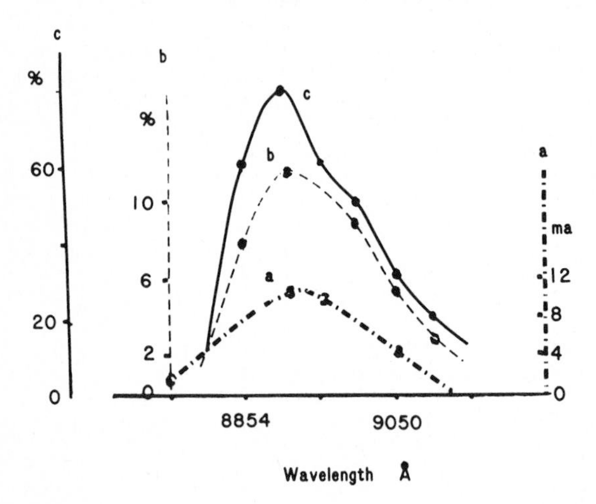

Fig. 6 The effect of the optical pulse wavelength on the properties of the GaAs sample in the presence of transmitted LED light and applied pulse voltage is shown.

(a) The current in the sample caused by the optical pulse is shown for voltage pulse 50v and optical pulse intensity 6.3×10^{-7} Joules/pulse.

(b) The percentage absorption after the optical pulse is shown for voltage pulse 250v and laser pulse intensity 6.7×10^{-7} Joules/pulse.

(c) Reduction of voltage caused by the optical pulse generated current in the sample is shown. The applied voltage is 300v and pulse laser intensity 1×10^{-4} Joules/pulse.

would also expect that the photo carriers excited by the λ_2 radiation would be confined more to the surface region near the electrodes as λ_2 becomes shorter than 8900 Å due to the increased absorption coefficient. Thus the reduction of electroabsorption for the λ_1 radiation will occur only near the electrode surface. The electroabsorption at λ_1 below this surface layer may not be affected significantly. Hence there is a drop in cross modulation effect for shorter λ_2. However, we may have expected a slower dependence on wavelength than the results shown in figure 6.

In conclusion, we have demonstrated a new cross modulation effect in SI GaAs. A number of side effects have not yet been explained. This cross modulation effect may be utilized in spatial light modulators, and will have much faster response than Liquid Crystal Light Valve. However, to demonstrate the modulation effect we have used high voltage (up to 400v across 400 μm) and high optical power densities ($\lambda_1 \sim 50$ m watts/cm^2, $\lambda_2 \sim$ 1kwatt/cm^2). The applied voltage and light intensities are expected to be reduced by replacing the LED by a laser source operating near absorption edge. A further improvement is expected for semiconductor materials that have an abrupt absorption edge. This work was partly supported by AFSOR grant 80-0037.

References

1. P. W. Smith, "On the Physical Limits of Digital Optical Switching and Logic Elements", Bell Sys. Tech. J., 61, 1975 (1982).
2. D. A. B. Miller, S. Des Smith, Colin T. Seaton, "Optical Bistability in Semiconductors", IEEE J. of Quantum Electronics, 17, 313 (1981).
3. D. S. Chemla, D. A. B. Miller, P. W. Smith, A. C. Gossard, W. Wiegmann, "Room-Temperature Excitonic Nonlinear Absorption and Refraction in GaAs/AlGaAs Multiple Quantum Well Structures", IEEE J. of Quantum Electronics, 20, 265 (1984).
4. D. A. B. Miller, D. S. Chemla and T. C. Daman, A. C. Gossard, W. Wiegmann, T. M. Wood, "Novel Hybrid Optical Bistable Switch: The Quantum Well Self-Electro Optic Effect Device", Appl. Phys. Lett. 45, 13 (1984).
5. A. Migus, A. Antonetti, D. Hulin, A. Mysyrowicz, H. M. Gibbs, N. Peyghambarian, J. L. Jewell, "One-Picosecond Optical NOR Gate at Room Temperature with GaAs-AlGaAs Multiple-Quantum-Well Nonlinear Fabry-Perot Etalon", Appl. Phys. Lett. 46, 70 (1985).
6. G. E. Stilman, C. M. Wolfe, C. O. Bozier and J. A. Rossi, "Electroabsorption in GaAs and its application to waveguide detectors and modulators", Appl. Phys. Lett. 28, 544 (1976).

22-GHz Bandwidth InGaAs/InP PIN Photodiodes

J.E. Bowers
Holmdel Laboratories, AT&T Bell Laboratories, Holmdel, NJ 07733, USA
C.A. Burrus and R.S. Tucker
Crawford Hill Laboratories, AT&T Bell Laboratories, Holmdel, NJ 07733, USA

We describe the fabrication and characterization of an improved back-illuminated PIN mesa photodiode composed of InGaAs absorbing layers on an InP substrate. The measured 3 dB bandwidth is 22 GHz.

Fast GaAs Schottky photodetectors for visible and near-infrared applications have been demonstrated [1], and high-speed PIN structures using InGaAs/InP for the emerging lightwave communications wavelengths (1.3 - 1.6 μm) have been reported [2,3]. We describe here the fabrication and characterization of an improved InGaAs/InP PIN photodiode with a measured 3 dB bandwidth of 22 GHz.

The structure of the improved detector (Fig. 1 (a)) is similar to detectors described previously [2]. It is a2.5- μm thick layer of zinc-diffused p^+InGaAs on a 1.5- μm thick layer of n-InGaAs grown on a (100) n^+InP substrate. The background doping of the n InGaAs layer was less than $1X10^{15}$ cm^{-3}. The structure is back illuminated, and a 25- μm diameter wire contact is epoxied to the top of the metalized mesa. The capacitance of the unpackaged detector at -10 V bias is about 0.05 pF, as measured with a capacitance bridge. The $1/(2\pi RC)$ frequency for a 50 Ω load is 64 GHz. Although the dark currents of most devices made in this way have been quite low, often 10 nA at -10V bias, the highest frequency batch of devices described here has exhibited 400 nA leakage current at -10V. A commercial package for microwave detectors (HP 3330C) was modified to accept the back-illuminated detectors. The biasing circuit is shown in Fig. 1(b).

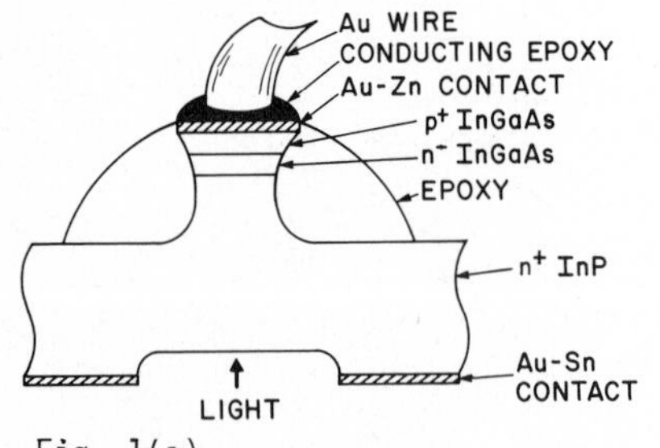

Fig. 1(a)
Schematic diagram of back-illuminated PIN InGaAs/InP photodetector.

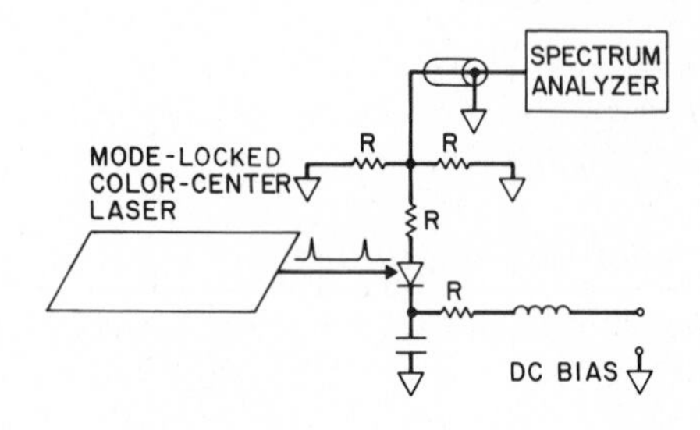

Fig. 1(b)
Experimental set-up and diagram of biasing network.

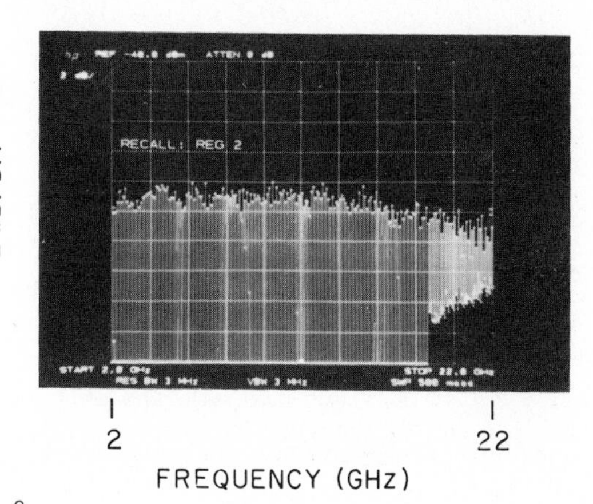

Fig. 2
Frequency response of photodector for excitation with 10-mW peak power, 10-ps FWHM pulses at 1.5 µm.

The frequency response of the detectors was determined by illuminating the diode with a 100-MHz train of 10-ps (FWHM) pulses at 1.5-µm wavelength (from a mode-locked color-center laser [4]) and measuring the rolloff with a spectrum analyzer connected directly to the detector package (Fig. 1(b)). The result is shown in Fig. 2 for optical pulses with a peak power of 10 mW. The response rolls off by just 3 dB at 22 GHz. Using the measured pulsewidth of 10 ps, we calculate that 2 dB of the rolloff is due to the finite frequency content of the laser source. This was confirmed by observing the degradation in the measured frequency response,which accompanied a slight detuning of the laser. Three detectors from this wafer were measured,and all had -3 dB frequencies in excess of 21 GHz. The -3 dB frequency did not decrease when the optical beam was laterally misaligned or defocused. The frequency response did degrade for peak optical powers greater than 20 mW.

The bias dependence of the -3 dB frequency is shown in Fig. 3. As the bias is increased from 0 to 10 V, the detector speed increases as (a) the n-InGaAs layer becomes increasingly depleted, reducing the capacitance and diffusion current, and (b) the electric field increases and the hole transit time decreases. The response is constant at 22 GHz from 10 to 20 V.

The influence of finite electron and hole transit times on the frequency response of InGaAs pin detectors was calculated for a field of 70 kV/cm, corresponding to a bias of 10 V across a 1.5 µm thick layer. The electron velocity in InGaAs at this field is $\sim 7 \times 10^6$ cm/s [5] and we assume a saturated hole velocity of $\sim 7 \times 10^6$ cm/s. The frequency responses are plotted in Fig. 4 for our layer thickness (1.5 µm) and layer thicknesses of 0.5 µm and 5.0 µm. Responses are drawn for three relevant wavelengths where

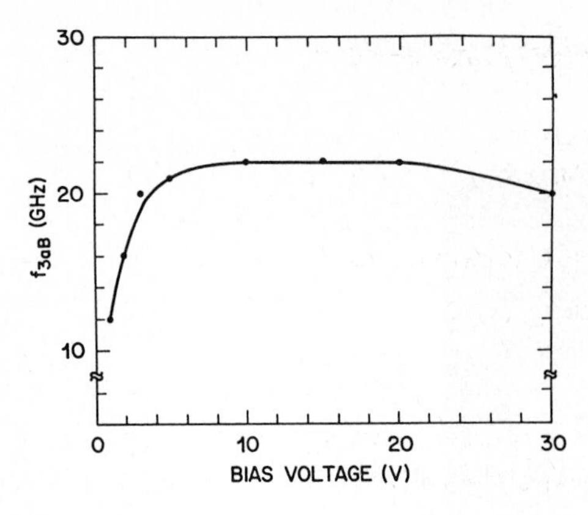

Fig. 3
Bias dependence of -3 dB frequency.

[6] α = 0.66 μm^{-1} (λ= 1.55 μm), α = 1.15 μm^{-1} (λ= 1.30 μm), and α= 1.8 μm-1 (1.06 μm). The 3 dB bandwidth limit due to transit time for our detectors is 26GHz. This supports our calculation that 2 dB of the measured rolloff at 22 GHz is due to the 10 ps color center laser pulsewidth.

The improved frequency response (22 GHz versus 12 GHz [2]) is due to (1) shorter hole and electron transit times resulting from the thinner (1.5 μm) n-layer in this detector, (2) lower doping in the n-layer, and (3) improved packaging.

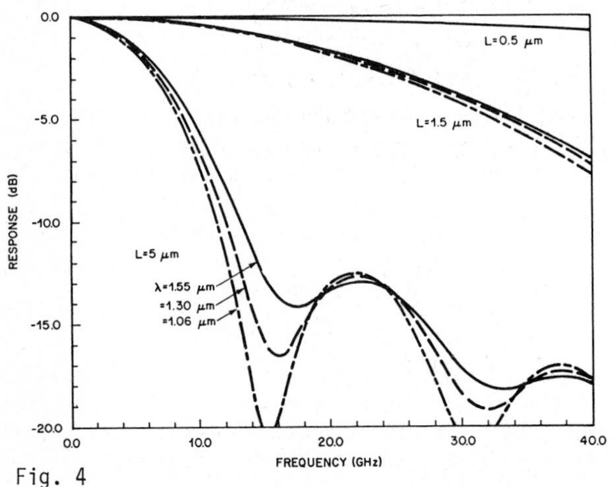

Fig. 4
Transit time-limited frequency responses of pin InGaAs detectors for several wavelengths and absorbing layer thickness.

We wish to thank L. F. Mollenauer for allowing us to use the color-center laser, S. Lorimer and A. Dentai for providing LPE wafers, and J. C. Campbell, S. K. Korotky, and D. M. Bloom for useful discussions.

References

1. S. Y. Wang and D. M. Bloom, Electron Lett. 19, 554 (1983).
2. T. P.Lee, C. A. Burrus, K. Ogawa, and A. G. Dentai, Electron Lett. 17,431 (1981)
3. K. Li, E. Rezek and H. D. Law, Electron Lett. 20, 196 (1984).
4. L. F. Mollenauer and D. M. Bloom, Optics Lett. 4 , 297 (1979).
5. T. H. Windhorn, L. W. Cook and G. E. Stillman, J. Electron. Mat. 11, 1065 (1982)
6. Yu-Ssu Chen, unpublished.

An Ultrafast Diffusion-Driven Detector

A.G. Kostenbauder and A.E. Siegman
Edward L. Ginzton Laboratory, Stanford University, Stanford, CA 94305, USA

I. Introduction

Numerous fast semiconductor photodetectors have been developed in the last decade. These devices have used short carrier lifetimes [1,2] or large electric fields and small geometries [3,4] to produce fast impulse responses. In contrast to these methods of making fast photodetectors, we describe a *photovoltaic* device whose operating speed is due to the diffusion of carriers. This device involves interfering light waves to form a grating of photoexcited carriers with a $\cos Kx$ dependence. Such a distribution will decay with a time-constant of $(D_a K^2)^{-1}$ where D_a is the material's ambipolar diffusion constant. Because K can be made very large, the response time can be made very short. For example, if the grating is formed in silicon, with optical beams of 1 μm wavelength having antiparallel k-vectors, then the diffusion-driven detector response time will be 0.27 psec.

II. Geometries for the Diffusion-Driven Detector

In its simplest form, the diffusion-driven detector consists of a slab of semiconductor with a transparent electrode on the front face, and an optically reflecting rear electrode. When such a structure is illuminated at normal incidence, the rear electrode produces counterpropagating optical signals in the semiconductor, which results in an electromagnetic energy density proportional to $1 + \cos 2knx$ where k is the (vacuum) optical k-vector and n is the material's index of refraction. The electrode spacing (i.e., the slab thickness) must be chosen so that a maximum of the energy density occurs at one electrode while a minimum occurs at the other.

Other geometries are possible; for example, one may use transparent electrodes on both faces and use external optics to produce an interference pattern in the semiconductor slab. These alternate geometries allow one to produce an arbitrary angle between the interfering beams,and thus produce energy densities proportional to $1 + \cos Kx$ where K can be varied between 0 and $2nk$. When varying K in this manner, one must only use values which produce a minimum of the energy density at one electrode and a maximum at the other.

III. Small Modulation Analysis

As mentioned above, the electromagnetic energy density inside a diffusion-driven detector is proportional to $1 + \cos Kx$, and as is well known, the sinusoidal part decays away very rapidly (by diffusion) leaving a spatially constant carrier distribution. This constant distribution is in general much longer lived, since it must decay away by recombination. Thus, at any given time during an incident optical pulse, we expect to have a relatively large spatially constant carrier distribution (which is the remnant of all of the pulse energy deposited in one carrier lifetime) in addition to a relatively small sinusoidal part (which results only from the part of the pulse which has arrived in one diffusion time constant). Under these conditions, we linearize the equations of motion [5] for the carrier densities about the large spatially constant values. These equations are easily solved for the carrier densities,which result from an optical impulse at $t = 0$:

$$n \doteq p \doteq \text{spatial constant} + \text{constant} \cdot (\cos Kx)e^{-t(1/\tau + D_a K^2)}.$$

When we account for displacement current in the semiconductor, the current generated under short circuit conditions is $q/\ell[nD_n - pD_p]_0^\ell$ where 0 and ℓ are the locations of the electrodes. Thus, with the electrodes placed as discussed in section II, there will be a current generated proportional to $2q/\ell(D_n - D_p)e^{-t(1/\tau + D_a K^2)}$. For most semiconductors, $1/\tau$ is much smaller than $D_a K^2$ and may be dropped from this expression. Thus, we see that the operating speed of such a detector is due to the rapid diffusive decay of sinusoidal carrier distributions. Since the carriers must move a distance of about $1/K$ to wash out the grating, and they have a velocity of about $D_a K$, we may interpret their ratio (i.e., $(D_a K^2)^{-1}$) as a transit time. Also, we stress that this current is a photovoltaic effect which exists in the absence of junction fields or an externally applied bias!

Also, we see that since $D_a = 2D_n D_p/(D_n + D_p)$, the quantum efficiency of a diffusion-driven detector is $(D_n^2 - D_p^2)/(D_n D_p K^2 \ell^2)$ which for GaAs ($D_n = 220 \text{ cm}^2/\text{s}$ and $D_p = 10 \text{ cm}^2/\text{s}$) and $K\ell = \pi$ is slightly larger than 2.

Finally, we note that our solution (based upon the diffusion equation) is only accurate when the predicted response times are > 1 ps. This is due to the fact that the achievement of response times < 1 ps requires grating peak-to-trough distances of about .1 μm which is in the regime of ballistic transport in GaAs. Also, when the response times become significantly less than 1 ps, the carriers will not have time to thermalize, and thus cannot be expected to move with the diffusion constant characteristic of thermalized carriers.

IV. Experimental Results

We have conducted some preliminary tests of a diffusion-driven detector. These tests were made on an InP sample with a 0.5 ns carrier lifetime and a 25 μm thick active region. The sample was excited by a 0.5 mW helium-neon laser whose output was focused to 55 μm. The sample was mounted on a translation stage so that the current could be measured as a function of the relative position of the gap and the laser spot.

The quantum efficiency (defined as electrons out per photon absorbed in the active region) under such conditions is:

$$\eta = 2\frac{x}{w}\frac{\tau(D_n - D_p)}{w\ell}$$

where x is the displacement of the spot from the center of the gap, τ is the carrier lifetime, w is the spot size, and ℓ is the device thickness. This result is in good agreement with the intuitively expected form for the current. When x/w is large, the number of photons entering the active region is exponentially suppressed, which dominates the linear increase in η and results in a small current. On the other hand, when x is zero, there are many photons entering the device, but symmetry forces the measured current (and thus η) to be zero. However, in the region where x/w is near ± 1 there is a large difference in the intensity at the electrodes, and thus a large generated current. Moreover, the direction of this current is tied to the sign of x and is responsible for η being less than zero when x is negative.

As shown in Fig. 1, the measured current is in excellent agreement with the current predicted by this expression. The experiment was repeated, using a 5 mW helium-neon laser focused to 80 μm and showed the linear scaling of current with power density.

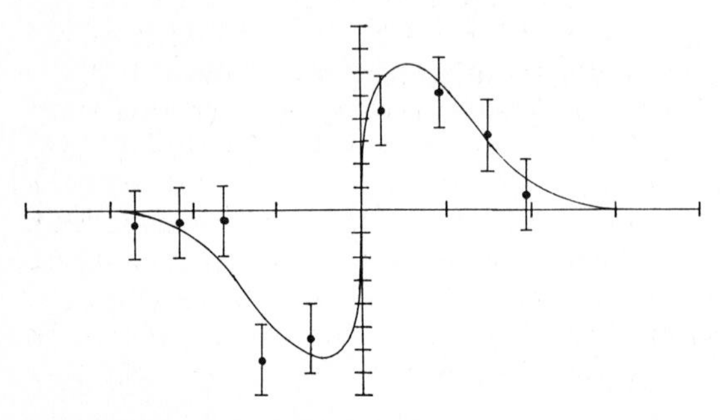

Fig. 1 Current versus position of the laser spot relative to the center of the 25 μm gap. Horizontal scale is 50 μm per division and the vertical scale is 2nA per division

V. Parasitic Effects

We now turn to the case where a diffusion-driven detector is connected to a transmission line of impedance Z_0. Two parasitic effects may now become prominent.

First, the large spatially constant carrier distribution in the device means that in addition to being a current source, the detector acts as a resistor. If the optical intensity becomes too high, this resistance will become comparable to Z_0 and the detector will cease to act as a current source. Moreover, if the resistance falls well below Z_0, the device acts as a $\epsilon(2KT/q)(D_n - D_p)/(D_n + D_p)$ volt voltage source, where ϵ is the fractional modulation of the carrier densities. This result implies that a diffusion-driven detector should operate in the regime where ϵ is large (i.e., $\epsilon \doteq 1$) which in turn implies that the carrier lifetime should not be too long, as this will cause ϵ to be very small. (See the beginning of section III).

Second, since the voltage between the contacts now varies in time, displacement current in the semiconductor slab appears in the form of a shunt capacitance. One of the key advantages of a diffusion-driven detector over a fast diode is that the thickness of the depletion region in a diode fixes both the transit time and the capacitance. This means that the thickness of the depletion region must be a compromise between being too small (achieving small transit times with unacceptably large capacitances), and being too large (resulting in an excessive transit time.) On the other hand, a diffusion-driven detector's capacitance is set by the thickness ℓ while the transit time depends only upon K. The fact that the two time constants depend upon physically independent parameters means that it is possible to construct a device which has a very low capacitance *and* a very short transit time.

VI. Conclusion

We have described a detector which is driven by diffusion, unlike other fast photodetectors. Our device does not rely on recombination or large electric fields in order to produce a fast impulse response. We believe that a diffusion-driven detector is superior to a fast diode, since it allows independent optimization of device capacitance and transit time.

Finally, we note that further theoretical investigation is needed since we have implicitly assumed equilibrium statistics in this analysis. This is acceptable when K is small, but is suspect when $K \simeq 2kn$ since we are predicting response times comparable to the carrier thermalization times. Also, when $K \approx 2kn$, the fringe spacing is about 0.1 μm which suggests that ballistic transport may become an important mechanism. Consequently, we feel that the diffusion-drived detector can serve both as a fast photodetector and as a tool for measuring fast processes in semiconductors.

References

1. A.M. Johnson, A.M. Glass, D.H. Olson, W.M. Simpson, and J.P. Horbison: *Appl. Phys. Lett.* **44**, 450 (1984).
2. R.B. Hammond, N.G. Paulter, R.S. Wagner, and T.E. Springer: *Appl. Phys. Lett.* **44**, 620 (1984).
3. S.Y. Yang and D.M. Bloom: *Elec. Lett.* **19**, 554 (1983).
4. B.H. Kolner, D.M. Bloom, and P.S. Cross: *Elec. Lett.* **19**, 574 (1983).
5. R.S. Muller and T.I. Kamins: *Device Electronics for Integrated Circuits,* (Wiley, New York, 1977.)

Picosecond Photoconductivity in Polycrystalline CdTe Films Prepared by UV-Enhanced OMCVD

A.M. Johnson, D.W. Kisker, W.M. Simpson, and R.D. Feldman
AT&T Bell Laboratories, Holmdel, NJ 07733, USA

Ultrafast photoconductive detectors are often built from semiconductors possessing a large density of structural defects, which function as effective trapping and recombination centers. One of the earliest demonstrations of this approach was made with high defect density, low mobility amorphous silicon [1,2]. The low carrier mobilities severly limit the sensitivity of amorphous semiconductor devices in most applications. Improvements in the sensitivity over the early devices have been made by using Schottky barriers and novel transmission line configurations in amorphous silicon [3] and by the radiation damage of crystalline semiconductors [4]. Another approach which has not been fully exploited is the use of polycrystalline semiconductors [5,6], which have a large defect density at the grain boundary and a high mobility within the grain. In this work, we have made the first measurements of picosecond photoconductivity in thin films of polycrystalline CdTe grown by photon-assisted Organometallic Chemical Vapor Deposition (OMCVD). Photoconductivity measurements were made on three samples of CdTe, intentionally prepared under different deposition conditions to controllably alter the electronic transport. The fastest of these photodetectors had a carrier relaxation time of 3.7 psec and an average drift mobility of 59 $cm^2/Vsec$.

The CdTe films were grown by a modified OMCVD technique. Normally, the low temperature limit of thermal deposition for these materials is about 380°C - 410°C [7]. Using UV light to enhance the reaction, CdTe films can be grown at temperatures as low as 250°C. By varying the deposition parameters such as temperature, stoichiometry, and substrate material we are able to change the crystallinity from 200 Å grain polycrystallites to single crystalline material. Further details of the deposition process are reported elsewhere [8].

A 2000 Å thick film of CdTe was deposited onto a 250 μm thick fused silica substrate at a deposition temperature of 250°C. This film was studied by transmission electron microscopy (TEM) in a Philips EM 400 operated at 120 kV. The film was removed from the substrate by a lift-off technique. Figure 1 shows a TEM bright field image of this film. The film contains small, equiaxed grains. The average grain size is 200 Å, although some grains are two to three times larger. Diffraction patterns of the film show no signs of preferred orientation.

The 200 Å grain film and supporting fused silica substrate were incorporated into a standard high-speed 50 Ω microstrip transmission line configuration [1].

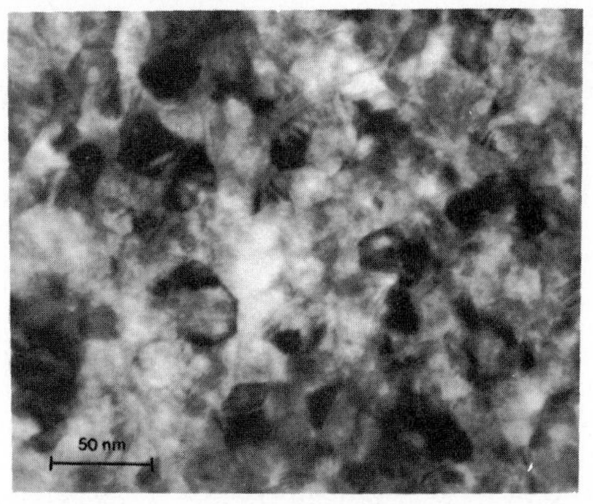

Fig. 1: TEM bright field image of a CdTe film grown at 250°C on a fused silica substrate. The substrate has been entirely removed. The average grain size of the film is 200 Å.

In this gap cell configuration a 16.4 μm gap between the aluminum electrodes served as the active area of the photodetector. The sampling oscilloscope limited response of this detector to optical pulses having a wavelength of 595 nm and a duration of 300 fsec, from a synchronously modelocked dye laser pumped by the compressed second harmonic of Nd:YAG [9], is displayed in Fig. 2. The excitation energy was 39 pJ and the bias was set at 60 V. For this bias level the dark resistivity was determined to be 3×10^7 Ω-cm. From the scope response we obtained a mobility-relaxation time product ($\mu\, t_r$) of 2.2×10^{-10} cm^2/V. The carrier relaxation time was determined by the now standard high-speed electronic correlation technique [10]. In the case of a single type of photoconductor, we measure a second order autocorrelation function, $G^2(t)$, which contains the desired information about the temporal response of the photoconductor as a function of optical delay, t. By fitting the decay of the autocorrelation function to an exponential, we obtain a carrier relaxation time t_r of 3.7 psec for this sample, clearly indicating the high-speed capability of fine-grain polycrystalline materials. The full-width at half-maximum (FWHM) of the autocorrelation function was 7.2 psec. Combining the independent measurements of μt_r and t_r on the same sample results in an average drift mobility of 59 $cm^2/Vsec$.

The next sample of CdTe was again grown on a fused silica substrate, but at the elevated deposition temperature of 350°C in order to increase the grain

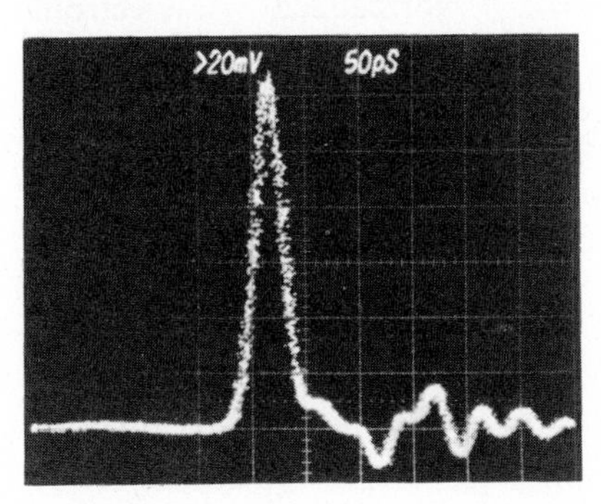

Fig. 2: Sampling oscilloscope response of 200 Å grain polycrystalline CdTe on fused silica photodetector to 300 fsec pulses from a synchronously modelocked dye laser operating at a repetition rate of 100.498 MHz. The FWHM is 35 psec.

size. This film had a thickness of approximately 2 μm. For this photodetector we had an electrode separation of 15.4 μm, an excitation energy of 29 pJ, a 45 V bias, a dark resistivity (45 V) of 2×10^6 Ω-cm, and a μt_r product of 9.2×10^{-10} cm^2/V. The sampling oscilloscope response of this detector had an identical temporal response to the previous detector (FWHM = 35 psec). The electronic autocorrelation measurement of this photoconductor resulted in a carrier relaxation time of 7.2 psec and a FWHM response of 10.4 psec. This larger grain polycrystalline CdTe sample had an average drift mobility of 128 cm^2/Vsec.

A third photodetector was made from a 1 μ thick film of CdTe grown on (100) semi-insulating GaAs. The film was determined to be *single crystal* using X-ray diffraction. For this photodetector we had an electrode separation of 15 μm, an excitation energy of 34 pJ, a bias of 10 V, and a μt_r product of 6.1×10^{-9} cm^2/V. The GaAs substrate prevented an accurate determination of the dark resistivity of the CdTe film. The sampling oscilloscope response of this single crystal detector had a FWHM of 70 psec, which was considerably slower than the polycrystalline detectors. From the scope response we obtained a μt_r product of 6.1×10^{-9} cm^2/V. From the electronic autocorrelation function of this photodetector we obtained a value of 54 psec for the carrier relaxation time and a FWHM response of 94 psec. The resultant average drift mobility was 113 cm^2/Vsec. The non-ohmic aluminum contacts can account for the reduced mobility of this detector, when compared to the larger grain polycrystalline detector. The effects of space charge and dielectric relaxation will reduce the photoconductivity, resulting in an artificially lowered mobility and a lower carrier relaxation time [11].

Figure 3 is a plot of the autocorrelation measurements [10] of the three CdTe photodetectors and the GaAs substrate, normalized to unity, at zero delay,

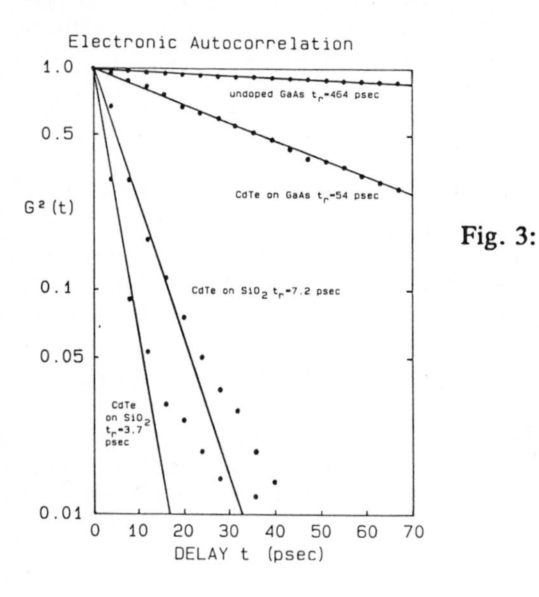

Fig. 3: Normalized electronic autocorrelation functions of three CdTe photodetectors, prepared under various deposition parameters described within the text, as a function of delay. The electronic autocorrelation of the semi-insulating GaAs used for substrate material is also displayed.

for ease of comparison. It is evident from this figure that the temporal response of the photoconductors can be varied by changes in the deposition parameters. The larger grain polycrystalline CdTe resulted in a longer carrier relaxation time and an increased average carrier drift mobility. A transport model frequently used in polycrystalline silicon [12] estimates an effective carrier mobility that consists of the sum of two terms, 1) transport within a grain of finite size with a carrier mobility equal to that of single crystal bulk material, and 2) transport within the disordered grain boundary layer of finite thickness and a carrier mobility representative of the disordered material. Thus, for small grains, the effective mobility will be dominated by the mobility of the disordered grain boundary layer. As the grain size is increased, one can expect the carrier relaxation time to increase because of the longer distance to the trapping sites at the grain boundaries. An increase in the effective carrier mobility is also expected because of the decreasing volume fraction of disordered grain boundary material to single crystal bulk material as the grain size in increased.

In many instances, the use of polycrystalline semiconductors as ultrafast photoconductors has been inhibited by the very large dark conductivity. In some cases one had nearly metallic conduction across the grain boundaries. Because of its low dark conductivity the OMCVD CdTe is ideal for ultrafast photodetectors. We have demonstrated a device with a tunable temporal response by the manipulation of the chemical vapor deposition parameters. These detectors have a speed and sensitivity that surpasses that of silicon-on-sapphire detectors that have been damaged by high energy oxygen ions [4]. One of the disadvantages of radiation-damaged photodetectors is the finite implant depth of the ions. Thus, for long wavelength devices, optical absorption in undamaged regions of the device can lead to long photocurrent tails, a problem not encountered with the polycrystalline materials. With the incorporation of Hg ($Hg_{1-x}Cd_xTe$) we will also have a detector with a tunable spectral response well into the infrared portion of the spectrum.

[1] D. H. Auston, P. Lavallard, N. Sol, and D. Kaplan, Appl. Phys. Lett. *36*, 66 (1980).

[2] A. M. Johnson, D. H. Auston, P. R. Smith, J. C. Bean, J. P. Harbison, and A. C. Adams, Phys. Rev. B*23*, 6816, (1981).

[3] A. M. Johnson, A. M. Glass, D. H. Olson, W. M. Simpson, and J. P. Harbison, Appl. Phys. Lett. *44*, 450 (1984).

[4] P. R. Smith, D. H. Auston, A. M. Johnson, and W. M. Augustyniak, Appl. Phys. Lett. *38*, 47 (1981).

[5] A. P. DeFonzo, Appl. Phys. Lett. *39*, 480 (1981).

[6] W. Margulis and W. Sibbett, Appl. Phys. Lett. *42*, 975 (1983).

[7] J. B. Mullin, S. J. C. Irvine, and D. J. Ashen, J. Cryst. Growth *55*, 92, (1981).

[8] D. W. Kisker and R. D. Feldman, Proc. 2nd International Conference on II-VI Compounds, Aussois, France, 1985. To be published in the J. Cryst. Growth.
[9] A. M. Johnson and W. M. Simpson, J. Opt. Soc. Am. B2, 619 (1985).
[10] D. H. Auston, A. M. Johnson, P. R. Smith, and J. C. Bean, Appl. Phys. Lett. *37*, 371 (1980).
[11] D. H. Auston, IEEE J. Quantum Electron. *QE-19*, 639, (1983).
[12] D. P. Joshi and R. S. Srivastava, J. Appl. Phys. *56*, 2375 (1984).

High-Speed Internal Photoemission Detectors Enhanced by Grating Coupling to Surface Plasma Waves[1]

S.R.J. Brueck, V. Diadiuk, T. Jones[2], and W. Lenth[3]
Lincoln Laboratory, Massachusetts Institute of Technology,
Lexington, MA 02173, USA

The use of gratings to couple optical energy into surface plasma wave (SPW) modes that are confined to a metal-dielectric boundary has been extensively explored [1]. With optimal choices of the grating period and profile, nearly 100% coupling efficiency has been achieved in wavelength ranges where the metal is normally almost totally reflecting. We report the use of this coupling to enhance the quantum efficiency of internal-photoemission detectors in which hot carriers, generated in a metal film by optical absorption, are collected by photoemission over a Schottky barrier into a semiconductor. This class of detectors is of interest as a result of its uniformity, high-speed potential, long wavelength response with binary rather than quaternary III-V compounds, and compatibility with integrated circuit fabrication techniques [2-4]. For metal-silicide/Si systems, which have been the most thoroughly explored, techniques such as back-illumination and double-pass insulator-metal overlayer structures already provide quite high absorption (> 40%) at infrared wavelengths. Surface texturing has also been used to enhance the responsivity of metal-insulator-metal tunnel junction detectors and of metal-(a-Si)-metal detectors [5,6].

The response speed of this class of detectors is generally limited by the RC time-constant associated with the Schottky barrier depletion region and by the carrier transit time across this depletion region; relaxation processes within the metal and the semiconductor are in the subpicosecond range and do not limit detector response. High-saturation-velocity III-V semiconductors can be used to minimize transit time effects. Here we report measurements of the angular response functions, limiting response speeds and quantum efficiencies of Au-(p-InP) internal-photoemission detectors fabricated on substrates textured with shallow gratings to enhance the coupling into the SPW modes of the metal film.

Gratings with a period of 2 μm and a line/space ratio of ~ 1/3 were fabricated photolithographically on p-InP substrates. Electron-beam-evaporated Au films were deposited to form the Schottky barriers. Both film thickness and groove depth were varied to find the optimum coupling conditions. Detectors were defined by removing most of the Au film, leaving circular active areas with diameters ranging from 67 to 150 μm. Thicker, 25-μm diameter, Au contact dots were evaporated within the active areas. The ohmic back contact was thermally evaporated AuMg alloyed at 420 C for 10 s.

---

[1]This research was supported by the Department of the Air Force, in part with specific funding from the Air Force Office of Scientific Research.
[2]Present address: Southern University, Baton Rouge, Louisiana 70813.
[3]Present address: IBM Research Laboratory, San Jose, California 95193.

Figure 1 shows the detector response as a function of the angle of incidence for cw 1.15-μm irradiation of a detector with 50-nm deep grooves and a 30-nm thick Au film. This curve was taken for TM-polarized incident light with the orientation of the grating grooves perpendicular to the electric field vector. The peaks in the responsivity at incident angles of 26, 8 and 45 degrees correspond to the m = 1, 2 and 3 surface plasma wave coupling resonances, respectively. The order of the resonance m corresponds to the multiple of the fundamental grating wavevector involved in the scattering. At the resonant coupling angles, more of the incident energy is coupled into the surface plasma wave mode at the air-Au interface, is absorbed in the metal, and contributes to the detector response. The maximum responsivity, at the peak of the m = 1 resonance, is a factor of 30 greater than the responsivity for an untextured Au-InP detector fabricated simultaneously with the grating-enhanced detector. Details of this angular responsivity pattern and its variation with grating parameters are given elsewhere [7].

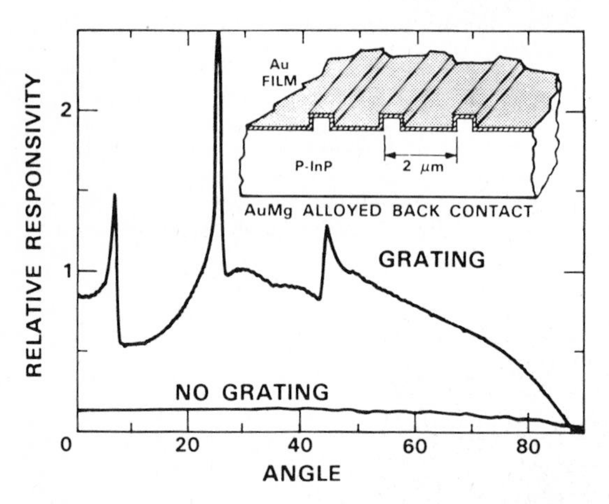

Fig. 1. Au-InP detector response vs incident angle for TM-polarized 1.15-μm radiation. The lines of the etched surface grating (see inset) were oriented at right angles to the plane of incidence. The grating parameters were: period 2 μm, groove depth 50 nm, Au film thickness 30 nm.

The speeds of these devices were limited by RC effects involving capacitance of the depletion region and the resistance of the back contact. This is illustrated in Fig. 2 which shows the RC time-constant (solid line) evaluated from the measured forward bias resistance (400 Ω) and the measured Schottky barrier capacitance vs reverse bias voltage for a detector with a 125-μm diameter active area and a hole concentration of 1×10^{16} cm^{-3}. The dashed curve is an extrapolation of the RC limit to higher bias voltages. Also shown are device response times measured using 150-ps duration mode-locked pulses at 1.06 μm. There is good agreement between these two measurements.

It is also interesting to compare the response time of the same detector for irradiation at 1.06 μm, where the detection mechanism is SPW-enhanced internal photoemission, and for above bandgap irradiation at 570 nm, where electron-hole pair generation within the depletion region dominates the responsivity. This comparison is shown in Fig. 3 for a 20-V reverse bias, where the response to 150-ps duration 1.06-μm pulses is shown on the left

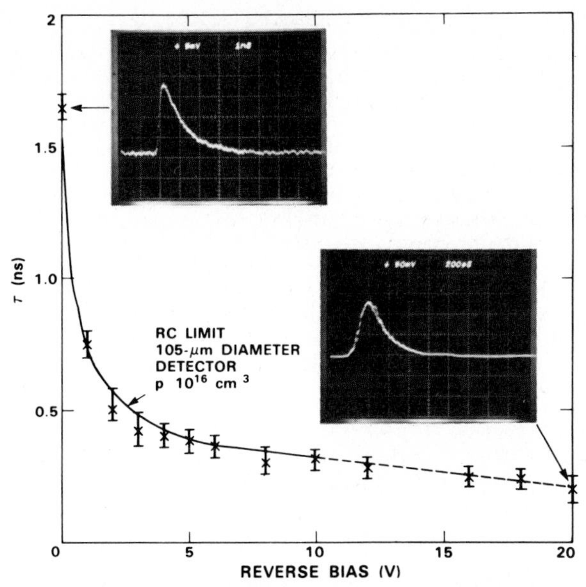

Fig. 2. Measured RC-time-constant limit (solid curve) and temporal response times (points) for a 125-μm diameter detector

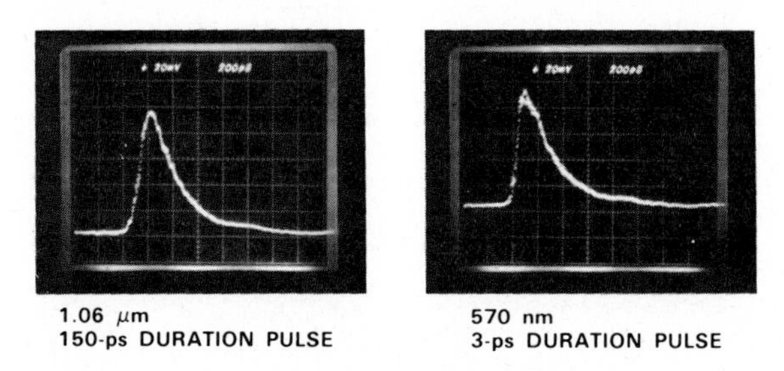

Fig. 3. Detector response at a 20 V reverse bias for below bandgap 1.06-μm radiation (left, 150-ps pulse duration) and for above bandgap radiation at 570 nm (right, 3-ps pulse duration)

side and the response to 3-ps 570-nm pulses is on the right. The risetimes of the signals vary as a result of the differing laser pulsewidths; the falltimes are identical and reflect the detector RC time-constant.

The quantum efficiency at the peak of the m = 1 resonance (cf. Fig. 1) for 1.06-μm irradiation was 0.4% at zero bias for heavily doped detectors ($p = 1 \times 10^{18}$ cm^{-3}). This increased to over 2% at reverse biases of 3 V probably as a result of both field-induced tunneling and electric-field depression of the Schottky barrier. Comparable zero-bias quantum efficiencies were observed for the lighter-doped faster detectors; however, as expected for these much lower field devices, these efficiencies were independent of reverse bias.

Improvements in both the detector response speed and the quantum efficiency will enhance the utility of these internal photoemission detectors.

The RC-limited decay times can be reduced by decreasing the capacitance using smaller areas and lower carrier concentrations (longer depletion widths). There is a lower limit to the detector area set by the SPW mean-free path. For smaller detectors, the enhancement due to the SPW coupling is reduced. These effects were observed using 67-μm-diameter detectors, which had decay times of less than 100 ps, but whose angular response patterns showed much reduced coupling and overall lower responsivities. Improvements in the detector time response can also be achieved by decreasing the back contact resistance; experiments using Zn skin diffusions to improve the back contact resistance are in progress.

The detector quantum efficiency should be improved by illuminating the detector through the substrate, and coupling to the SPW mode at the Au-InP boundary. Hot carriers would then not be required to diffuse across the metal film before collection. In addition, thicker Au films with better coupling to the SPW would be possible. For the present detectors, where the metal film thickness is limited to a hot-carrier diffusion length, optimal coupling is not achieved, and less than 10% of the incident radiation is coupled into the surface mode. Higher spatial frequency gratings (period ~ λ/n where n is the InP refractive index) are required to insure optimum coupling to this SPW.

References:

1. G. N. Zhizhin, M. A. Moskalova, E. V. Shomina and V. A. Yakovlev, in Surface Polaritons, V. M. Agranovich and D. L. Mills, Eds. (North-Holland, New York, 1982).
2. F. D. Shepherd in SPIE Proceedings 443: Infrared Detectors, W. L. Wolfe, Ed. (SPIE, 1984), p. 42.
3. W. F. Kosonocky and H. Elabd, ibid, p. 167.
4. R. C. McKee, IEEE Trans. Electron Devices, ED-31, 968 (1984).
5. A. M. Glass, P. F. Liao, D. H. Olson and L. M. Humphrey, Opt. Lett. 7, 575 (1982).
6. A. M. Glass, P. F. Liao, A. M. Johnson, L. M. Humphrey, D. H. Olson and M. B. Stern, Appl. Phys. Lett. 44, 77 (1984).
7. S. R. J. Brueck, V. Diadiuk, W. Lenth and T. Jones, Appl. Phys. Lett. (May 15, 1985, to be published).

Submicron-Gap Photoconductive Switching in Silicon

Ghavam G. Shahidi, Erich P. Ippen, and John Melngailis
Department of Electrical Engineering and Computer Science and Research Laboratory of Electronics, Massachusetts Institute of Technology, Cambridge, MA 02139, USA

In recent years there has been an increased interest in fast photoconductive switches: switches based on the use of a light pulse to modulate the resistivity of a semiconducting gap in a high-speed transmission line. To make such a switch fast, one has to restore the gap resistivity to its dark value as soon as possible after the light pulse is turned off. Thus, the object is to reduce the lifetime of carriers in the gap. In previous work with silicon, this has been achieved by increasing the carrier recombination rate through the use of radiation damaged or amorphous materials [1-2]. This introduction of recombination centers, however, also reduces carrier mobility and thus switch sensitivity. An alternative technique, the one used in this work, is to sweep the carriers out of the gap with an applied electric field.

For our experiments, we fabricated gaps as narrow as 0.2 μm on 1 μm thick silicon on sapphire with a sum electron and hole mobility of 470 cm^2/Vsec. By operating these switches in the sweep-out mode, we were able to increase the switch speed from a lifetime-limited response of 700 psec down to 27 psec.

Preliminary observations of the effects of sweep-out could be made directly with a sampling scope. Fig. 1a shows the observed, lifetime-limited, response (about 700 psec FWHM) of a wide gap switch. As the bias is increased or as a narrower gap is used, the response becomes faster. Fig. 1b shows that the response of a 0.2 μm gap can no longer be resolved with the sampling scope.

For better temporal resolution, we studied switch response using the cross correlation technique of Auston illustrated in Fig. 2. Switch (A)

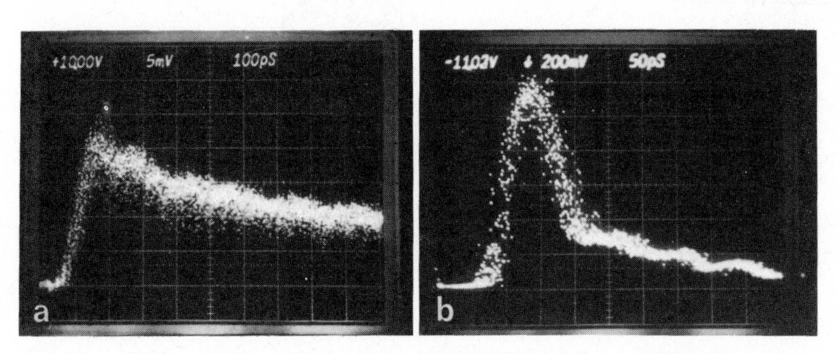

Fig. 1: (a) Lifetime-limited response of a wide gap detector.

(b) Scope limited of a 0.2 μm gap detector in sweep-out.

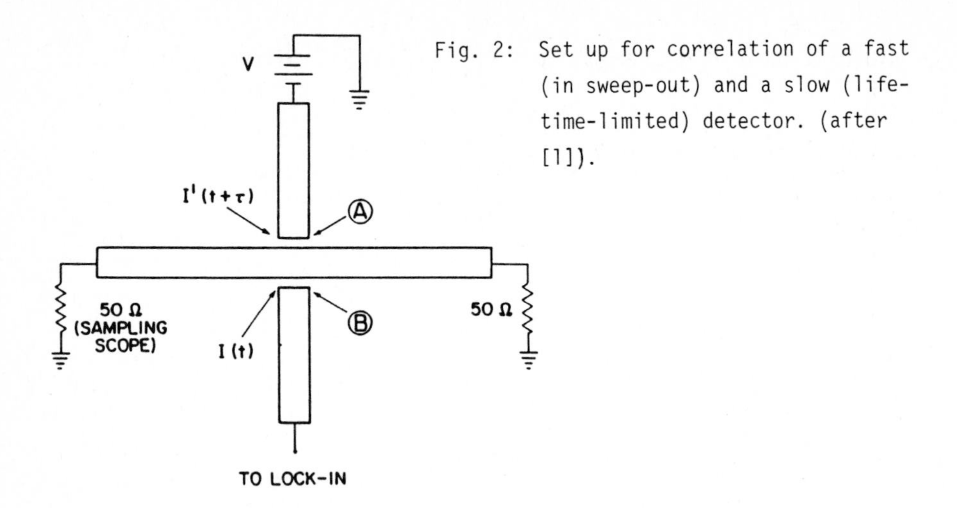

Fig. 2: Set up for correlation of a fast (in sweep-out) and a slow (lifetime-limited) detector. (after [1]).

is operating in the sweepout regime; its response is fast and variable with applied voltage. Switch (B) is operating in the lifetime-limited mode. Correlation of the responses of these two switches is obtained simply by varying the delay between their respective optical excitations. It will be asymmetric with respect to zero delay, with the decay time on each side proportional to one of the speeds.

Experimental results for two different gaps are shown in Fig. 3. Fig. 3a is a result of correlating two 14 μm gap detectors, when one of them is influenced by sweepout. As the bias is increased to 15 volts, recovery time decreases to 146 psec. Increasing the voltage further does not improve the speed, because carrier velocity is saturated at this point. Fig. 3b shows the result of correlating a 0.5 μm gap with a 5 μm gap, 1 mm away. The fastest response obtained in this case is 27 psec.

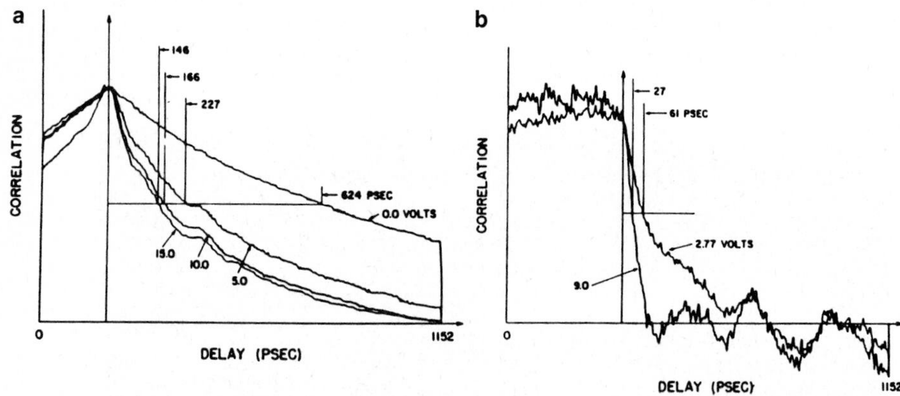

Fig. 3: (a) is result of correlation of two 14 μm gap detectors.

(b) is result of correlation of a 5 μm gap a 0.5 μm gap, 1 mm away.

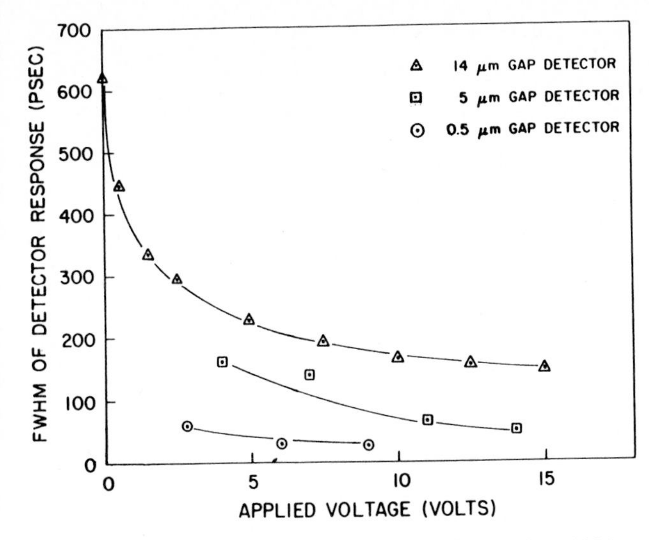

Fig. 4: Switch speeds vs. applied biases for different size detectors

Figure 4 summarizes our measurements for devices with gaps of 14, 5 and 0.5 μm at different applied biases. The maximum speeds of the 14 μm and 5 μm gaps scale as expected according to dimension. But the 0.5 μm gap does not become as fast (5 psec) as might be expected from the same scaling. The reasons for this are probably the generation of carriers far from the gap due to optical diffraction, the relatively large absorption depth (2.5 μm) of silicon, and the confinement of the applied electric field to the surface. Gap capacitance is still expected to be negligible. These problems can be ameliorated somewhat by using a direct-gap semi-conductor with a small absorption depth at the laser wavelength.

An especially interesting result of our work with sub-micron gap detectors is the demonstration of their excellent sensitivity. As gap size is reduced, we are able to focus less and less of the light into the gap. Nevertheless, the efficiency of voltage switching continues to increase. For a fixed optical excitation density, the gap "on" resistance decreases linearly with dimension. The sensitivity therefore increases linearly with decreasing gap dimension. The "off" resistance remains large enough ($\sim 10^5$ ohms) to keep the dark current negligible.

An example of this improved sensitivity is shown in Fig. 1b: With 11 applied volts and 310 pJ incident pulse energy focused to a diameter of 10 μm, we were able to switch more than 1 volt across a 0.2 μm gap detector.

In summary: we have fabricated and studied submicron gap detectors using silicon on sapphire. By making use of carrier sweep-out the speed of response was increased from 700 psec (lifetime limited) to 27 psec. At the same time an increase in switch sensitivity has been observed.

This work was supported in part by a grant from the Joint Services Electronics Program under DAAG 29-83-K-0003.

References

[1] P.R. Smith, D.H. Auston, A.M. Johnson, and W.M. Augustyniak, Appl. Phys. Lett. 38, 47 (1981).

[2] D.H. Auston, P. Lavallard, N. Sol and D. Kaplan, Appl. Phys. Lett. 38, 66 (1980).

Hertzian Dipole Measurements with InP Photoconductors

P.M. Downey and J.R. Karin
AT&T Bell Laboratories, 4D-411, Holmdel, NJ 07733, USA

In a recently described measurement scheme [1], the picosecond photoresponses of two radiation-damaged photoconductors were correlated by using these detectors as Hertzian dipoles, i.e., radiating and receiving dipole antennas. As in earlier microstrip circuit correlation experiments [2], one photoconductor is DC biased and acts as an electromagnetic pulse generator in response to excitation by a short optical pulse. The second photoconductor is biased by the electromagnetic transient from the first photoconductor, and functions as a sampling gate when excited by a second, delayed optical pulse. By monitoring the current induced in the circuit of the second photoconductor as a function of the delay between the optical pulses, the correlated photoresponse of the two circuits is measured. In Hertzian dipole experiments, a transient electromagnetic pulse from the pulse generator is radiated freely through a bulk dielectric to the second, receiving photoconductor. This results in a high pass coupling scheme, as opposed to the low pass transfer characteristics of microstrip interconnections.

In this paper we present Hertzian dipole measurements which differ from previously described dipole measurements [1] in that we used InP photoconductors [3] arranged in a face-to-face configuration (see inset, Fig. 1). Each photoconductor consisted of a gold electrode (40 μm wide) interrupted by a 10-μm gap, on helium-ion-bombarded Fe:InP. The two photoconductors were spaced 5-10 mm apart, and were excited by subpicosecond optical pulses from a CPM ring dye laser [4]. For some of the measurements, a circular aperture cut in a piece of aluminum foil was placed between the photoconductors. This aperture limits the field of view of the receiving photoconductor, and operates as a high-pass filter on the radiated field from the pulse generating photoconductor.

Figure 1 shows the results obtained from correlating the photoresponses of two InP photoconductors with a 1-mm diameter aperture between them. The correlation trace can be characterized by a 1/e risetime of 0.5 ps and a FWHM of 1.2 ps. The risetime can be attributed primarily to the decay time of the photoexcited carriers in the receiving photoconductor, and corresponds to a bandwidth of 300 GHz. As we will show, the short duration of the correlated response is a result of the differentiating transmission characteristics of the circular aperture.

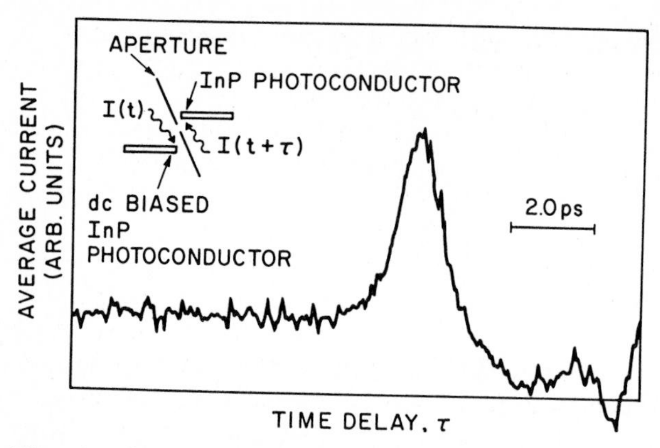

Fig. 1: Cross-correlated photoresponse of two helium ion bombarded InP photoconductors with a 1 mm (diam.) aperture between them. The sample configuration is illustrated in the inset.

We present in Fig. 2 the correlated photoresponse of the same two samples, collected both with a 2-mm diameter aperture and with no aperture at all. The high-frequency oscillation ($\nu \approx 0.15$THz) which dominates the data collected with the aperture is due to a Fabry-Perot-like resonance in the photoconductor substrates. These oscillations are also evident in the data collected without the aperture; however, the peaks in the oscillations occur at longer time delays, indicating that the peaks arrive at the receiver later when the aperture is not present than when it is present. This phenomenon can also be explained by the differentiating transmission function of the circular aperture.

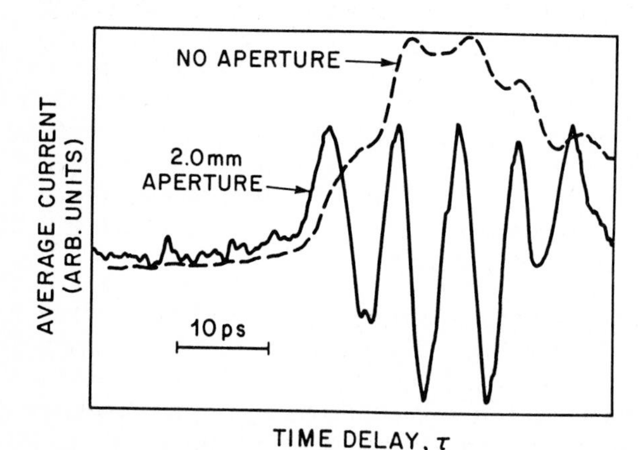

Fig. 2: Cross-correlated photoresponse of two helium ion bombarded InP photoconductors, both with and without a 2-mm (diam.) aperture between them. (Vertical scale is 8x larger for data taken without the aperture.)

The spectral transmission function of a circular aperture can be understood by considering the vibration spiral for Fresnel half-period zones of a circular opening [5]. A circular aperture of radius s can be characterized by a wavelength $\lambda_0 = s^2 (a+b)/ab$, for which the aperture transmits only the first half-period zone of radiation from a source a distance a in front of the aperture to a point a distance b behind the aperture. For frequencies less than c/λ_0 (where $c =$ speed of light), only a fraction of the first half-period zone will pass through the aperture. For these lower frequencies the transmitted electric field can be described by a phasor of amplitude $2\sin(2\pi\nu/\nu_0)$ and angle $2\pi\nu/\nu_0$, where $\nu_0 = 4c/\lambda_0$. Taking into account the one-quarter period phase advance for Huygens' secondary wavelets, we arrive at the following expression for the transmission function of a circular aperture at frequencies less than ν_0 :

$$T(\nu) = 2\sin(2\pi\nu/\nu_0)\exp i[(\pi/2)-(2\pi\nu/\nu_0)] \qquad (1)$$

At higher frequencies this approximation breaks down, because the vibration curve is a spiral rather than a circle.

Filtering the radiated electric waveform with this transmission function is equivalent to convolving the radiated electric waveform in the time domain with $\delta(t)-\delta(t-2/\nu_0)$. In other words, if $f(t)$ is the cross-correlated photoresponse of the two photoconductors without an aperture between them, $f(t)-f(t-2/\nu_0)$ is what we'd expect to measure when the aperture is in place. We have carried out this operation on the data collected without an aperture (Fig. 2) by shifting the data 1.3 ps to the right and subtracting it from the unshifted data. Although the result of this operation (Fig. 3) does not exactly reproduce the data collected with the aperture, the positions of the relative maxima and minima of Fig. 3 all coincide in relative time-delay with the same features in Fig. 2.

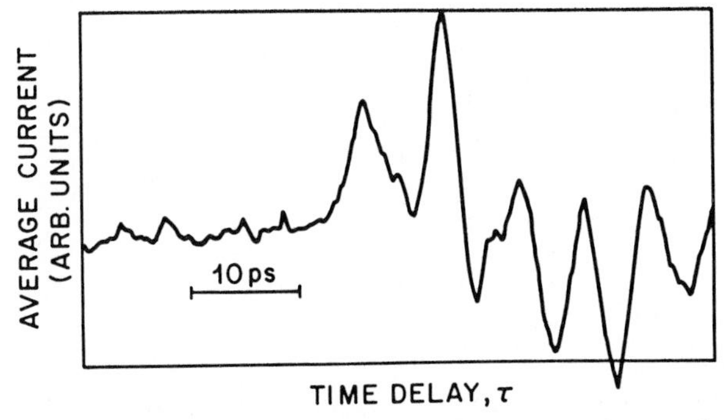

Fig. 3: The data from Fig. 2 (collected without the aperture) after shifting it to the right by 1.3 ps, then subtracting it from itself.

The positions of the maxima and minima are quite sensitive to the value used for the time-shift. Since this value corresponds to $2/\nu_0$, by choosing the time-shift so that the maxima and minima coincide in relative time delay, we have measured ν_0. Furthermore, the measured 1.3 ps is in agreement with the other physical parameters of our experiment; when the data in Fig. 2 was collected, each of the photoconductors was situated approx. 3mm from the aperture, so $\lambda_0 = 670\mu m$, and $2/\nu_0 = 1.1$ ps.

In summary, we have presented Hertzian dipole measurements carried out in a face-to-face configuration. The bandwidth of this measurement is in excess of 300 GHz, and limited by the relaxation time of photoexcited carriers in the photoconductors. We have demonstrated the utility of this measurement scheme by studying the spectral transmission characteristics of a circular aperture.

References

1. D. H. Auston, K. P. Cheung and P. R. Smith: Appl. Phys. Lett. 45, 284 (1984).
2. P. R. Smith, D. H. Auston, A. M. Johnson and W. M. Augustyniak: Appl. Phys. Lett. 38, 47 (1981).
3. P. M. Downey and B. Schwartz: Appl. Phys. Lett. 44, 207 (1984).
4. R. L. Fork, B. I. Greene and C. V. Shank: Appl. Phys. Lett. 37, 371 (1980).
5. F. Jenkins and H. White: Fundamentals of Optics, McGraw-Hill, New York 1976) p 386.

Pulse Waveform Standards for Electro-Optics

Robert A. Lawton
Electromagnetic Fields Division, National Bureau of Standards, 325 Broadway, Boulder, CO 80303, USA

The maturing of the semiconductor industry has resulted in the need for faster methods for the measurement of pulse waveforms. The measurement requirements are fast outstripping the capability of integrated circuit testors. The fastest commercial sampling oscilloscope today has a step response transition duration (risetime) of 20 picoseconds. For the measurement of single transients, available time resolution is somewhat less. This time resolution is not adequate to support the measurement of the faster semiconductor devices being developed,such as gallium arsenide logic gates whose switching times have been estimated to be 12 picoseconds [1]. Measurement of such transitions requires a significant advance in measurement technology. A prime candidate for making the required quantum leap in time measurement resolution is the electro-optic sampler [2]. This sampler has a demonstrated resolution of 0.4 picoseconds and a theoretical resolution limit much shorter than that.

Much work remains to be done, however, to make this an accurate means of measuring pulse parameters. An important step in that direction would be the development of a reference waveform standard,by means of which the pulse waveform measurement accuracy of the Automatic Pulse Measurement System [3] (APMS) at the National Bureau of Standards (NBS) can be applied to the electro-optic sampler. Pulse measurement capability on the APMS has been refined over a number of years to include time base and voltage (vertical) scale calibrations, and deconvolution by stable techniques to remove the distortion caused by both the measurement system impulse response [4] and time jitter [5]. An example of the recent application of these techniques to the measurement of a tunnel diode steplike waveform gave the following results. The observed step had a transition duration of 30 picoseconds. The same waveform after removal of the system distortion had a transition duration of 14 picoseconds, and the fine structure which had been deemphasized by the system distortion was restored.

The most promising way to transfer the accurate measurement capability of the APMS to the electro-optic sampler is by the use of a pulse transfer standard. That is a pulse generator that can be measured on the APMS to determine the shape of the pulse generator output waveform. The same pulse generator can then be measured on the electro-optic sampler, and the pulse generator output waveform can then be deconvolved from the electro-optic waveform measurement to determine the distortion of the electro-optic sampler.

Three commercial pulse generators have been investigated as possible transfer standards: a tunnel diode; a complete pulse generator with variable pulse duration, repetition rate etc.; and a comb generator. The tunnel diode pulse generator was not usable due to its low repetition rate and the fact that a cavity dumper was not readily available to drive the electro-optic sampler.

Measurements were made on both the APMS and the electro-optic sampler of the complete pulse generator since it could be operated at a fast enough repetition rate to be triggered by a signal derived from the laser. However, the transition duration (900 picoseconds as measured on the APMS) was too long to be measured by one scan on the electro-optic sampler. As a result, several scans were made for various values of fixed electronic delay and the resulting photographs of the output were combined in a composite photograph,from which the transition duration was estimated to be 750 picoseconds.

The comb generator turned out to be the most promising,in that good signal to noise ratio was obtained and the most significant parts of the waveform were contained in one scan of the sampler. Measurements made with the electro-optic sampler yielded a pulse duration of 290 picoseconds. Measurements have not yet been made with the APMS but measurements made with similar electrical sampling instruments yielded a pulse duration of 200 picoseconds. Much work remains to be done in examining the output stability and sensitivity to variations in power and frequency of the input sinewave. Nevertheless, this generator does have the potential of being a very compact and convenient transfer standard for transferring the accuracy of the APMS to the electro-optic sampler.

The author would like to acknowledge the help of Dr. Gerard Mourou and Kevin Meyer of the Laser Energetics Laboratory in Rochester, N.Y. in making measurements of the pulse generator waveforms with their electro-optic sampler,and William L. Gans of the National Bureau of Standards in Boulder, Colo. for providing the data that was used in the deconvolution example.

1. G. Nuzillat, "GaAs IC Technology for Ultra-High Speed Digital systems", URSI XXIst General Assembly digest and lecture, Florence, Italy, Aug. 28 - Sept. 5, 1984, p. 496.

2. J.A. Valdmanis, G. Mourou, and C.W. Gabel, "Subpicosecond Electrical Sampling", Proceedings of the SPIE, Vol. 439, Aug. 24-26, 1983, p. 142.

3. N.S. Nahman, J.R. Andrews, W.L. Gans, M.E. Guillaume, R.A. Lawton, A.R. Ondrejka and M. Young, "Application of Time Domain Methods to Microwave Measurements", IEE Proc., Vol. 127, Pt. H, No. 2, April, 1980, p.

4. R.A. Lawton, S.M. Riad, J.R. Andrews, "Pulse and Time Domain Measurements", IEEE Proc., Nov. 1985 (to be published).

5. W.L. Gans, "The Measurement and Deconvolution of Time Jitter in Equivalent-Time Waveform Samplers", IEEE Trans. Instr. & Meas., Vol. IM-32, March 1983, p. 126.

High-Speed Optoelectronic Track-and-Hold Circuits in Hybrid Signal Processors for Wideband Radar[1]

I. Yao, V. Diadiuk, E.M. Hauser, and C.A. Bouman
Lincoln Laboratory, Massachusetts Institute of Technology, P.O. Box 73, Lexington, MA 02173, USA

A high-speed track-and-hold circuit based on picosecond InP optoelectronic switches has been developed with a 260-MHz analog bandwidth, 100-MHz sampling rate, and better than 0.2% accuracy. This circuit, a surface-acoustic-wave convolver, and a charge-coupled-device matrix-matrix-product chip are being incorporated in a hybrid radar signal processor for processing radar signals with bandwidths up to 200 MHz.

1. Introduction

The signal processing task for radar systems with large instantaneous bandwidth and wide range coverage stresses the throughput rate of conventional digital processors. Instead, the approach of using combinations of analog signal processing techniques for processing wideband pulse-Doppler radar signals offers the potential of a compact, high-throughput processor. A hybrid analog signal processor for processing radar signals with bandwidth up to 200 MHz is being developed [1]. One of the crucial components required is the high-speed track-and-hold (T/H) circuit. This circuit is needed to capture and freeze the high-speed signals in the processor, such that they can be further processed.

InP optoelectronic switches have shown switching times ranging from nanoseconds to tens of picoseconds and on-off ratios exceeding six orders of magnitude [2] and therefore are well matched to the high-speed T/H application in the hybrid radar signal processor. The advantages of using the optoelectronic switch over the conventional diode bridge are its inherent optical isolation and its zero dc offset. This paper describes the high-speed T/H circuit [3] based on picosecond optoelectronic switches specifically developed for the hybrid radar signal processor cited earlier. In the subsequent sections, the processor architecture will first be introduced. Details of the optoelectronic T/H circuit and results of a demonstration processor will then be presented.

2. Processor Architecture

Three types of analog signal processing devices are incorporated in the proposed hybrid radar signal processor. They are: (1) a surface-acoustic-wave (SAW) convolver [4] to perform programmable pulse compression for radar signals with 200-MHz instantaneous bandwidth in order to provide target range information with 0.75-m resolution, (2) optoelectronic T/H circuits [3] to perform the range gating function and the buffering of sampled data into the Doppler processor, and (3) charge-coupled-device

[1]This work was supported by the Department of the Army and the Defense Advanced Research Projects Agency.

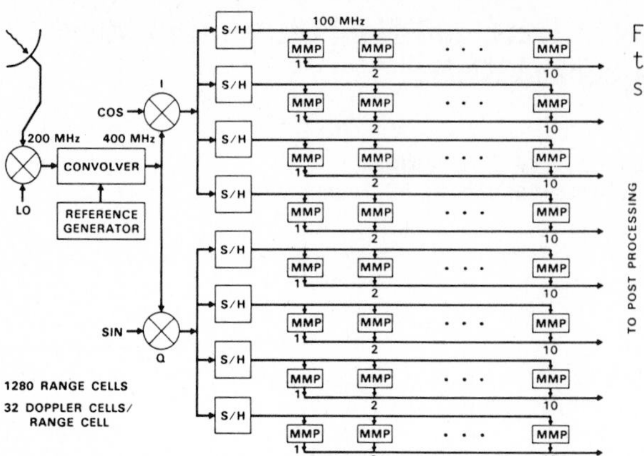

Fig. 1 Analog portion of the hybrid signal processor architecture

(CCD) matrix-matrix-product (MMP) chips [5] to perform Doppler Fourier analysis in order to provide target velocity information.

Figure 1 shows the block diagram of the analog portion of the proposed signal processor. It consists of one SAW convolver and its reference waveform generator, eight optoelectronic T/H circuits, and eighty CCD-MMP chips. The radar return signal with 200-MHz bandwidth is first pulse-compressed in the SAW convolver. Because of the frequency doubling effect in the convolver, the bandwidth of the compressed pulse is 400 MHz. After video demodulation, this signal is split into in-phase (I) and quadrature (Q) channels. There are 4 optoelectronic T/H circuits operating at a 100-MHz sampling rate in each channel to provide a 800-MHz effective sampling rate. Since the CCD-MMP chips are to be clocked at 10 MHz, the output of each T/H feeds 10 CCD-MMP chips. However, because the input stage of each CCD-MMP chip is capable of capturing data samples in 10 ns, no additional T/H circuits are necessary. The CCD-MMP chips perform Doppler processing on each range bin based on the discrete Fourier transform (DFT) algorithm to derive the target velocity information. Finally, the Doppler processor output feeds the digital post processor. This architecture would provide 1280 range bins by 32 Doppler bins in real time and would therefore be useful for search, track, and imaging applications.

3. Optoelectronic Track-and-Hold Circuits

In the hybrid radar signal processor, optoelectronic T/H circuits will buffer the analog signal path from a SAW convolver-based pulse compressor to a Doppler processor using CCD-MMP chips. The performance requirements of the T/H in this system call for a 200-MHz analog bandwidth, 10-ns sampling period, and better than 1% accuracy.

This T/H circuit is created by connecting a hold capacitor and an output buffer amplifier to the optoelectronic switch as shown in Fig. 2. The GaAs diode laser shown in the figure is used to generate optical command pulses for controlling the switch. The JFET source follower shown is the output buffer amplifier and its input gate capacitance is used as the hold capacitor. With a switch on-resistance of 100 Ω and a hold capacitor of 2.5 pF, this circuit is designed for sampling analog signals with bandwidths exceeding 200 MHz at a sampling rate as high as 100 MHz.

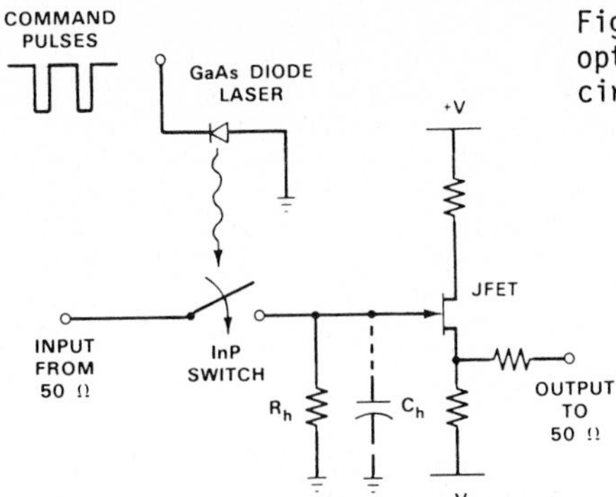

Fig. 2 Circuit diagram of the optoelectronic track-and-hold circuit

The most important parameters of a T/H circuit are speed and accuracy. The speed of the circuit specifies the analog bandwidth, and is determined by the turn-off time of the switch and the charging time constant of the capacitor-switch combination during the track mode. The accuracy of the circuit is affected by the residual charge left over from the prior sampled value and by the charge leakage to or from the current sampled value through undesirable parasitics.

The InP optoelectronic switch used was fabricated on a semi-insulating InP substrate, formed by introducing Fe doping during crystal growth. It demonstrated a transient-response time constant of about 300 ps. Although the measured dc dark resistance of the switch is on the order of 100 MΩ, experimental studies consistently show significant input-to-output leakage through the switch for times as long as hundreds of nanoseconds after an optical sampling pulse. The magnitude of this leakage is dependent on input-to-output biasing, switch geometry, and switch contacts. This effect is thought to be caused by deep trapping levels other than the desired recombination centers and/or nonohmic contacts.

A SPICE simulation of an equivalent circuit which includes parasitics predicts, and experimental results confirm, that this optoelectronic T/H circuit has an analog bandwidth of 260 MHz with 4% input-to-output leakage. Since the circuit met the 200 MHz bandwidth requirement in a radar signal-processor application but fell short of the goal of 1% accuracy, a circuit configuration using two identical T/Hs cascaded in series can be employed. The SPICE-simulated and experimental results both indicate significant improvement of input-to-output leakage to less than 0.2% while the bandwidth is maintained. Figure 3 shows the actual result

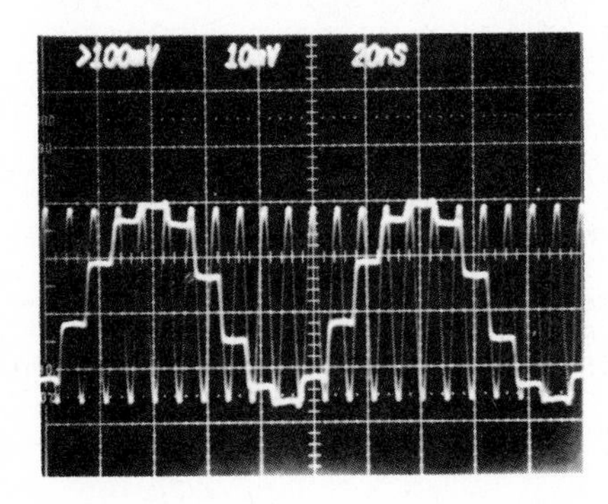

Fig. 3 Experimental result of track-and-hold circuit for 110-MHz sine-wave input being sampled at a 100-MHz rate

of the cascaded version of the T/H circuit for a 110-MHz sine-wave input being sampled at a 100-MHz rate. The staircase-like output waveform has a beat frequency of 10 MHz and it clearly shows the circuit tracking and holding the input waveform with good fidelity. The optoelectronic T/Hs incorporated in the radar signal processor require simultaneous sampling of both I and Q signal channels. A dual-channel optoelectronic T/H circuit has been implemented on a miniature optical bench using conventional optics to collect the light output from both facets of a single laser, thereby minimizing the relative timing jitter between the two channels. Future extension to the optical fiber coupling of the control signal path would further reduce the circuit size.

4. Demonstration Processor

Under development is a demonstration system using a thinned version of the processor as shown in Fig. 1 for proving the validity of the concept. In this thinned processor, one SAW convolver, two optoelectronic T/H circuits, and two CCD-MMP chips will be incorporated. All analog elements exist, and currently this demonstration system has been integrated up through the optoelectronic T/H circuits. Figure 4 shows a range-Doppler plot of two targets, one stationary and one moving, produced by the demonstration system in which the Doppler processing was simulated in a host computer.

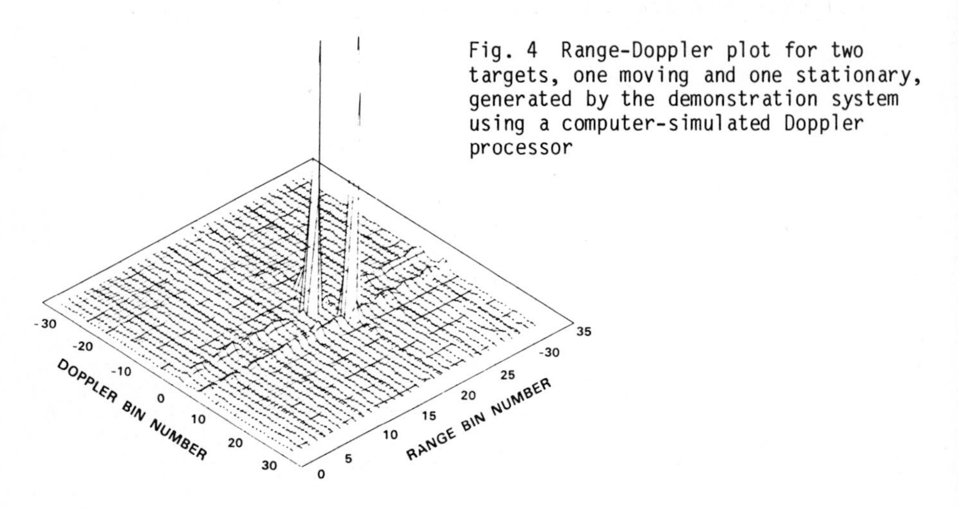

Fig. 4 Range-Doppler plot for two targets, one moving and one stationary, generated by the demonstration system using a computer-simulated Doppler processor

5. Conclusions

The properties of fast switching time, high on/off conductance ratio, and optically isolated command pulse make optoelectronic switches well-matched to high-speed T/H applications. In a cascaded configuration, an optoelectronic T/H circuit based on picosecond InP optoelectronic switches has demonstrated an analog bandwidth of 260 MHz and an accuracy better than 0.2%. Although it is designed specifically for the hybrid radar signal processor, the applicability of this circuit is general. If reasonable sensitivity is maintained and the on/off ratio is improved for switches demonstrating transient times less than 100 ps, T/H circuits with analog bandwidths exceeding 1 GHz can be achieved.

References

1. I. Yao, E. M. Hauser, C. A. Bouman, G. T. Flynn, and J. H. Cafarella, "Wideband Radar Signal Processor Based On SAW Convolvers," 1984 IEEE Ultrasonics Symposium, Dallas, Texas, 1984. Proceedings to be published.

2. A. G. Foyt and F. J. Leonberger, "InP Optoelectronic Switches," in Picosecond Optoelectronic Devices, edited by C. H. Lee. New York: Academic Press, to be published, 1984.

3. C. H. Cox, III, V. Diadiuk, I. Yao, F. J. Leonberger, and R. C. Williamson, "InP Optoelectronic Switches and Their High-Speed Signal Processing Applications," Proc. Soc. Photo-Opt. Inst. Eng., vol. 439, pp. 164-168, 1983.

4. I. Yao and S. A. Reible, "Wide Bandwidth Acoustoelectric Convolvers," 1979 Ultrasonics Symposium Proceedings. New York: IEEE, 1979, pp. 701-705.

5. A. M. Chiang, R. W. Mountain, D. J. Silversmith, and B. J. Felton, "CCD Matrix-Matrix-Product Parallel Processor," Tech. Dig. 84 Int. Solid-State Circ. Conf. New York: IEEE, 1984, pp. 110-111.

Optoelectronic Modulation of Millimeter Waves in a Silicon-on-Sapphire Waveguide

Chi H. Lee, Aileen M. Yurek+, M.G. Li, E.A. Chauchard, and R.P. Fischer
Electrical Engineering Department, University of Maryland,
College Park, MD 20742, USA

1. Introduction

Recently a great deal of attention has been given to a general class of switching and gating devices based on the photoconductivity effect using picosecond optical pulses. The photoconductivity effect is a long wavelength limit of a more general phenomenon, in which the complex dielectric constant of a semiconductor is modified by the introduction of an optically induced electron-hole plasma. A millimeter-wave propagating in a semiconductor waveguide experiences both attenuation and phase-shift when an electron-hole plasma is optically induced in the waveguide.

In this paper we report on the performance of a dielectric waveguide made of silicon on sapphire (SOS). The high speed modulation of the millimeter-wave in the SOS is obtained using the pulse train from a modelocked YAG laser. A short (< 2 ns) individual millimeter-wave pulse produced by the SOS is used to probe the behaviour (density and transport properties) of optically induced carriers in bulk Silicon.

2. Modulation of Millimeter-waves in a SOS Waveguide

Experimental technique

The optically controlled millimeter-wave device consists of a rectangular semiconductor waveguide with tapered ends to allow efficient transition of

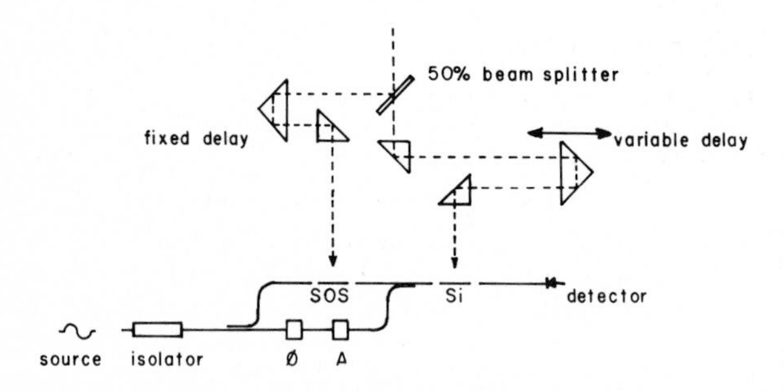

Fig. 1. Experimental Set-up. source: 94 GHz, ϕ: phase shifter, A: attenuator

+ Present Address: Naval Research Lab, Washington, DC

millimeter-waves to and from a conventional metal waveguide [1]. The experimental set-up shown in fig. 1 has been described in detail elsewhere [2]-[3]. The millimeter-wave bridge is initially (without laser illumination) balanced by adjusting the variable attenuator and the mechanical phase shifter so that there is no signal at the output. When an optical pulse illuminates the SOS (no light on the silicon waveguide), an electron-hole plasma is generated causing phase-shift and attenuation of the millimeter-wave signal. The bridge becomes unbalanced and a signal appears at the output. This signal depends strongly upon the laser intensity, which determines how much plasma is generated, and upon the laser wavelength, which determines the initial plasma depth. Different waveguides will also produce profoundly different millimeter-wave signals.

In the case of the SOS, it is expected that the plasma will be confined in a thin layer of epitaxially grown silicon. Thus, a high density but thin plasma layer may be maintained resulting in a small dynamic insertion loss. To obtain the modulation of the CW millimeter-wave signal, a frequency doubled pulse train of the mode locked YAG laser is used. Frequency doubling is required since the light at a wavelength of 532 nm (unlike that at 1.06 μm) is very efficiently absorbed in the thin layer of silicon [4]. Each pulse of the pulse train has a duration of 30 to 40 ps.

Results and Discussion

It is clear from fig. 2 (obtained without light on the bulk silicon waveguide) that the millimeter-wave output is mimicking the incident optical pulse train, resulting in a modulation bandwidth of about 1 GHz at a repetition rate of 200 MHz. The bandwidth of the scope does not allow an accurate measurement of the millimeter-wave pulse shape, but we can observe that the pulse is asymmetric with a faster rise time than fall time. A rise time equal to that of the laser pulse ($\sim$ 20 ps) is expected. The rapid decay (< 2 ns) of the millimeter-wave is attributed to the short lifetime of the free carriers in the silicon epitaxial layer. Strong surface recombination at the silicon-air and silicon-sapphire interfaces may also contribute to the rapid decay.

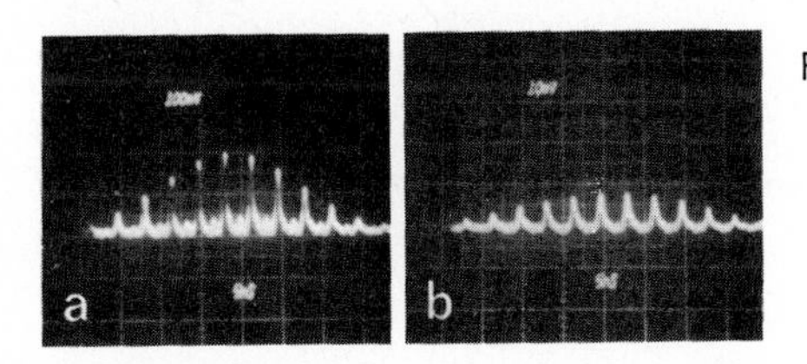

Fig. 2. Millimeter-wave pulse train (b), Incident optical pulse train from a frequency doubled mode-locked YAG laser (a). Oscilloscope sweep speed: 5 ns per division. The traces are bandwidth-limited by the oscilloscope.

With a single picosecond light pulse, we can generate a single ultrashort millimeter-wave pulse, which is used to probe the behaviour of carriers in bulk silicon.

3. Application: probing the carriers' behaviour in bulk silicon

Experimental technique

In bulk silicon, the electron-hole plasma can be generated using either infrared light at 1.06 μm or frequency doubled light. [2] In the first case the light penetrates deeply (α^{-1} $\sim$ 1 mm) generating a volume plasma in the bulk. It is thus possible to obtain an instantaneous (i.e. as fast as the rise time of the laser pulse) switch-off of the millimeter-wave. It persists after illumination until the excess carriers recombine (10-20 μs). Since the switch-off is instantaneous, we can determine which position of the prism along the delay line corresponds to the meeting at the silicon waveguide of the millimeter-wave pulse from the SOS and the light pulse. If the light pulse arrives first, the millimeter-wave pulse will not propagate through. If it arrives after, the millimeter-wave pulse will propagate through. The time during which there is only partial cut-off gives a measurement of the duration and shape of the millimeter-wave pulse. When the silicon is illuminated by 532 nm wavelength light, the absorption depth is only 1 μm [4] resulting in a thin initial layer of plasma. As shown by previous theoretical studies (fig. 3) [2], the conditions required to observe strong attenuation of the millimeter-wave would only be fulfilled after diffusion of the plasma into the bulk. We thus expect to observe some attenuation only if the silicon waveguide is illuminated by the laser pulse prior to the arrival of the millimeter-wave pulse. The model used is based on a "rectangular" distribution of plasma, and therefore does not take into account the instantaneous attenuation due to the exponential tail of the initial distribution of plasma.

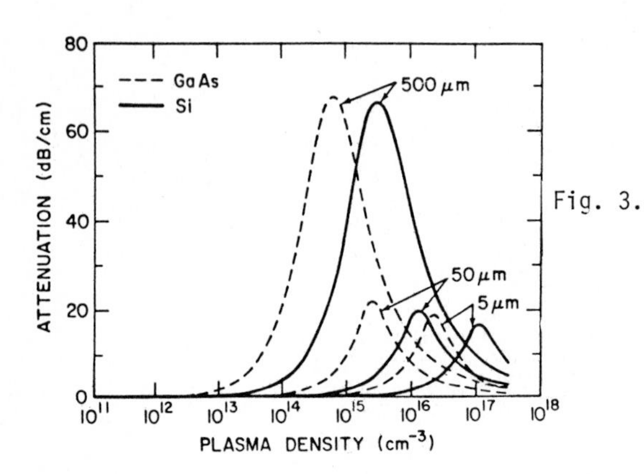

Fig. 3. Comparison of the attenuation of an E^y_{11} mode propagating at 94 GHz in a 2.4 x 1.0 mm^2 waveguide for Si and GaAs plotted with respect to plasma density. The three sets of curves correspond to plasma depths of 500 μm, 50 μm and 5 μm respectively.

The experiment consists of varying the time delay between the millimeter-wave pulse and the light pulse incident on the silicon waveguide using a sliding delay. For each delay we measure the portion of the millimeter-wave pulse that propagates through the silicon waveguide.

Results and Discussion

The curve showing attenuation versus delay (fig. 4) reveals that we achieve an attenuation of more than 4.5 dB with 1 μJ per pulse of 1.06 μm wavelength light. Assuming that the 20 ps laser pulse rise time is negligible compared to the millimeter-wave pulse duration, the curve can be deconvoluted by simply plotting its derivative versus time-delay. The millimeter-wave pulse so obtained has a duration of 1.5 ns. Its flat-top shape indicates a saturation of the phase-shift produced by the SOS as the number of carriers created increases. This can also be seen on fig. 2 : the height of the millimeter-wave pulses is not proportional to that of the laser pulses. This effect of saturation has not been explained previously.

The preliminary measurements obtained with 532 nm wavelength light show that we can also achieve total or partial attenuation of the millimeter-

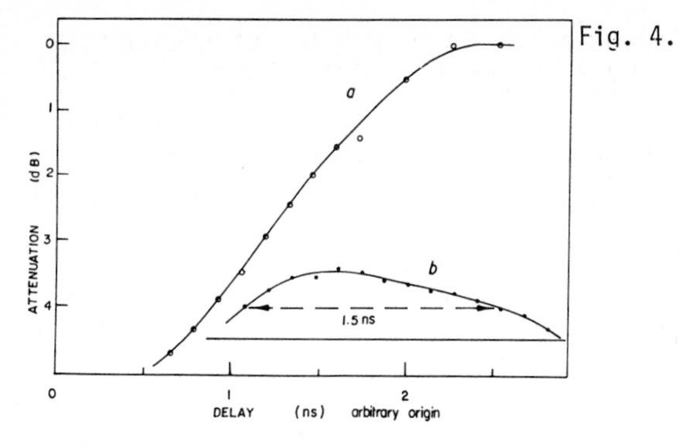

Fig. 4. a:portion of the millimeter-wave pulse propagating through the silicon waveguide versus time delay. Average infrared energy: 10^{-7}J. b: derivative of curve (a) giving the millimeter-wave pulse shape.

wave pulse depending on the intensity of light. We observe a small delay (< 300 ps) with respect to the infrared curve,indicating a qualitative agreement with the theoretical model. According to the curves shown on fig. 3, this small delay implies very fast diffusion speeds. The interpretation of the results is complicated by the existence of some instantaneous attenuation from the tail of the initial exponential distribution of plasma.

4. Conclusion

We have demonstrated the picosecond speed capability at the millimeter-wave spectral region using a dielectric waveguide made of silicon on sapphire. We have also analyzed, with a time resolution of 50 ps, the attenuation of the ultra-short millimeter-wave pulses in a bulk silicon where an electron-hole plasma has been optically induced.

5. Acknowledgement

This work was supported in part by the U. S. Army Research Office and the National Science Foundation.

References

[1] Chi H. Lee, Aileen M. Yurek, M. G. Li and C. D. Striffler: 9th International Conference on Infrared and Millimeter-Waves, Osaka, Japan, Oct. 22-26, 1984, pp. 495-498, Conference Digest.
[2] Aileen M. Yurek (Vaucher): Ph.D thesis, University of Maryland 1984.
[3] Chi H. Lee: Chapter 5 in Picosecond Optoelectronic Devices, edited by Chi H. Lee, Academic Press, 1984.
[4] Loferski Proc. IEEE 63, 669, 1963.

Direct DC to RF Conversion by Impulse Excitation of a Resonant Cavity

Ming G. Li, Chi H. Lee, A. Caroglanian, and E.A. Greene+
Department of Electrical Engineering, University of Maryland,
College Park, MD 20742, USA
C.Y. She* and P. Polak-Dingels
Laboratory for Physical Sciences, 4928 College Avenue,
College Park, MD 20740, USA
A. Rosen
RCA Laboratories, Princeton, NJ 08549, USA

Conversion of energy from DC to RF has been demonstrated with an impulse-excited coaxial resonant cavity using picosecond optoelectronic switching techniques. A single pulse is capable of generating more than a hundred RF cycles with an energy conversion efficiency greater than 50%. Frequencies up to 3 GHz have been generated, and CW oscillations at 1.6 GHz have been obtained.

1. Introduction

Direct DC to RF conversion with an optoelectronically switchable frozen-wave generator has recently been demonstrated using picosecond optical pulses [1,2]. A sequential waveform consisting of two and one half cycles has been obtained, with a voltage conversion efficiency better than 90% [2]. There are, however, some drawbacks associated with the frozen-wave device. One is that a large number of switches must be activated simultaneously, and that each switch must operate at close to 100% efficiency in order to form a good pulse sequence. These requirements are rather difficult to meet, because of the inevitable differences in the fabrication of each switch. Also, the simultaneous closure of several switches demands a large amount of laser energy.

2. Theory and Experiment

In this paper, we report on our recent study of DC to RF conversion using a single optoelectronic switch and a coaxial resonant cavity (Fig. 1). An optoelectronic switch generates an electrical impulse which contains a large number of frequency components. The coaxial cavity, which is closed at one end and open at the other, acts as a filter and will only resonate at the frequency determined by the inner conductor length (Fig. 2). The configuration of the cavity requires that the voltage be zero at the closed end, and the current in the inner conductor be zero at the open end. Because the length of the cavity is a quarter wave-length, the current is a maximum at the closed end and the voltage is largest at the open end. The electrical impulse is coupled into the closed end of the cavity via a loop antenna. The resulting magnetic field induces currents in the center conductor. A voltage probe, located in the region of maximum electric field, is used to couple the RF signal out of the cavity. The damping rate of the sine wave depends on the coupling coefficient and the loaded quality factor (Q) of the cavity.

+Mailing address: Department of Physics, University of Maryland, College Park, MD 20742.
*On leave from Colorado State University, Fort Collins, Colorado 80523.

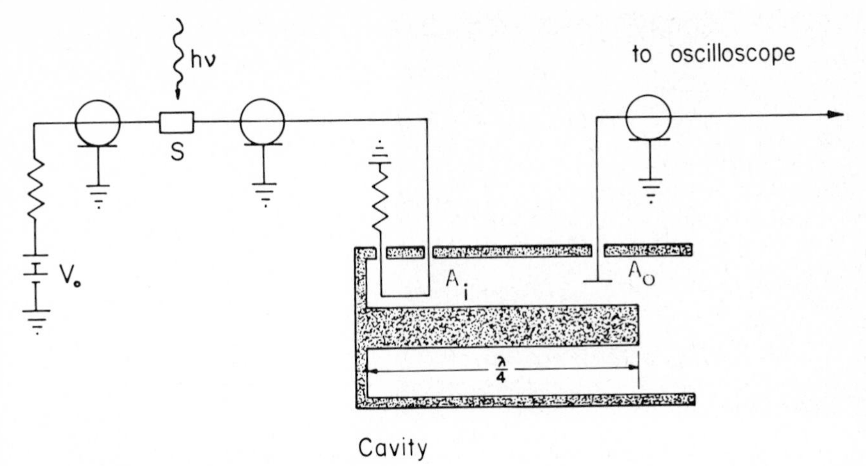

Figure 1: Experimental arrangement. V_o is the switch bias voltage, S is the semiconductor switch, A_i and A_o are antennae and the cavity is a coaxial structure with resonant wavelength λ.

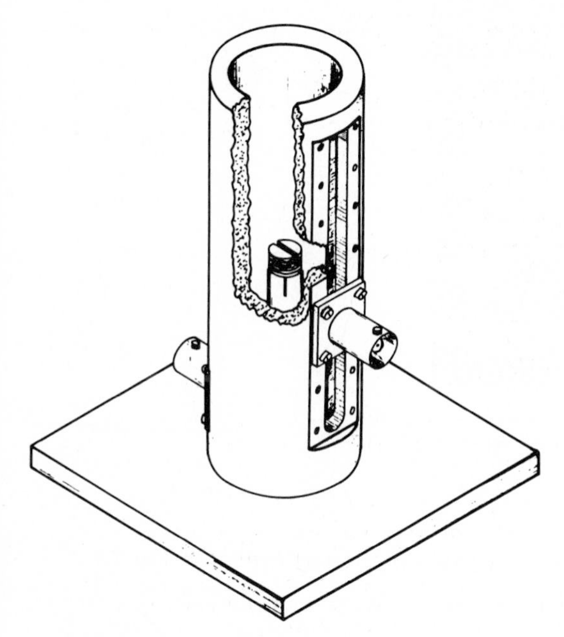

Figure 2: Cutaway view of the resonant cavity. The theoretical frequency range of this design is 0.75 to 3.0 GHz.

Two different optoelectronic switches were used: Cr doped GaAs and Fe doped InP. These switches are capable of producing electrical impulses much shorter than the period of the RF signals. In the first experiment, a switch, with a bias voltage of 93 V, was illuminated with a single pulse of 532 nm green light from a frequency-doubled, Q-switched and mode-locked Nd:YAG laser (30 psec, 400 μj). Using the resulting electrical impulse to excite a coaxial resonator, we obtained a signal at 300 MHz consisting of more than 100 cycles. The output waveform, not shown here, indicates that the oscillation

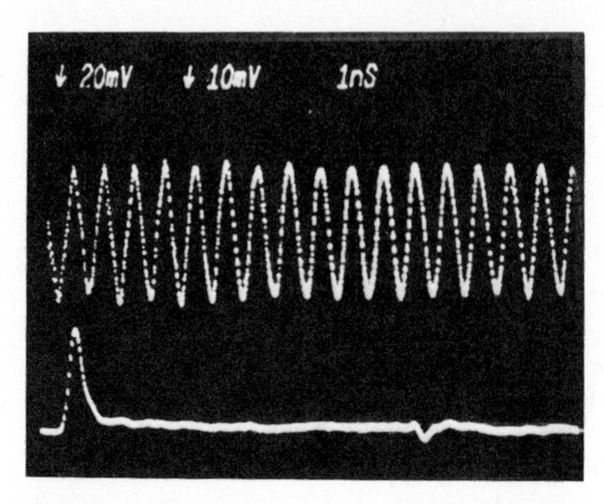

Figure 3: The lower trace shows the mode-locked laser pulse. The upper trace shows the CW 1.6 GHz RF signal created from the laser pulse train. Note that the amplitude remains nearly constant.

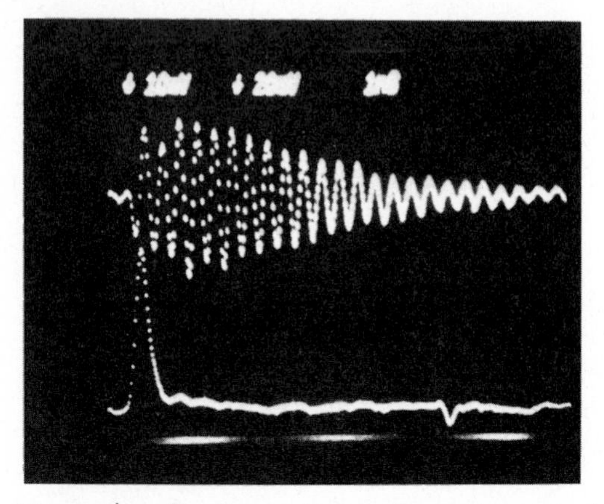

Figure 4: The lower trace shows the mode-locked laser pulse. The upper trace shows the 2.8 GHz RF output of the resonant cavity when excited with a single pulse.

builds up within two cycles. We were able to convert a single electrical pulse into an RF signal with a total energy efficiency greater than 50%.

This technique was extended to generate a CW RF signal by using a CW mode-locked doubled Nd:YAG laser (4 nj, 100 MHz repetition rate). The CW electrical pulse train was coupled to the resonator shown in Fig. 2. For these experiments, the bias voltage for the switch was approximately 20 V. To obtain an oscillation with near constant amplitude, it was necessary to : (1) tune the cavity by means of the inner screw to an integer multiple of the repetition frequency of the laser, (2) match the length of the transmission line between the semiconductor switch and the input antenna with that of the laser cavity, and (3) obtain a high Q for the cavity by reducing the input and output coupling coefficients. Sustained single frequency oscillation is achieved at the expense of conversion efficiency. Figure 3 shows the results of this technique at a frequency of 1.6 GHz. The highest frequency generated with our resonator was 2.8 GHz (Fig. 4). Due to the high losses of our cavity at this frequency, we were unable to prevent a high damping rate. In order to sustanin a high frequency oscillation,either the Q of the cavity could be increased, or a multiplexing technique could be used to increase the repetition rate of excitation of the cavity.

3. Conclusions

We demonstrated the conversion of DC to RF energy in a coaxial resonator. This approach was shown to be an alternative to the frozen wave generator. The design of the resonant cavity limited the RF frequency to 2.8 GHz. Higher frequencies are possible with a suitable resonator. This work was supported in part by the National Science Foundation, the Air Force Office of Scientific Research and the Laboratory for Physical Sciences.

References

1. J. M. Proud, Jr., and S. L. Norman: "High-Frequency Waveform Generation Using Optoelectronic Switching in Silicon.", IEEE Trans. Microwave Theory Tech., Vol. MTT-26, pp. 137-140, 1978.

2. C. S. Chang, M. C. Jeng, M. J. Rhee and Chi H. Lee, A. Rosen and H. Davis: "Direct DC to RF Conversion by Picosecond Optoelectronic Switching.", 1984 IEEE MTT-S International Microwave Symposium Digest, pp. 540-541; also in the Proceedings of SPIE - The International Society for Optical Engineering on "Optical Technology for Microwave Applications", Vol. 477, pp. 101 - 104, 1984.

Kilovolt Sequential Waveform Generation by Picosecond Optoelectronic Switching in Silicon

C.S. Chang, M.J. Rhee, and Chi H. Lee
Electrical Engineering Department, University of Maryland,
College Park, MD 20742, USA
A. Rosen and H. Davis
RCA Laboratory, Princeton, NJ 08549, USA

We demonstrate the generation of a sequential waveform by picosecond optoelectronic switching in silicon. A periodic pulse of two and one half cycles, with a peak to peak voltage of 850 volts, and an aperiodic step pulse are generated.

1. Introduction

Several methods for generating the sequential waveform have been demonstrated [1,2]. A frozen wave generator, simple and effective device [2], consists of several segments of transmission lines connected in series by means of switches. When the switches are closed simultaneously, the "frozen wave" within the segments of the transmission line will be released sequentially. We have reported and demonstrated [3] the generation of a picosecond rise time sequential pulse of two and one half cycles (250 MHz) by using three silicon switches. However, due to the limited high voltage handling capability of the switches, previous experiments [1,2,3] cannot be easily extended to the kilovolt regime.

Intrinsic silicon has been widely used as an ultrafast optoelectronic switch for producing nanosecond square pulses. However, due to the thermal runaway phenomenon, a silicon switch with a gap of 2 mm cannot withstand a DC bias voltage higher than several hundred volts across the gap [4,5]. Two methods were suggested to overcome this problem. The first method suggests the use of a DC bias with the switches at cryogenic temperature [6]. The second method proposes the use of a pulse bias [4,7]. In this report, we will describe the extension of our previous work [3] by using pulse bias. The frozen wave generator also has the capability of generating any kind of aperiodic waveform in addition to the periodic sequential waveform. The schematic diagram of the frozen wave generator used in this work is shown in Fig. 1. It consists of three charged lines and three silicon switches. The three charged lines have lengths of L_1, L_2, and L_3, producing pulse lengths of L_1/v,* L_2/v, and L_3/v, respectively. Three switches of 0.25 mm gaps are

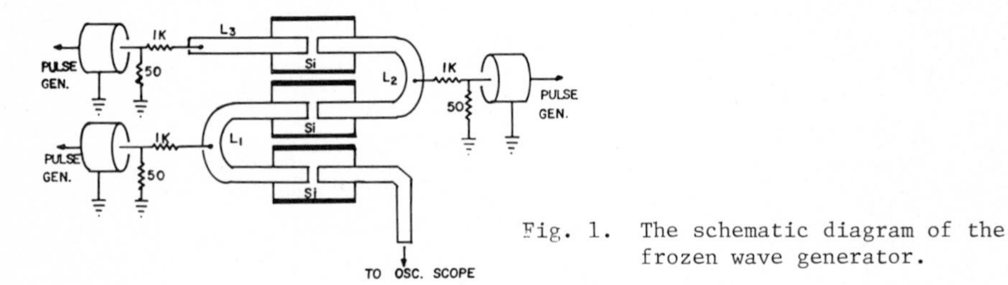

Fig. 1. The schematic diagram of the frozen wave generator.

*where v is the velocity of wave traveling in the transmission lines.

fabricated on a silicon wafer, and they are separated in order to prevent coupling between them. The pulse power supply employed in this experiment consists of a high voltage charged capacitor, a krytron switch, and an output transformer. The transformer has three secondary windings to produce both positive and negative polarity outputs. A typical output pulse has a ~200 ns pulse duration and a 500 to 3000 volts amplitude across a 50 ohm load for both polarities.

2. Experiment and Results

A pulse train is generated by a mode-locked Nd:YAG laser with a wavelength of 1.06 μm. A single picosecond excitation pulse (35 ps) is selected from a pulse train by a Pockel's cell. A photodiode and a variable delay unit are employed to synchronize the charging pulse and the single pulse from the laser.

A single switch is first tested in a charged line configuration. The pulse charging voltage is varied up to 1300 volts, and a square pulse of up to 600 volts is obtained. The upper limit of charging voltage is chosen in order not to exceed the surface breakdown of an electric field of 5×10^4 v/cm, as reported earlier [7].

In this paper, we will discuss both the periodic and aperiodic waveform generation. For the periodic waveform generation, the three charged line segments have lengths of 50 cm, 50 cm, and 25 cm, which correspond to pulse lengths of 2.4 ns, 2.4 ns, and 1.2 ns, respectively. This frozen wave generator produces a pulse of two and one half cycles with a period of 4.8 ns. Three switches are coated with clear epoxy which makes them possible to increase the pulse charging voltage up to ± 900 volts. A single picosecond laser pulse of total energy ~20 μJ is used to activate the three switches simultaneously. The resultant output waveform, as shown in Fig. 2, has a pulse of two and one half cycles (200 MHz), a period of 4.8 ns, and an amplitude of 850 V_{p-p}. For the aperiodic waveform generation, three different DC bias levels are applied to each of the three charging cables of different lengths. A series of experiments has been conducted. For example, in Fig. 3, it is shown that the frozen wave generator with bias levels of 35, 68, and 100 volts, which applied to the three charging cables of lengths 50 cm, 50 cm, and 25 cm, a resulting pulse consists of steps of square pulse with the step duration of 2.4 ns is obtained. Then, by applying three different pulse bias voltages of 400, 800, and 1200 volts, respectively, to all the transmission lines, and using three charging cable lengths of 50cm, 50 cm, and 71 cm, steps of square pulse is observed, as shown in Fig. 4. A computer-

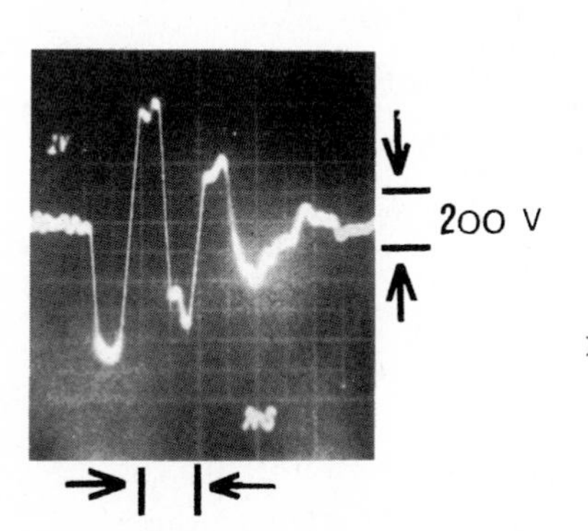

Fig. 2. The Oscillogram of the sequential wave. L_1=50cm, L_2=50cm, and L_3=25cm. The Peak-to-peak voltage is 850 volts.

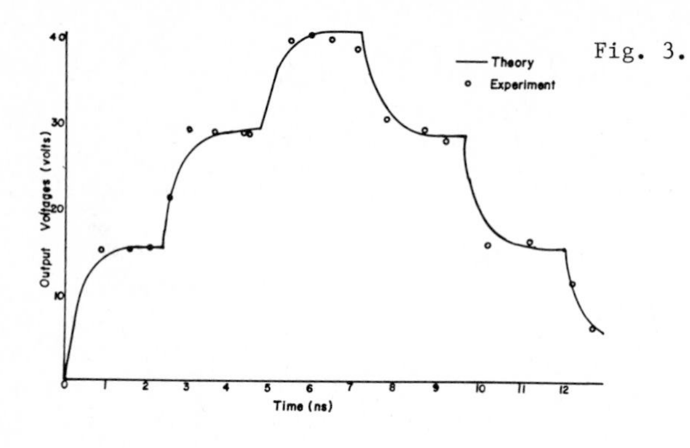

Fig. 3. The generated step waveform. The solid curve is a computer simulated waveform. The dots are experimental points.

program SPICE 2 is also employed to simulate the aperiodic waveform generation for both cases. In the DC bias case, three parameters of the resistance of the three switches are assumed to be 16 Ω, 4 Ω, and 10 Ω. Figure 3 shows both the experimental result and theoretical calculation. The first three steps show voltage amplitudes of 15, 29, and 40 volts, respectively. In the pulse bias case, three values of switch resistance are assumed to be 27 Ω, 4 Ω, and 2 Ω. The voltages of the first three steps are 156, 312, and 468 volts, respectively.

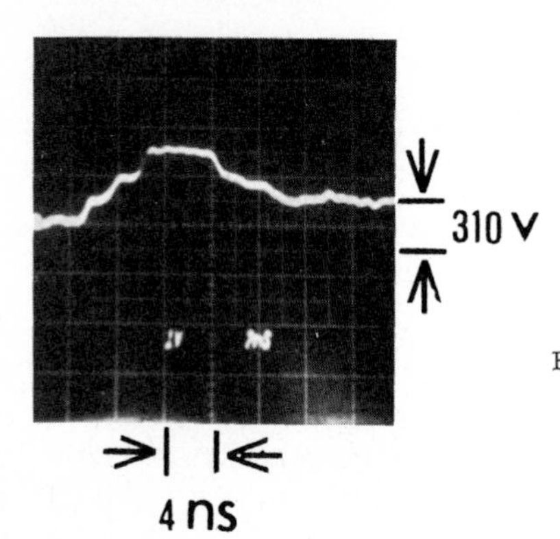

Fig. 4. High voltage step waveform generated with the pulsed charging voltages. The peak voltage is 470 volts.

3. Conclusion

In conclusion, we have demonstrated a method of generating kilovolt sequential waveform and aperiodic waveform by applying pulse bias and DC bias to a frozen wave generator. This method may possibly generate sequential waveform with higher voltages if the surfaces of the silicon switches are properly protected. However, the pulse bias which is applied across the gap of the silicon switch can extend only to the limit set by the intrinsic bulk breakdown ($E \sim 3 \times 10^5$ v/cm). It therefore suggests that in the future experiments, a design utilizing switches with wider gaps will be necessary for the generation of multikilovolt sequential waveform.

4. Acknowledgement

This work was supported by the Air Force Office of Scientific Research.

5. References

1. H. M. Cronson, "Picosecond-Pulse Sequential Waveform Generation," IEEE Trans. Microwave Theory Tech., Vol. MTT-23, pp. 1048-1049, 1975.
2. J. M. Proud, Jr., and S. L. Norman, "High-Frequency Waveform Generation Using Optoelectronic Switching in Silicon," IEEE Trans. Microwave Theory Tech., Vol. MTT-26, pp. 137-140, 1978.
3. C. S. Chang, M. C. Jeng, M.-J. Rhee, Chi H. Lee, A. Rosen, and H. Davis, "Sequential Waveform Generation by Picosecond Optoelectronic Switching," Picosecond phenomena, June 1984.
4. A. Antonetti, M. M. Malley, G. Mourou, and A. Orszag, "High-Power Switching with Picosecond Precision: Applications to High Speed Kerr Cell and Pockets Cell," Opt. commun., Vol. 23, No. 3, pp. 435-439, Dec. 1977.
5. G. Mourou and W. Knox, "High Power Switching with Picosecond Precision," Appl. Phys. Lett., 35(7) pp. 492-495, Oct. 1979.
6. M. Stavola, M. G. Sceats, and G. Mourou, "Picosecond Switching of a Multikilovolt DC Bias with Laser Activated Silicon at Low Temperatures," Opt. commun., Vol.34, No. 3, pp. 409-412, Sept. 1980.
7. P. Lefur and D. H. Auston, "A Kilovolt Picosecond Optoelectronic Switch and Pockel's Cell," Appl. Phys. Lett., Vol. 28, No. 1, pp. 21-23, Jan. 1976.

Femtosecond Nonlinearities of Standard Optical Glasses

J. Etchepare, I. Thomazeau, G. Grillon, A. Migus, and A. Antonetti
Laboratoire d'Optique Appliquée, Ecole Polytechnique- Ecole Nationale Supérieure de Techniques Avancées, F-91120 Palaiseau, France

We have used the optical Kerr effect technique to measure and resolve on a femtosecond time-scale the nonlinear properties of several inorganic glasses. We demonstrate first, the ability of standard transparent optical glasses in performing relatively large, very high speed, nonlinearities through the whole visible spectral range. We point out, next, a very different behavior of colored (semiconductor-doped) glasses, where a non-instantaneous process is superimposed on a pulse-limited temporal response.

The experimental set-up has been described earlier [1]. It consists of a two wavelength configuration, one wavelength for excitation (w_e = 620 nm) and the other one for probing, w_t. Gigawatt peak power pulses of 100 fsec FWHM at 620 nm are obtained at a 10 Hz repetition rate after amplifying the output of a colliding pulse mode-locked dye laser. The probe pulse is selected at a specific wavelength from a continuum of light generated in a water cell, through a monochromator located after the interacting region. The experiment provides a direct measurement of the transmission as a function of time for a sample located between crossed polarizers.

We have reproduced, on Fig. 1, the time-resolved transmission of a 1.5 mm thick piece of SF59 Schott glass, as compared to that of a cell of same length filled with liquid CS_2. The actual measured SF59 Kerr kinetics is identical to the apparatus function, implying that the response may be attributed to a unique process, with a characteristic relaxation time less than 100 fsec. All the other transparent glasses studied present a very similar temporal behavior, the relative magnitudes of the signals are also independent of the probe pulse wavelength.

Quantitative measurements of the Kerr constants in glasses have been performed using liquid CS_2 as a calibration standard [2]. Taking into account the relationship
$n_{2B} = \frac{48\pi}{\lambda n} \chi^{(3)}_{eff.}$, one can then have access to the nonlinear susceptibility coefficients. We have plotted on Fig. 2 the values of the Kerr constants n_{2B} of several binary lead silicate glasses with respect to their PbO mole content. The following correlation is evident: the nonlinearity increases with increasing

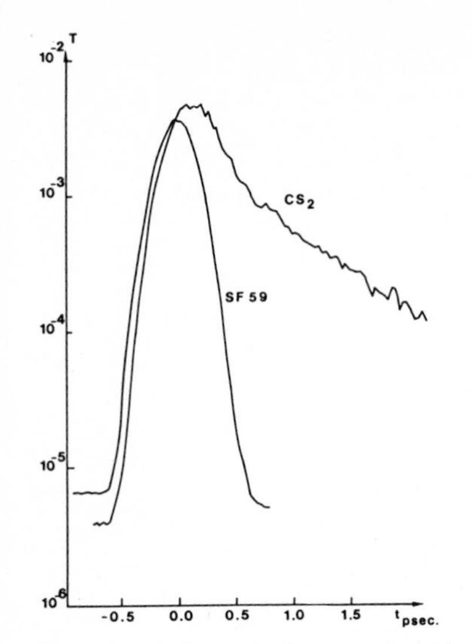

Fig.1 Kinetics of the transmission of a 1.5 mm thick piece of SF59 Schott glass, and of a 1.5 mm cell of liquid CS_2

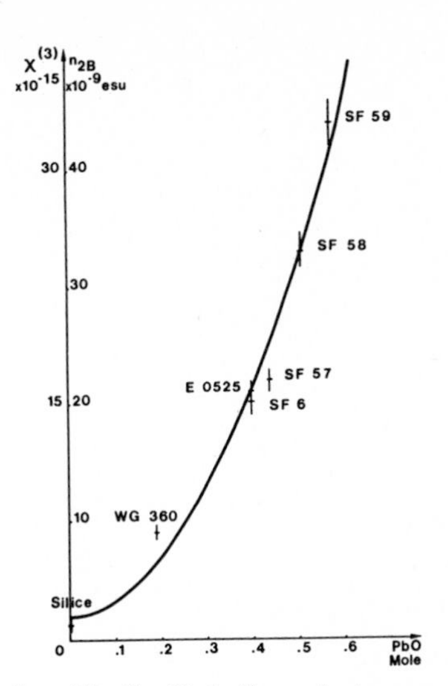

Fig.2 Relation between the Kerr constant n_{2B} and PbO mole content in binary lead silicate glasses.

lead concentration. As a matter of fact, the nonlinearity due to the existence of Pb^{2+} ions, is connected with their high polarizability, owing to the presence of two $6s^2$ electrons in their external shell and the high value of their ionic radius. More precisely, the Kerr constants are quadratically dependent on the PbO mole content, with a correlation factor r of 0.99; on the opposite; we recall that the linear refractive indices of these glasses are directly proportional to PbO concentration (r=0.99). We have shown, as stated in detail elsewhere [3], that the optical nonlinearities in lead silicate glasses increase with the linear index values,and their dispersion in the frequency range of interest; an excellent agreement has been found between our measured values and those calculated by using an empirical expression, by N. L. BOLING et al. [4], which connects the electronic nonlinear refractive index n_2 with n_d (linear d-line refractive index) and ν_d (Abbe number).

The RG645 Schott colored filter exhibits an optical nonlinearity drastically dependent on the probe wavelength (Fig. 3). This behavior is characteristic of semiconductor doped glasses, based on the CdS_xSe_{1-x} ternary compounds [5]: in order to obtain an optimal nonlinear efficiency with these materials, one can match the laser wavelength and the semiconductor band gap and absorption by simply choosing a glass filter of suitable composition and thickness. Their relative magnitude notwithstanding, two temporal different processes are clearly

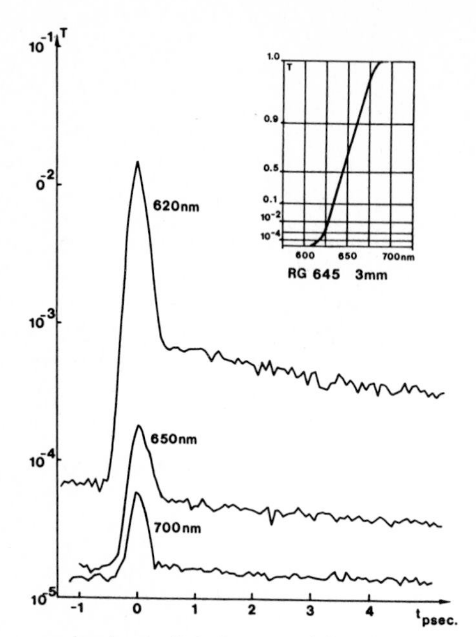

Fig.3 Kinetics of the Kerr signal in a 3 mm thick RG645 slab for 3 different values of the probe pulse. The inset gives the transmission of the colored filter in the spectral range of interest.

resolved, at all values used for the probe pulse wavelength. One process is limited by the pulse width, while to the other corresponds a relaxation time which increases with the probe wavelength value. At w_t = 620 nm, the non-instantaneous process(τ = 9 psec) is characterized by an important nonlinear susceptibility: $\chi^{(3)}_{eff.}(-w_t, w_t, w_p, -w_p) = 2.10^{-13}$ esu, whereas for the instantaneous one $\chi^{(3)}_{eff} = 4\ 10^{-14}$ esu . The overall susceptibility is nevertheless several orders of magnitude smaller than that estimated by R. K. JAIN et al. on similar glasses by DFWM with 10 nsec pulses [5]. This discrepancy is still a puzzle, but is tentatively attributed either to an only preresonnant effect in our case or to the influence of other nonlinear phenomena in the DFWM experiments.

As a conclusion, we point out a potentially better efficiency of lead silicate glasses, compared to semiconductor-doped colored filters; for performing nonlinear materials: although characterized by smaller nonlinearities, indeed lead to responses in the subpicosecond time regime and independent of the optical pulse wavelength in the whole visible spectral range.

1. J. Etchepare, G. Grillon, I. Thomazeau, J. P. Chambaret, and A. Orszag, Ultra Fast Phenomena IV., D. H. Auston, K. B. Eisenthal, Ed , Springer Verlag (1984)
2. Y. A. Volkova, V. A. Zamkoff and N. V. Nalbanoff, Opt. Spectros., 30 556 (1971)
3. I. Thomazeau, J. Etchepare, G. Grillon and A. Migus, Optics Lett., to appear
4. N. L. Boling, A. J. Glass and A. Owyoung, IEEE J. Quantum Electron., **QE14** 601 (1978)
5. R. K. Jain and R. C. Lind J. Opt. Soc. Am. **73**(5), 6475 (1983)

Part VI

Cryoelectronics

High-Speed Analog Signal Processing with Superconductive Circuits*

Richard W. Ralston
Lincoln Laboratory, Massachusetts Institute of Technology,
Lexington, MA 02173, USA

1. Abstract

A new technology utilizing superconductive circuits to provide wideband analog signal processing is described. The function of a canonical signal processor is reviewed, followed by a description of the structure and performance of recently implemented superconductive tapped delay lines and convolvers. The technology has two key features: low-loss and low-dispersion electromagnetic striplines, which provide tapped delay on compact substrates; and superconductive tunnel junctions, which provide efficient low-noise mixing and high-speed sampling circuits. As a demonstration of the utility of this new class of superconductive devices, an ultrawideband chirp-Fourier spectrum analyzer has been developed, which provides a 60-channel analysis of a 2.4-GHz band in 40 ns. Projections based on these results indicate that superconductive technology should support analog device bandwidths of at least 10 GHz with time-bandwidth products up to 1000. Such circuits will provide real-time signal-processing functions with the digital equivalent of up to 10^{12} operations/sec. In addition to pure analog structures, superconductive technology supports high-speed digital components and thus offers the potential for integrated analog/digital signal processors of greater efficiency, speed and processing gain than would be achievable by either approach in isolation.

2. Introduction

The purpose behind the processing of signals is the improvement of the signal-to-noise or -interference ratio to a level at which detection and other decisions can be made with high accuracy. Signal processing is usually accomplished by a combination of pre-processing, which involves linear techniques, and post-processing, which involves nonlinear or conditional techniques. During the last decade there has been a trend toward all-digital implementations. This has been due primarily to the accuracy, stability and flexibility of digital systems. The tendency has been to place the analog-to-digital (A/D) converter at the front end just behind the sensor. But in many emerging applications which stress real-time processing of signals [1], such as pulse-compression radar, spread-spectrum communication and spectral analysis systems, the computational rate required is not compatible with projected digital components. The appropriate performance can be most efficiently realized

*This work was sponsored by the Department of the Army, the Defense Advanced Research Projects Agency, and the Department of the Air Force, in part under a specific program for the Air Force Office of Scientific Research.

by doing the pre-processing in the analog domain and placing the A/D converter just ahead of a digital post-processor, thereby creating a hybrid processor.

Most of the analog pre-processing structures are variations of the transversal filter indicated schematically in Fig. 1. A tapped delay line provides temporary storage and sampling at τ_n intervals for the signal. The samples are multiplied by weights w_n and coherently combined in space or time. Electromagnetic delay media offer great bandwidth (up to a few tens of gigahertz) but the high electromagnetic propagation velocity makes difficult the attainment of large delays. To achieve reasonable delays (10-1000 ns) in a small volume, the transmission lines must be reduced in size. This results in an increased transmission loss and precludes the use of normal conductors. We have exploited the low loss of superconductive transmission lines to realize transversal filters as miniature tapped electromagnetic delay lines. These lines form the basis for constructing planar integrated circuits incorporating mixers, resonators, switches or any of the many rather highly developed superconductive devices.

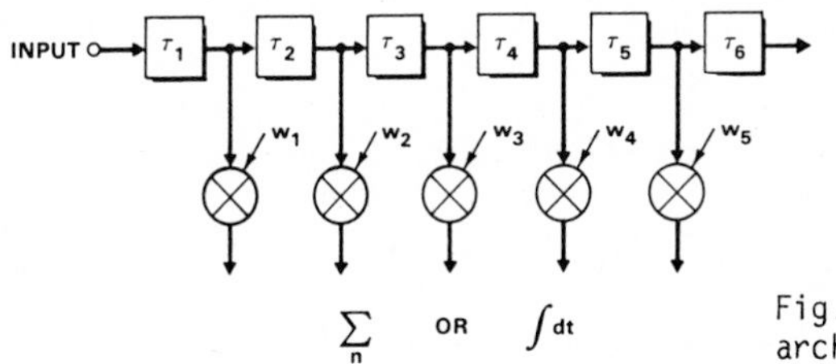

Fig. 1. Transversal filter architecture.

A method by which a pre-processor can enhance the signal-to-noise ratio (S/N) is with a matched filter. The filter impulse response is chosen equal to the conjugate of the signal to be processed. Noise or interference which does not match the impulse response suffers the CW insertion loss of the filter as it passes to the output. The matched signal, however, has its energy collapsed coherently and an impulse is generated at the output. Functionally, the filter forms the autocorrelation of the signal. The width of the central lobe of the correlation is inversely proportional to the bandwidth (B) of the waveform. The correlation peak power rises to a level above the noise at the output in direct proportion to the number of information cycles of the waveform gathered coherently in the filter.

The improvement in S/N is termed processing gain, and in an ideal filter is given by the time-bandwidth (TB) product of the waveform. It is typically in the range of 20 to 30 dB, or 100 to 1000 cycles of information combined. Tight control of amplitude- and phase-vs-frequency characteristics is required. For the full processing gain to be realized, the dynamic range of the filter must substantially exceed (>10 dB) the TB product.

Two superconductive devices which provide matched filtering at multigigahertz bandwidth are described below. The first, a tapped delay line, has a fixed response; the second, a convolver, is electronically reprogrammable. The former is appropriate for pulse compression in millimeter-wave or laser radar, the latter for demodulation in spread-spectrum communication.

3. Superconductive Tapped Delay Line

The basic device cross-section is shown in Fig. 2(a). A coupled pair of superconductive striplines of width w is separated by a variable distance s. The striplines are surrounded by two layers of low-loss dielectric of thickness h which are coated with superconductive ground planes.

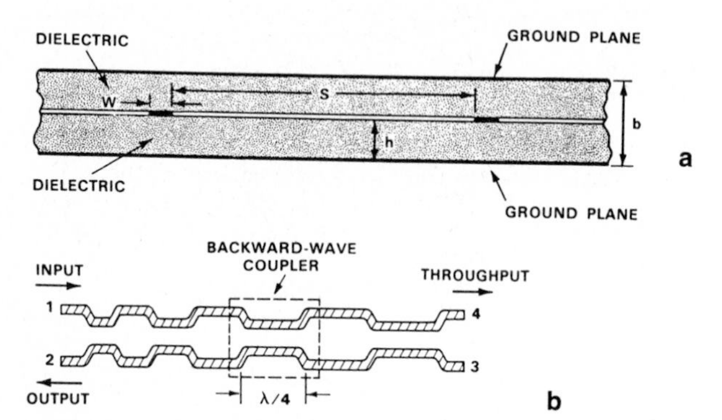

Fig. 2. (a) Cross-section of the coupled stripline structure and (b) chirp filter formed by cascading backward-wave couplers.

The two lines are coupled by a cascaded array of backward-wave couplers as shown in the schematic of Fig. 2(b). Each coupler has a peak response at frequencies for which it is an odd number of quarter-wavelengths long. If the length of the coupler is proportional to the reciprocal of distance down the line, then the resulting structure has a local resonant frequency which is a linear function of delay. In other words, it is a chirp filter with a linear group-delay-vs-frequency (or quadratic phase) relationship. The strength of each coupler is controlled by varying the line spacing s to give a desired amplitude weighting to the response.

The delay line shown schematically in Fig. 2(b) would be difficult to fabricate, being typically 1-mm wide and 3-m long. To utilize round substrates with a minimum number of small-radius bends, the delay line is wound up into a quadruple-spiral configuration, with all four electrical ports brought to the wafer edge [2].

The lower wafer (cf. Fig. 2(a)) carries a niobium ground plane on its lower surface and the niobium stripline pattern on the upper surface. The upper wafer subsequently clamped against the spiral carries a single niobium ground plane. The niobium films are deposited to a thickness of 3000 Å by RF sputtering and patterned by reactive ion etching [3]. The two substrates are 5-cm-diameter, 125-μm-thick silicon wafers. The use of high-resistivity silicon provides a low-loss dielectric substrate at 4.2 K [4].

Figure 3(a) shows the measured and predicted magnitudes of the frequency response of a Hamming-weighted filter. The device was designed to have a 4-GHz center frequency, 2.32-GHz bandwidth, 37.5-ns dispersion, and 10-dB insertion loss. A matched device of opposite chirp slope and flat amplitude response was also produced.

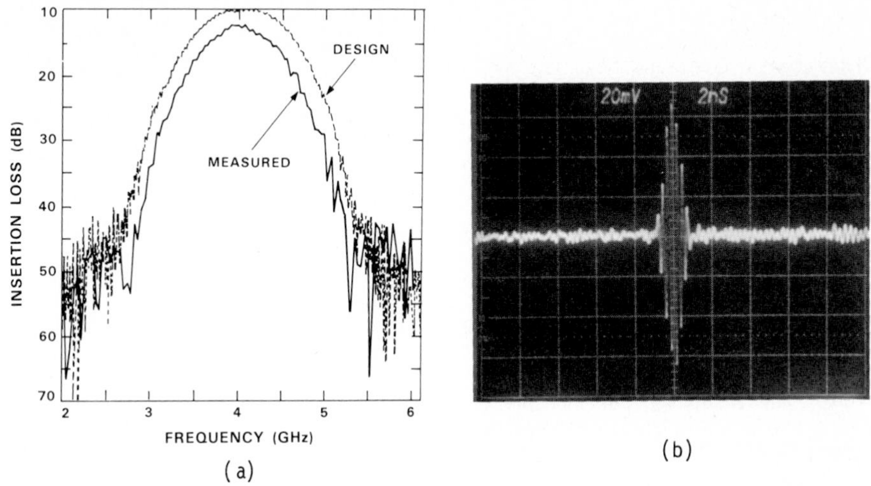

Fig. 3. (a) Insertion loss and (b) compressed-pulse output of a Hamming-weighted filter.

The fidelity of the phase and amplitude characteristics of these filters is borne out in the pulse-compression performance shown in Fig. 3(b). Relative side-lobe levels of -26 dB were obtained when using the flat- and Hamming-weighted devices as matched filters, and -32-dB levels were obtained when the latter device was used as its own matched filter. Devices free of phase and amplitude errors would produce -41-dB side lobes.

4. Superconductive Convolver

Operation of the superconductive convolver [5] is shown schematically in Fig. 4. A signal s(t) and a reference r(t) are entered into opposite ends of a superconductive delay line. Samples (delayed replicas) of the two counterpropagating signals are taken at discrete points by proximity taps weakly coupled to the delay line. Sampled energy is directed into junction ring mixers. The mixing products are spatially integrated by summing in multiple nodes connected to short transmission lines. The mixer ring with

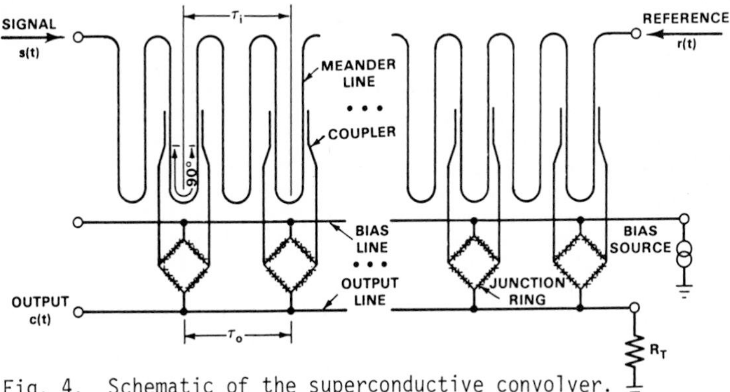

Fig. 4. Schematic of the superconductive convolver.

quadrature feed is chosen to enhance the desired cross product (signal times reference), while suppressing undesired self products and higher order terms. Because both the s(t) and r(t) spatial patterns are moving, there is a halving of the time scale at the output; that is, the center frequency and bandwidth are doubled. If the reference is a time-reversed version of a selected waveform, then the convolver in effect becomes a correlator, and functions as a programmable matched filter for that waveform.

Efficient mixing is obtained using the very sharp nonlinearity produced by tunneling of electrons in a superconductor-insulator-superconductor junction. The ring structure has a series array of four tunnel junctions in each of its four legs. The junctions consist of a niobium base electrode, a 30-Å thick niobium oxide tunnel barrier and a lead counterelectrode [3]. Junction areas are defined by 4-μm-diameter windows etched in a silicon monoxide insulation layer. Unlike logic circuits, which employ the Josephson current due to tunneling of superelectrons for mixing, the junctions are biased at the onset of quasiparticle (normal electron) tunneling because of the larger RF impedance available.

The convolver circuit is fabricated on a 125-μm-thin, 2.5 x 4 cm sapphire substrate with a niobium ground plane deposited on the reverse side. The central region of the device consists of a 14-ns meander delay line with a 50-ohm characteristic impedance. Twenty-five proximity tap pairs, located along opposite sides of the delay line, sample the propagating waveforms and direct the sampled energy into a corresponding number of junction-ring mixers. The resultant mixing products are collected and summed by two 15-ohm transmission lines located near opposite edges of the rectangular substrate. Each end of the output transmission line has a tapered line section which transforms the characteristic impedance of the line to a standard 50 ohms. The desired outputs from the two transmission lines are then summed externally with a microwave combiner. During device assembly, a second sapphire substrate with another niobium ground plane is placed against the delay line region of the first substrate to form a stripline circuit.

The real-time output of the convolver with CW input tones gated to a 14-ns duration and entered into signal and reference ports is shown in Fig. 5(a). The envelope of the convolver output has a triangular shape, as expected for the convolution of two square input envelopes. The trailing side lobes are associated with reflections in the measurement set and spurious signals in the device. The maximum output power level of the device is -58 dBm, and is set by the saturation of the mixer rings as determined by the number of junctions used in each leg.

In wideband measurements, input waveforms consisting of a flat-weighted upchirp and a complementary downchirp were applied to the signal and reference ports of the convolver. The waveforms were generated by two superconductive tapped-delay-line filters. The waveforms had chirp slopes of about 62 MHz/ns and were effectively truncated to bandwidths of about 0.85 GHz by the 14-ns-long interaction length of the convolver. The resultant output waveform shown in Fig. 5(b) has a bandwidth of about 1.7 GHz. Use of flat-weighted chirps should yield a (sin x)/x response with a null-to-null width of 1.2 ns and peak relative side lobes of -13 dB. A null-to-null width of 1.5 ns was observed with excessively high -7 dB side-lobe levels. These distortions are attributed primarily to mixer products produced from undesired leakage of input signal onto the output line and inadequate balance in the taps. Further engineering is required to provide adequate dynamic range.

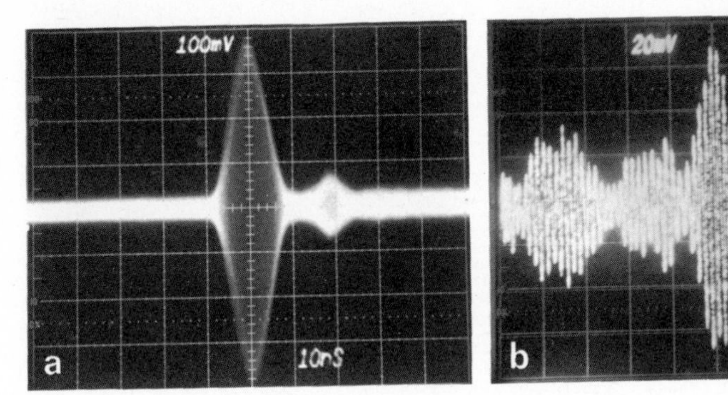

Fig. 5. Output waveform of convolver with input of (a) two-gated CW input tones and (b) two linearly chirped waveforms.

5. Superconductive Spectrum Analyzer

Fourier analysis is certainly an important signal processing function. Here too, the constituent elements of the signal are coherently combined by means of the Fourier algorithm for enhanced S/N. Real-time applications include Doppler analysis in moving-target and imaging radar and power spectral analysis in frequency-shift-keyed communication and electronic warfare. The standard Fourier integral can be written as

$$H(\mu t) = \exp(-i\mu t^2/2) \cdot \int [h(\tau)\exp(-i\mu\tau^2/2) \cdot \exp(i\mu(t-\tau)^2/2)d\tau.$$

The quadratic phase terms are linear-frequency chirps with slope $\pm\mu$. It is seen that the time function h(t) is Fourier analyzed by the following steps: multiply with a chirp, convolve with a chirp of opposite slope and post-multiply with the original chirp. The analog hardware realization of this transform is shown schematically in Fig. 6. Three chirp lines are required for full amplitude and phase analysis; the post-multiply chirp is neglected for power spectral analysis. Essentially, the transformer maps the spectral channels into time bins. The analysis bandwidth is equal to the convolving compressor bandwidth and the number of channels is the TB product of that device. The pre-multiply chirp is required to be twice the T and B of the compressor.

Figure 7 shows the compressor output in response to seven sequential CW tones from 3400 to 4000 MHz in 100-MHz increments. The system responds to RF inputs from 2.8 to 5.2 GHz with an amplitude uniformity of $\pm$1.2 dB. From Fig. 7 the dispersion rate of the analyzer is measured to be 61 MHz/ns, in good agreement with the designed device chirp slope of 61.9 MHz/ns.

The width of the peaks at the -4 dB level is 0.7 ns, implying a two-tone frequency resolution of 43 MHz. For comparison, the expected resolution, which is the inverse of the compressor dispersion multiplied by a factor to account for the broadening caused by the Hamming weighting, is 39.2 MHz.

The compressor output amplitude is a linear function of the RF input amplitude for input power levels up to 3 dBm. Input CW signals of -43 dBm can be resolved from the noise floor set by the thermal noise of the room-temperature electronics, implying a dynamic range over the noise floor of 46 dB.

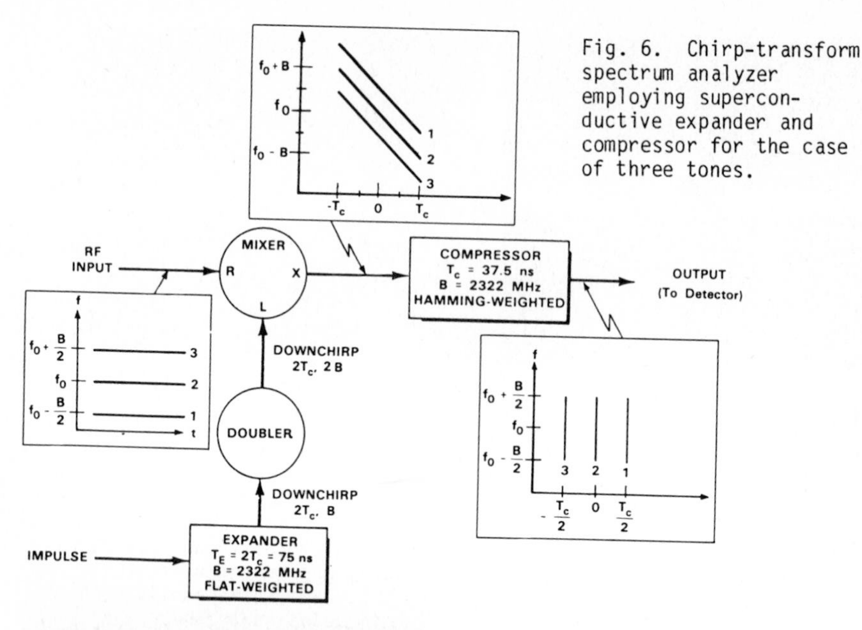

Fig. 6. Chirp-transform spectrum analyzer employing superconductive expander and compressor for the case of three tones.

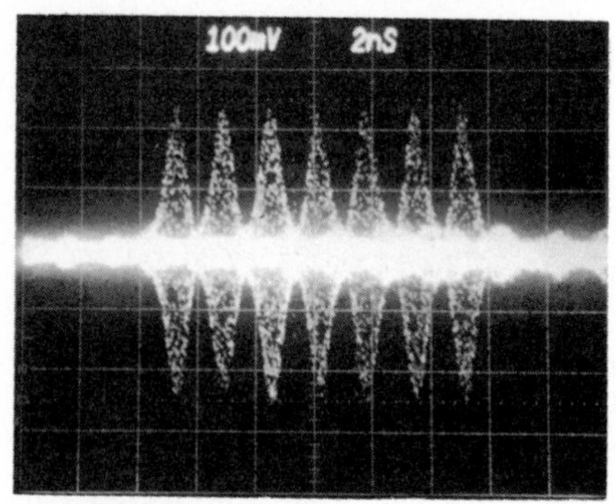

Fig. 7. Multiple exposure of sampling-scope output for successive single-tone inputs of 4000 MHz (leftmost pulse) to 3400 MHz (rightmost pulse) in 100-MHz increments. The input power level is -7 dBm.

6. Analog vs Digital Comparison

It is instructive to compare analog and digital processing. A reasonable basis is the Fourier transform. The number of points N in the transform is equal to the TB product. The transform requires the order of N^2 operations as a direct computation. Substantial savings are possible by using the fast Fourier transform (FFT). For simplicity the radix-2 FFT is assumed. Under the constraint of real-time processing, the digital rate required for the FFT is directly proportional to the signal bandwidth, i.e., at least 5 B $\log_2(TB)$.

Correlation in time of signal against a reference is equivalent to multiplying in the frequency domain. Thus for a fixed filter response, two transforms are required, one to transform the signal into the frequency domain, where it is multiplied by a pre-computed weighting function, and a second to convert back to the time domain. For programmable filtering, a forward transform of the changing reference is also required. This model for digital processing yields the equivalent computational rates of projected superconductive pulse compressors (B = 4 GHz, T = 250 ns) and convolvers (B = 2 GHz, T = 100 ns) as nearly 10^{12} operations/sec, orders of magnitude beyond existing or projected digital machines.

There is a growing interest in signal-processing architectures which combine analog and digital technologies in order to exploit the best features of each while overcoming their individual limitations. This trend is based on the realization that digital and analog techniques are complementary in wideband applications. Thus analog processing can be used for high-speed computation in combination with digital control and post-processing circuitry to extend the flexibility, improve the accuracy, and deliver large (up to 60 dB) signal-processing gain [1].

7. Conclusion

These superconductive devices have, with only a few years of development, surpassed many of the performance goals of competing signal-processing devices within the optical and magnetostatic technologies.

The hybrid approach to signal processing presents a significant opportunity for superconductive electronics. Both the hybrid architecture and superconductive device physics have natural applications at ultrawide bandwidths. The hybrid approach stresses compatibility of analog and digital elements; superconductive technology promises to support a full family of elements with compatible interfaces. Additionally, optoelectronic or cold semiconductor devices might be employed in combination with superconductive pre-processors. The hybrid approach offers a means by which to achieve substantial signal-processing capability with modest circuit density. It is the author's opinion that a spectrum analyzer will be the first practical superconductive processor.

8. Acknowledgements

The author thanks R. S. Withers, A. C. Anderson, J. B. Green, S. A. Reible and J. H. Cafarella for critical contributions to analog superconductive technology which are included in the paper. He thanks K. A. Lowe for preparing the manuscript.

9. References

1. R. W. Ralston: "Signal Processing: Opportunities for Superconductive Circuits," IEEE Trans. Magn. (in press)

2. R. S. Withers, A. C. Anderson, J. B. Green, and S. A. Reible: "Superconductive Tapped Delay Lines for Microwave Analog Signal Processing," IEEE Trans. Magn. (in press)

3. S. A. Reible, "Reactive Ion Etching in the Fabrication of Niobium Tunnel Junctions," IEEE Trans. Magn. MAG-17, 303 (1979)

4. A. C. Anderson, R. S. Withers, S. A. Reible, and R. W. Ralston: "Substrates for Superconductive Analog Signal Processing Devices," IEEE Trans. Magn. MAG-19, 485 (1983)

5. S. A. Reible, "Superconductive Convolver with Junction Ring Mixers," IEEE Trans. Magn. (in press)

Picosecond Sampling with Josephson Junctions

Peter Wolf
IBM Zurich Research Laboratory, CH-8803 Rüschlikon, Switzerland

Introduction

Sampling oscilloscopes are the instruments with the best time resolution for the investigation of electrical waveforms. Their operation principle is the same as that of the stroboscope. For sampling a waveform, a fast switch is required to take a sample of the waveform, and a short pulse to operate the switch. By scanning the waveform point by point with a variable time delay, it can be reconstructed on a recording device. Josephson junctions are well suited for this purpose because they are not only picosecond switches but have a sharp switching threshold, too. Today, time resolution of Josephson samplers approaches 2 ps.

We shall first describe the operation of Josephson samplers, and review the work published. Then, sampling pulse generators, means of delay generation, the room-temperature instrumentation, and the time resolution will be discussed. Finally, the future prospects, especially for better time resolution and as a general-purpose instrument, will be addressed.

1. Josephson Samplers

1.1 Principle of the Sampling Operation

The Josephson junction is a superconducting tunneling device which operates at temperatures of some K. The junction essentials necessary to understand sampler operation are collected in the Appendix. The schematic diagram of a Josephson sampler is shown in Fig. 1a. To the sampling gate, a Josephson junction, three signals are applied: (i) The signal to be measured; (ii) a short pulse generated with the sampling pulse generator, and (iii) a dc current.

As Fig. 1b shows, the dc current is adjusted so that the sum of the three currents just reaches the junction threshold at which the junction switches to the voltage state. This is detected by an external voltage detector. By means of an external feedback loop, the adjustment can be made automatically. To sample the whole waveform, by means of a variable time delay, the sampling pulse is moved timewise along it. As is obvious from this figure, the adjusted dc current is then a replica of the waveform.

The three currents are coupled to the junction galvanically, however, these currents are frequently sent through control lines and then influence the junction magnetically.

1.2. Realized Josephson Samplers

The first use of Josephson junctions in waveform analysis was as a variable threshold gate. It was possible to trace out fairly fast waveforms provided they were monotonically rising. This was extensively used to measure transients in Josephson circuits [1]. The best resolution achieved was about 9 ps [2].

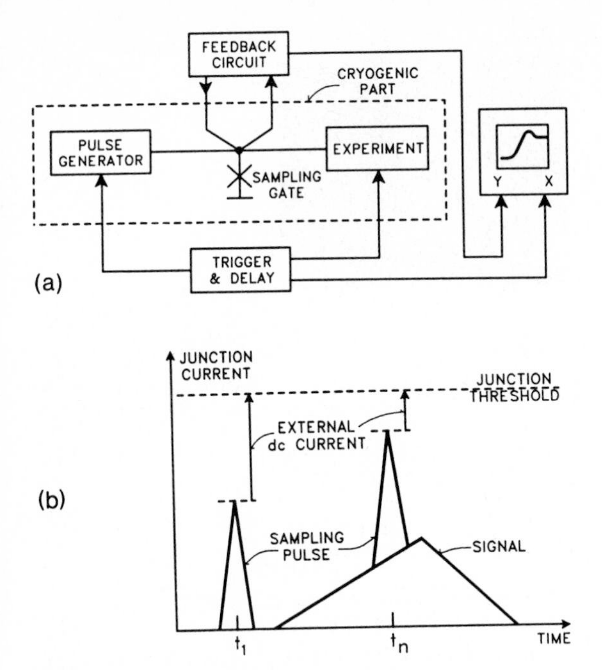

Fig. 1. Sampler operation. (a) Block diagram of a Josephson sampling system. (b) Principle of the sampling process

The next important step, due to FARIS, was the invention of a method to sample general waveforms, including a sampling pulse generator [3]. This first Josephson sampler was realized with three-junction interferometers. As a signal, the short pulse generated by another Faris pulser was used, which at half amplitude was 26 ps wide. TUCKERMAN [4] introduced the feedback circuit to keep the sampling gate automatically at its switching threshold. He was able to record fairly complicated waveforms. The time resolution was about 12 ps. Both samplers generated the time delay for the sampling pulse by a variable transmission line.

An electronically adjustable delay stage built with Josephson junctions was introduced by HARRIS et al. [5]. The electronic delay allowed a higher sampling recording rate which gave a flicker-free display of the waveform sampled on an oscilloscope. This sampler also used faster junctions which resulted in a time resolution of about 8.5 ps.

All these samplers used magnetically-controlled junctions for the sampling gate. The inductance of the control lines and the junction turn-on delay limit their time resolution. In general, better resolution is possible with direct injection, the method shown in Fig. 1a. It was suggested by FARIS [6], and realized experimentally first by AKOH et al. [7,8]. The time resolution of their system approached approximately 6 ps.

Another direct-coupled sampler using junctions with very high Josephson current density and consequently very high switching speed were described by WOLF et al. [9]. They achieved a time resolution approaching 2.1 ps, the best reported so far for a Josephson sampler. A waveform oscillographed by them is reproduced in Fig. 2a. They also introduced a simple method of electronic delay.

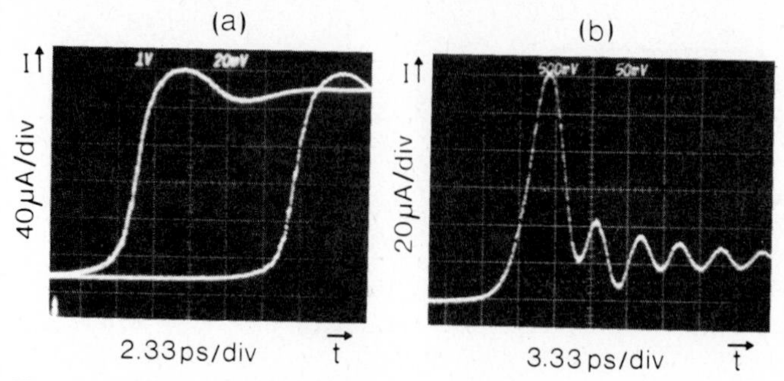

Fig. 2. Measured waveforms [9] (a) Output pulse of a Josephson interferometer which switches to the voltage state. Double exposure with one waveform shifted by 10 ps with a variable air line. The 10%-90% risetime is 2.1 ps. (b) Output of a Faris pulser. The width at half amplitude is 3.7 ps

Josephson samplers have been used extensively to measure Josephson-junction phenomena and the behavior of Josephson circuits.

1.3. Sampling Pulse Generation

Fig. 3 shows the simple circuit first proposed and built by FARIS [3] to produce the spike-like sampling pulse. The Faris pulser consists of a junction which is here magnetically controlled (a single junction or an interferometer) and has a maximum Josephson current I_0. To it, a junction with a smaller Josephson current I_s and a small resistor are connected in series. The output current is sent to the sampling gate which in its unswitched state has a negligible impedance. The circuit operates as follows. First, to the larger junction a dc bias current is applied,which is kept smaller than I_0 so that the junction stays in its superconducting state. Then a control current is applied, strong enough to switch the junction to its voltage state which is reached in a very short time. The voltage over the junction generates a rapidly rising current through the resistor and the small junction until its threshold current I_s is reached. Then this junction switches to the voltage state, too, and transfers most of the current out of the series connection very rapidly. In this way, the short sampling pulse is obtained. The small resistor prevents bias current from being diverted into the small junction during bias set up, and it also damps circuit oscillations. Theoretically, the pulse width is about equal to the junction switching time T_j (Appendix). An example of a measured sampling pulse is shown in Fig. 2b [9]. At half amplitude it is 3.7 ps wide.

Another method of generating a sampling pulse is to switch an interferometer between its vortex states [10,11]. The resulting pulse can be narrower than that achieved with Faris pulsers. On the other hand, vortex pulsers are harder to optimize.

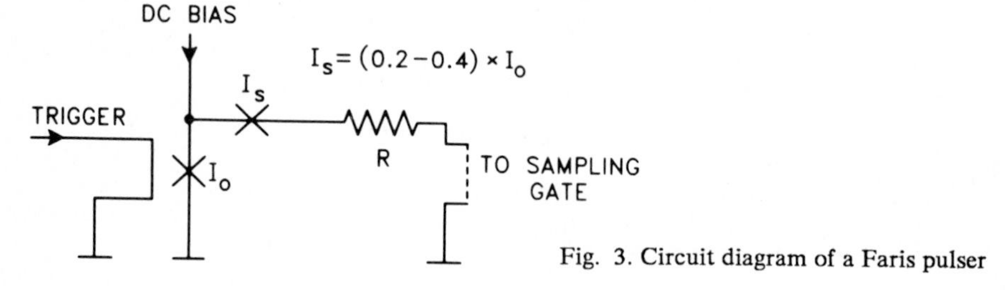

Fig. 3. Circuit diagram of a Faris pulser

2.4. *Delay Generation*

All samplers need a delay stage allowing the sampling pulse to be moved in a controlled fashion timewise along the waveform to be measured. The simplest and most precise delay devices are variable air lines,which are commercially available and have a delay resolution and accuracy of 0.1 ps or better. They can be hand operated or motor driven, and to generate a timebase for an oscilloscope or a recorder, a potentiometer can be coupled with them [3,4].

The aforementioned methods necessarily allowed slow recording only. This was overcome by introducing an electronic delay stage [5]. The output voltage of a switching interferometer is applied to an L-R network. Depending on the time constant $\tau = L/R$, the current in the network rises slowly. It is fed to one control line of another interferometer, usually the input stage of the Faris pulser. To a second control line, a variable bias current is applied. If the bias current is low then the Faris pulser will switch late, because the current from the L-R network has to reach a high value. If the bias current is high, then it will switch early. By using a low-frequency sawtooth signal for the bias current, a variable delay can be achieved. Delay ranges of several 100 ps have been made, and, with sawtooth frequencies of more than 30 Hz, a flicker-free display of the signal sampled on an oscilloscope is possible. The disadvantage of this method is that the delay is a nonlinear function of the bias, and therefore calibration is necessary, for which a variable air line can be used.

An even simpler method without an extra delay stage is possible [9]. Instead of generating a ramp on-chip with an L-R network, the rising edge of the external trigger pulse is used. With commercial pulsers, convenient risetimes in the range of 0.3-1 ns can be obtained. The method can be simplified further by injecting the trigger pulse and the low-frequency sawtooth into the same control line. As with the on-chip delay stage, a calibration of the delay with a variable air line is necessary, and the delay is nonlinear when the rising ramp of the trigger pulse is not linear. Furthermore, the transmission path from the trigger pulse generator to the Josephson chip has to be rather free of discontinuities to maintain a smooth rising ramp.

1.5. *Sampling-Gate Feedback Electronics*

As already mentioned, the purpose of the sampling gate feedback electronics is to provide a bias current which always keeps the sampling gate at its threshold. This bias current is the replica of the sampled waveform. A simple version is shown in Fig. 4 [5]. As a voltage

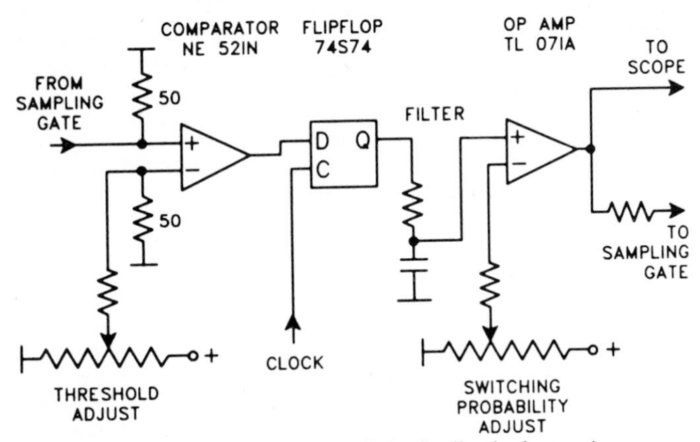

Fig. 4. Schematic circuit diagram of the feedback electronics

detector, a high-speed comparator NE 521 is used. It can sense the sampling-gate output of 2.5 mV without a preamplifier. A potentiometer allows adjustment of the comparator threshold set to about 1 mV. The output of the comparator with TTL logic levels is fed to a D-flipflop and clocked into it by a pulse positioned slightly after the trigger for the Faris pulser. In this way, the short signal from the sampler is stretched over the whole sampler repetition time. The output of the flipflop is applied to an R-C filter with a time-constant determined by the recording rate of the sampled signal, typically some ms. Since the sampling-pulse repetition frequency can be high, typically some MHz, this filter averages the flipflop output strongly. The filtered voltage is applied to an op-amp, the output of which is fed back to the sampling gate and also to the vertical input of an oscilloscope. At the inverted input of the op-amp, the switching probability of the sampling gate can be adjusted by a potentiometer. Usually, a value of 50% is chosen. Because of the D-flipflop, the switching probability adjustment is independent of sampling rate and duty cycle of the sampling-gate output.

The sensitivity of a sampler depends mainly on the noise in the feedback electronics and less on the intrinsic noise of the sampling gate. Because the sampling repetition frequency can be high compared to the recording bandwidth, effective averaging is possible, consequently giving a good sensitivity. Values of about 1 μA have been found for a recording bandwidth of 1 Hz and sampling rates of 4 MHz. Sensitivity can certainly be improved with better designs.

A source of coherent noise which cannot be averaged out is inductive crosstalk of the external pulses, especially the trigger pulses into the sampling gate. By careful design of the chip and the chip holder this noise can be avoided.

1.6. Time Resolution

Because Josephson junctions are quite nonlinear devices, most theoretical investigations of samplers and their time resolution have been carried out with computer simulations. Besides using correct circuit models for the sampling gate and the pulser, one has also to take into account the interconnection parasitics. Recent examples of simulations can be found in [11,12]. In a well-designed sampler with low parasitics, the resolution is limited by the sampling gate and not by the pulser.

Recently, it has been shown by VAN ZEGHBROECK that a good insight into the resolution limits can be obtained with a linearized circuit model of the sampling gate [13]. Essentially, the sampling gate behaves as a low-pass structure with a cut-off frequency given by the junction plasma frequency ω_p (see Appendix). Therefore, the signals applied to the gate are filtered. A further smoothing of the response takes place because the gate needs time to switch to the voltage state, a process also influenced by the plasma frequency. Accordingly, improvements in time resolution of Josephson samplers depend on how much the plasma frequency of the sampling gate can be increased.

Another effect adversely affecting resolution is time jitter. It is important when the sampler and/or the signal source are triggered by external trigger pulses, which usually have risetimes in the ns range. Experimentally, it was found that the slewing rate of the external trigger signal has to be sufficiently high to avoid jitter problems [5,9]. Jitter is caused by external thermal and other noise signals which, superposed on the trigger signal, cause random shifts of the trigger time. A faster rise of the trigger signal decreases this switching uncertainty. Values of 20 μA/ps have been found to be satisfactory to avoid jitter in a system with a 2 ps resolution.

2. Future Improvements

In this section, we shall briefly discuss the prospects of the Josephson sampler as a general-purpose instrument, especially for room-temperature operation, and how time-resolution can be improved.

As shown in the Appendix, the plasma frequency $\omega_p \sim (I_0/C_j)^{0.5}$ depends on the junction capacitance C_j and the maximum Josephson current I_0. This ratio can be improved by decreasing the thickness t of tunnel insulator. I_0 increases exponentially with decreasing t but C_j increases only with 1/t. On the other hand, the current cannot be increased indefinitely, since the junctions become less and less hysteretic. Sufficient hysteresis is necessary for a sharp switching threshold. Junctions with a Josephson current density of about 30 kA/cm^2 and a plasma frequency of the order 250 GHz have allowed samplers to be built with 2 ps resolution [9]. Similar junctions with current densities approaching 100 kA/cm^2 still showing good hysteresis have been made [14]. From this, it follows that a doubling of the plasma frequency should be possible with available technology, which should result in a time resolution of about 1 ps. This, of course, would require an improved layout of the sampler circuits to avoid parasitic effects as much as possible. It seems that 1 ps is not a limit and, with a junction made of high T_c materials such as niobium nitride, subpicosecond resolution should be obtainable in the future.

So far, the Josephson samplers have mostly been used to observe switching phenomena of devices on the same chip in the same cryogenic environment. For using a Josephson sampler for room-temperature events, one has to guide the signals with a sufficiently long transmission line into the cryostat, which has hardly ever been done up to now.

The dominant loss mechanism in transmission lines is skin losses, the influence of which on pulse propagation is well known [15]. Usually, one would choose lines which in the pass-band, about 300 GHz for a 1 ps pulse, have no higher-order modes. A teflon-insulated 50-Ω copper line with an inner conductor diameter of 0.125 mm would fulfill this requirement. But the length allowed for a 10%-90% risetime of 1 ps is only 2 cm, much too short for a helium cryostat. However, the situation can be remedied. First, by using oversized lines, e.g., a 3.5-mm air line has a 10%-90% risetime of 1.7 ps for a length of 50 cm. This is long enough for specially designed cryostats. Such a line has a cut-off frequency for higher-order modes of about 38 GHz. By careful design of the line, it should be possible to avoid the excitation of higher-order modes which travel slower than the fundamental one. A second and additional measure may be to subtract the influence of line attenuation by Fourier transformation. For this, the attenuation has to be known from measurements. Today, even small computers can perform such transformations essentially in real time.

In summary, the techniques described should allow connection of a sampler from its liquid-helium to a room-temperature environment without degrading picosecond resolution severely.

It is possible to simplify the present Josephson samplers to three junctions and one resistor only. No control lines are needed, which simplifies fabrication technology considerably.

3. Conclusions

Josephson samplers are simple systems which today have reached a time resolution close to 2 ps. Further improvements are possible which should allow resolutions below 1 ps. It should also be feasible to use them for room-temperature measurements without a big loss in resolution. Present sampler structure can still be simplified. Accordingly, a Josephson sampler with picosecond resolution, which would be adequate for most investigations of electrical circuits, should be considerably simpler and cheaper than comparable optical samplers.

4. Appendix

Here, we briefly summarize properties of Josephson junctions, in so far as they are important for sampler applications. For more information we refer to the literature [16,17].

As Fig. 5a shows, a Josephson tunnel junction consists of two superconducting electrodes separated by a tunnel insulator only some nm thick. The I-V characteristics in Fig. 5b consist of two branches. The superconducting or Josephson branch, caused by tunneling of Cooper pairs, is limited to a maximum Josephson current I_0 determined mainly by the thickness of the tunnel insulator and the junction area. If I_0 is exceeded, the junction will switch to another branch, the voltage state, given by the tunneling of unpaired electrons. The voltage V_g at the vertical part of this branch is some mV. As is shown in the figure, the junction has a hysteresis. It occurs only when the junction capacitance C_j is large enough. For samplers, hysteretic junctions are required.

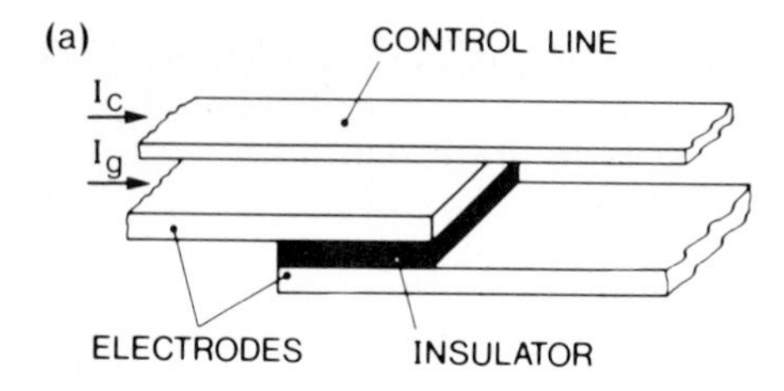

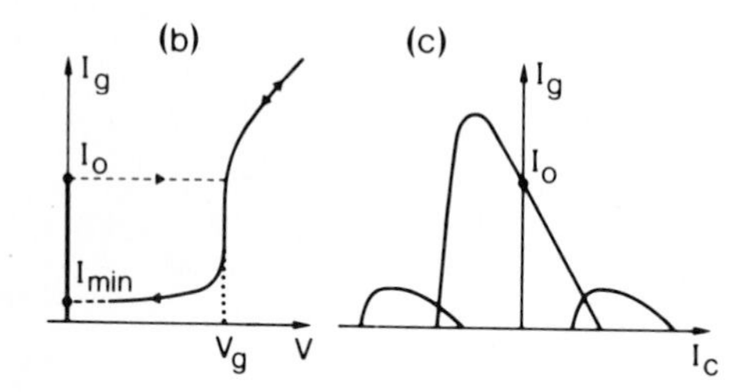

Fig. 5. Josephson junction:
(a) Structure. (b) I-V characteristics.
(c) Control characteristics

As shown in Fig. 5c, the value of I_0 can be altered by a magnetic field. The field is generated by a current I_c sent through a control line on top of the junction. The switching time from the Josephson to the voltage state is approximately $T_j = C_j V_g / I_0$. In the Josephson state, the Josephson branch behaves like an inductor, which together with the capacitance leads to an eigenfrequency of the junction, the plasma frequency ω_p:

$$\omega_p = \left(\frac{4\pi q I_0}{h C_j}\right)^{0.5}$$

with q and h being the electronic charge and Planck's constant, respectively.

References

1. H.H. Zappe: IEEE J. Solid State Circuits **SC-10**, 12 (1975)
2. C.A. Hamilton, F.L. Lloyd, R.L. Peterson, J.R. Andrews: Appl. Phys. Lett. **35**, 718 (1979)
3. S.M. Faris: Appl. Phys. Lett. **36**, 1005 (1980)
4. D.B. Tuckerman: Appl. Phys. Lett. **36**, 1008 (1980)
5. R.E. Harris, P. Wolf, D.F. Moore: IEEE Electron Device Lett. **EDL-3**, 261 (1982)
6. S.M. Faris: U.S. Patent 4,401,900, filed Dec. 20, 1979; issued Aug. 30, 1983
7. H. Akoh, S. Sakai, A. Yagi, H. Hayakawa: Jpn. J. Appl. Phys. **22**, L435 (1983)
8. S. Sakai, H. Akoh, H. Hayakawa: Jpn. J. Appl. Phys. **22**, L479 (1983)
9. P. Wolf, B.J. Van Zeghbroeck, U. Deutsch: IEEE Trans. Magn. (in press)
10. H. May: Ph.D. Thesis, University of Stuttgart (1981)
11. H.A. Kratz, W. Jutzi: IEEE Trans. Magn. (in press)
12. H. Hafdallah, P. Crozat, R. Adde: IEEE Trans. Magn. (in press)
13. B.J. Van Zeghbroeck: J. Appl. Phys. (in press)
14. D.F. Moore, P. Vettiger, T. Forster: Physica **108 B+C**, 983 (1981)
15. R.L. Wigington, N.S. Nahman: Proc. IRE **45**, 166 (1957)
16. T. Van Duzer, C.W. Turner: *Principles of Superconductive Devices and Circuits* (Arnold, London 1981)
17. A. Barone, G. Paterno: *Physics and Applications of the Josephson Effect* (Wiley, New York 1982)

Transmission Line Designs With a Measured Step Response of 3 ps Per Centimeter

Charles J. Kryzak*
Hypres, Inc., Elmsford, NY 10523, USA

Kevin E. Meyer and Gerard A. Mourou
Laboratory for Laser Energetics, University of Rochester,
Rochester, NY 14623, USA

In order to effectively exploit the extreme speed and sensitivity potential of both the electro-optic and superconducting technologies, proper generation, coupling, and propagation of signals with bandwidths in excess of 100 Gigahertz is required. In this report we present results of an experimental investigation, supported by theoretical predictions, on a transmission structure which satisfies the ultra wide bandwidth requirements. The experiments, which incorporated a new variation of an electro-optical sampling technique [1], investigated the generation of picosecond electrical pulses, an impedance matching tapered coupling configuration, and the propagation of the pulses down the transmission line. The technique allows the measurement of the waveforms at any point along the transmission line while minimizing dispersional effects associated with the sampling. This resulted in an accurate determination of pulse degradation, i.e., step response, per unit length of transmission line. The investigation supports the conclusion that transmission lines and connectors capable of supporting bandwidths in excess of 100 Gigahertz, while allowing relatively dispersionless propagation over distances exceeding a centimeter, are realizable.

The transmission line structure chosen, for its high bandwidth capabilities was the inverted microstrip [2]. By employing thin, low dielectric constant substrates, a static effective dielectric constant near that of air could be achieved. This results in a propagation characteristic approaching a homogeneous, air dielectric, transmission line, i.e., near dispersionless T.E.M. mode propagation. Similarly, the low static effective dielectric constant implies a larger strip width to height above ground plane ratio to achieve a desired characteristic impedance. This has the added benefit of reducing the net skin resistance and thus the attenuation.

The inverted transmission line configuration has nearly the same topology as the open microstrip, except that the roles of the two dielectrics i.e., air and substrate, are interchanged. The reported expressions governing the frequency dependence of the phase velocity, characteristic impedance, and attenuation will be applicable if the finite thickness of the dielectric substrate is taken into account [3,4,5]. Also, the task of supporting risetimes of 5 ps or less requires the strip width to be scaled to about 75 mils or less so as to keep the transmission structure from becoming too radiative, i.e., maintain $\lambda/2 \geqslant$ strip width.

Figure 1 illustrates the pulse degradation mathematically predicted for a inverted microstrip. The model employed a 5.2 mil thick glass substrate,

‡Present address, Physics Dept., Rensselaer Polytechnic Institute, Troy, New York 12181

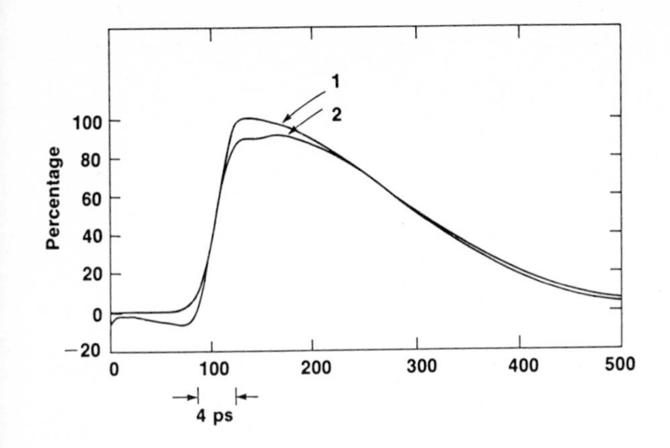

Fig.1. Propagation of electrical pulses in inverted transmission line. (1) input pulse 10-90% risetime ~3.99 ps (2) pulse after propagating 1.5 cm, risetime ~4.1 ps

a strip width of 75 mils, characteristic impedance of 50 ohms, and 5000 Å thick niobium strip and ground plane metalization. The curves are obtained by calculating the frequency dependent propagation constant, then multiplying by the frequency domain representation of the input pulse, and finally transforming the product back to the time domain.

From Fig. 1, it can be seen that over a distance of 1.5 centimeters only a slight pulse degradation is predicted. Also, an undershoot due to dispersion in combination with frequency insensitivity of the attenuation may be observed. The modelling does predict, however, that if radiative effects can be ignored and the large band width pulses properly coupled, the inverted structure described above will support pulses with risetimes less than 5 ps over distances exceeding one centimeter.

The electrical characteristics of inverted microstrips employing several different dielectric substrates were measured utilizing a variation of the tranverse linear electro-optic sampling technique, [6]. The sampling crystal, i.e., Pockel cell -$LiTaO_3$, which is placed between the strip and ground plane, remains an independent structure from the transmission line. This allows the $LiTaO_3$ crystal to be freely moved along the line so that the sampling may be performed within 1.5 mils of the crystals leading edge. Thus, one is able to electrically sample the pulse which has suffered only a small amount of dispersional degradation attributed to a microstrip on a $LiTaO_3$ substrate.

The picosecond signal generation was accomplished with a Cr-doped GaAs switch excited by a 120 femtosecond duration optical pulse from a C.P.M. ring dye laser, [1,6]. Each transmission structure tested was fitted with its own generator coupled to one end of the inverted microstrip. The coupling configuration included an impedance matching linear tapering of the strip and ground plane, Fig. 2 and 3, and an impedance matched non-tapered butt connection. Figure 4 shows a measured waveform at a point located at the end of a linear taper, i.e., 4 millimeters from the pulse source. The 4.4 ps risetime pulse, Fig. 4, was then used as a probe to investigate the step response of the inverted microstrip.

Figure 5 is a plot of the 10-90% risetimes verses distance propagated. Curves A', B, and C are the measured risetime degradations for inverted microstrips while curves A and D are calculated risetimes for an inverted microstrip and microstrip, respectfully. A and A' represent similar

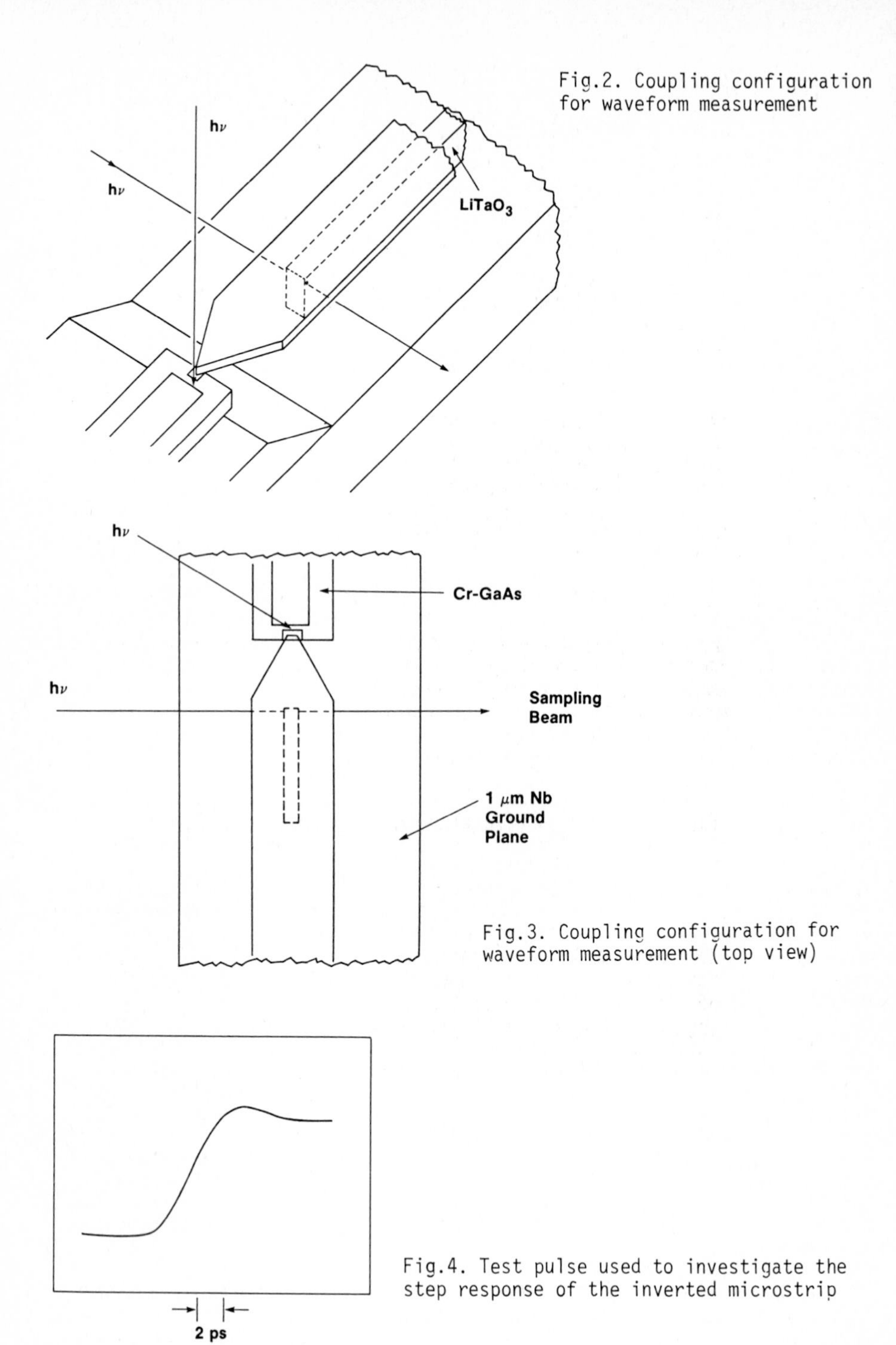

Fig.2. Coupling configuration for waveform measurement

Fig.3. Coupling configuration for waveform measurement (top view)

Fig.4. Test pulse used to investigate the step response of the inverted microstrip

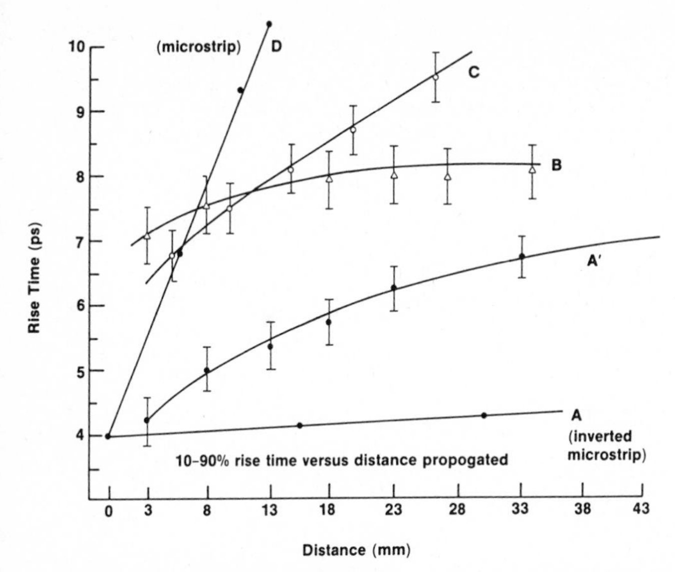

Fig.5. Dispersion characteristics of transmission lines

inverted structures, i.e. similar substrates (glass) and metalization (Nb), which employ a tapered coupling structure. B and C are inverted lines employing Kapton and RT Duroid as substrates, respectfully. Both have copper strip metalization, niobium ground planes and non-tapered couplers. Table 1 below summarizes the results.

From Fig. 5, one finds an inverted structure with a measured step response of 3ps per centimeter, i.e. curve A'. Curve B shows a measured

Table 1

	Line	Substrate, thickness,	Strip metalization	Connector	Step Response per centimeter
A	inverted	glass, 5.2 mils $\varepsilon_r \cong 4.2$	Nb	-	< 1 ps
A'	inverted	glass, 5.2 mils $\varepsilon_r \cong 4.2$	Nb	Tapered	2.98 ps
B	inverted	Kapton, 1 mil $\varepsilon_r \cong 3.3$	Cu	Non-tapered	< 2.5 ps
C	inverted	RT Duroid, 10 mils $\varepsilon_r \cong 2.2$	Cu	Non-tapered	6 ps
D	microstrip	Fused quartz 10 mils $\varepsilon_r \cong 3.78$	Nb	-	> 8.7 ps

response of less than 2.5 ps, i.e., > 100 GHz, except that the lack of a tapered coupler allowed only a 6.9 ps risetime pulse to be loaded successfully. Thus, it appears that inverted microstrips, designed similarly to those discussed (including a tapered coupling), will allow pulses with bandwidths of 100 Gigahertz and greater to be successfully loaded and propagated over distances in excess of a centimeter.

[1] Mourou, G. A., Meyer, K. E., Applied Physics Letters, Vol. 45, pp. 492, 1984.

[2] Schneider, M. V., Bell Sys. Tech. J, May-June, 1969.

[3] Yamashita, E., Katsuki, IEEE Trans. Microwave Theory Tech., Vol. MTTT-24, pp. 195-200, 1976.

[4] Woermbke, J. D., Microwaves, pp. 89-98, January 1982.

[5] Kuester, E. F., Chang, D. C., IEEE Trans. Microwave Theory Tech Vol. MTT-28, pp. 259, 1980.

[6] Validmanis, J. A., Ph.D. Thesis, Univesity of Rochester, 1983.

Development of a Picosecond Cryo-Sampler Using Electro-Optic Techniques

D.R. Dykaar[1,2], T.Y. Hsiang[1], and G.A. Mourou[2]
[1]Department of Electrical Engineering, University of Rochester, Rochester NY 14627, USA
[2]Laboratory for Laser Energetics, University of Rochester, 250 East River Road, Rochester, NY 14623, USA

1. Introduction

The first all-superconducting sampler, with a rise time of 26 ps, was reported by Faris in 1980 [1]. Recently, an improved version was developed by Wolf et al. with a 2.1 ps response [2]. In this approach, a Faris pulser [1] is used to superimpose a fast current spike on an unknown waveform. This sum signal is then applied to a sense junction, whose dc bias is adjusted to reach the switching threshold. By sweeping the fast pulse in time, the unknown signal can be replicated. The speed of these systems was limited by the response time of the superconducting Josephson junctions, although junctions with switching speeds on the order of one picosecond are possible.

Electro-optic sampling can also be used to study these high speed phenomena. Previously we reported on a picosecond electro-optic sampling system used to characterize Josephson devices [3]. A block diagram is shown in Figure 1. The sampler [4,5] was driven by a colliding-pulse mode-locked (CPM) laser, which produced two 100 fs FWHM pulses at 100 MHz. One pulse train was used to generate an electrical signal of adjustable height and width. This was accomplished using a photoconductive switch and a resistively-charged section of microstrip transmission line. In addition, a long, looping wire bond was used to connect the switch output to ground. This prevented dc charging of the output transmission line, while preserving the shape of the switch output.

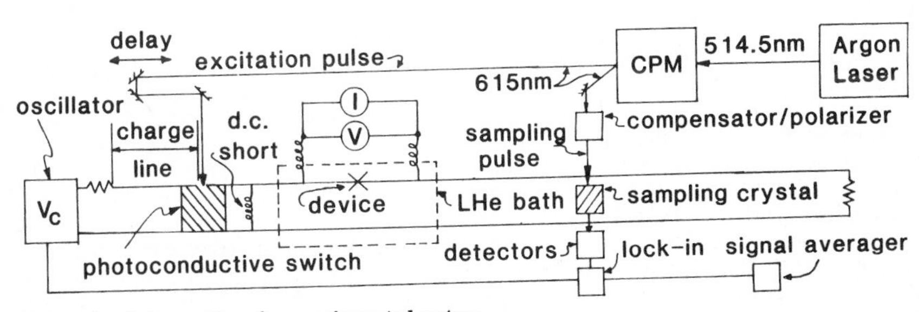

Figure 1. Schematic of experimental setup

This signal was used to excite the test device placed in the cryogenic environment. The resultant electrical transient was then sampled with the second CPM laser pulse train using a birefringent lithium tantalate crystal. By changing the relative delay between the optical excitation and sampling pulses, the temporal development of the electrical transient was recorded using conventional slow speed electronics. The limitations on the response of this system were found to be due to the several meters of cable required to move signals into and out of the cryogenic environment.

2. Cryo-Sampling

Clearly, to improve the response time of this system, the various cables must be eliminated. To accomplish this,the high-loss electrical paths were replaced with low loss optical paths. As shown in Fig. 2a, in the initial version of the cryo-sampler approximately half of the electrical path length has been eliminated by placing the sampling crystal in the cryogenic environment. Since the remaining electrical path was coaxial cable, the sampling geometry was also basically coaxial in nature. As shown in Fig. 2b, a cut coaxial cable was used in conjunction with a microstrip type of sampling crystal. This hybrid geometry is still limited by the high-frequency attenuation of the coaxial cables. The sensitivity is also limited, since most of the electric field still propagates in the cable and not in the sampling crystal. However, despite these limitations, a rise time of 16.4 was obtained as shown in Fig. 3.

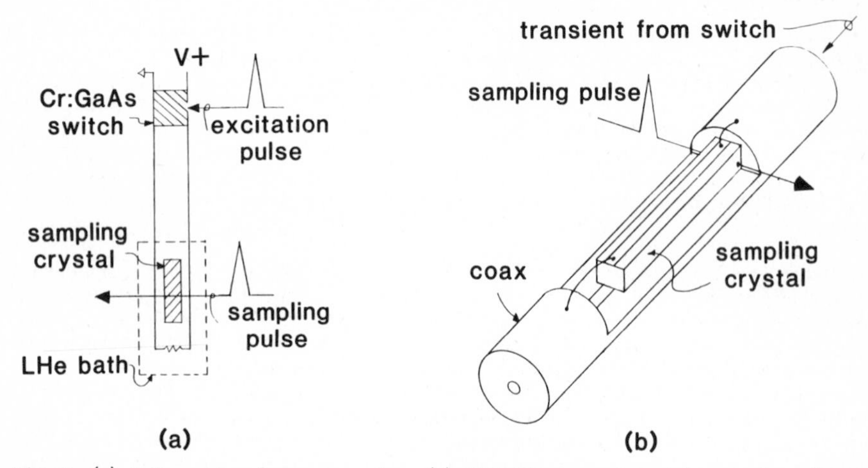

Fig. 2. (a) Schematic of cryo-sampler. (b) Hybrid coax-microstrip sampler. (Note: Spring clips for sampling crystal are omitted for clarity)

In the present version, the final impediments to high speed operation are being systematically removed. First, the coaxial cables have been eliminated. While several geometries of different transmission line have been tested, the present trend is toward using a coplanar geometry. A coplanar geometry sampler has been demonstrated with a 0.46 ps response time at room temperature [6]. In a coplanar transmission line, the spacing between the electrodes determines the cutoff frequency, so small photolithographically formed lines are being used. While it may seem that small lines would be more resistive, and hence more lossy, it appears that wider lines will cause the electric field to experience a larger dielectric mismatch, causing more dispersion. This is a change in the velocity of the signal as a function of frequency. It causes the higher frequencies to arrive later, degrading the signal rise time. Loss in these transmission lines however, is a result of skin effect attenuation of the high frequencies. In this case the rise time is still degraded, but the high frequencies are diminished, not delayed. Note that while the results are similar, the mechanisms are quite different.

Next, in the interest of robustness, a reflection mode sampling geometry was chosen. In this mode, a sampling crystal with a high reflection dielectric coating on one side is pressed down onto the circuit to be tested. A sampler operating in the reflection mode has been demonstrated to have a response of 0.75 ps at room

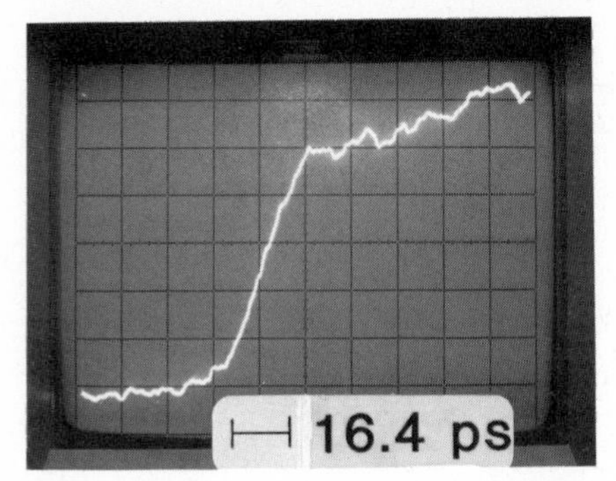

Figure 3. Cryogenic response of hybrid coax-microstrip sampler

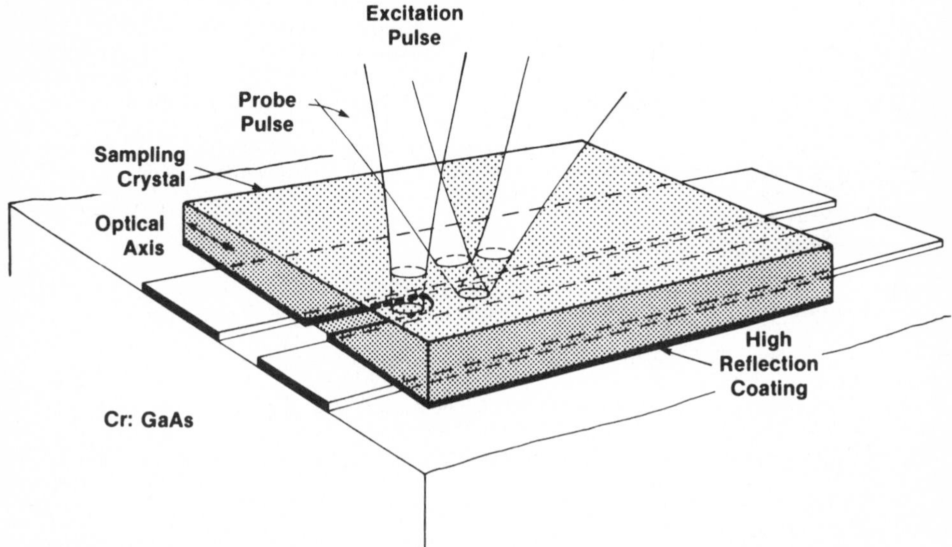

Figure 4. Integrated sampling structure. The excitation pulse is focused through a hole in the high reflection coating.

temperature [7]. This situation is shown in Fig. 4. By using a spring clip (not shown) to position the crystal, the problems of differential thermal contraction are avoided.

Finally, a "switchless" geometry has been used. As shown in Fig. 4, the switch gap has been eliminated, and replaced by the gap between the electrodes. In this geometry the electrical pulse is created by photoconductively shorting together the two transmission lines in the presence of a dc bias. Now the semiconductor switch is not defined a priori on the GaAs substrate using photolithography. Instead, the switch is defined optically in real time and in situ using the CPM laser pulse. This arrangement allows the electrical path length to be minimized by changing the position of the focused CPM laser beam. In addition, small movements of this focal spot can be used to maximize the electrical signal output. This enables the switch to be defined in a region where good ohmic contacts have been made between the transmission lines and the Cr:GaAs substrate.

This geometry has been tested at room temperature using Pb electrodes with a 10 μm line spacing and 100 μm line width. The electrical signal was generated just outside the crystal, and sampled just inside the crystal. The response time was measured to be 1.8 ps despite the low conductivity of the Pb transmission lines. By directing the excitation beam through a hole in the high reflection coating, it is possible to decrease the electrical path length even further, as shown in Fig. 4.

3. Conclusion

The picosecond electro-optic sampling technique has been used to characterize superconducting Josephson tunnel junctions. To improve system response, a cryo-sampler was developed with a 16 ps rise time. An integrated version with Pb electrodes had a room-temperature response of 1.8 ps.

4. Acknowledgement

This work was supported by the Laser Fusion Feasibility Project at the Laboratory for Laser Energetics which has the following sponsors: Empire State Electric Energy Research Corporation, General Electric Company, New York State Energy Research and Development Authority, Northeast Utilities Service Company, Ontario Hydro, Southern California Edison Company, The Standard Oil Company, and the University of Rochester. Such support does not imply endorsement of the content by any of the above parties. Additional support was provided by NSF Grant #ECS-8306607. One of us (D.R.D.) has an IBM Pre-Doctoral Fellowship.

5. References

1. S.M. Faris, Appl. Phys. Lett., Vol. 36, No. 12, 1005 (1980).
2. P. Wolf, B.J. Van Zeghbroech, V. Deutsch, to appear IEEE Trans. Magn., March, 1985.
3. D.R. Dykaar, T.Y. Hsiang, G.A. Mourou, to appear IEEE Trans. Magn., March, 1985.
4. J.A. Valdmanis, G.A. Mourou, C.W. Gabel, Appl. Phys. Lett., Vol. 41, No. 3, 211 (1982).
5. J.A. Valdmanis, G.A. Mourou, C.W. Gabel, IEEE J. Quantum Electron., QE19, 664 (1983).
6. K.E. Meyer, G.A. Mourou, Appl. Phys. Lett., Vol. 45, No. 5, 492 (1984).
7. K.E. Meyer, G.A. Mourou, Electron. Lett., submitted February 1985.

Picosecond Josephson Logic Gates for Digital LSIs

Jun'ichi Sone, Jaw-Shen Tsai, and Hiroyuki Abe
NEC Corporation, 1-1, Miyazaki 4-Chome, Miyamae-ku, Kawasaki, Kanagawa 213, Japan

1 Introduction

Because of its high-speed properties and low power-dissipation, the Josephson junction digital integrated circuits (JJ IC) have been intensively investigated especially for applications in high performance digital systems. The high-speed properties of the JJ IC can be attributed to the quantum-effect-dominating switching speed as well as the light-velocity signal transmission between gates. A wide gate operation margin, multiple logic function capability, and a small gate size are among other important factors in the evaluation of the JJ logic gates. The small gate size is, in particular, very important because it make it possible to densely pack the logic gates and to shorten the signal transmission line length.

Resistor Coupled Josephson Logic (RCJL) gate family, proposed by SONE [1], satisfies these requirements and is the promising candidate to be integrated in large scale digital systems. By the recently developed RCJL 4 b x 4 b multiplier, implemented in 5-μm lead alloy technology, a multiplication time as small as 280 psec has been demonstrated [2].

In this paper, the realized and projected picosecond gate speed of Resistor Coupled Josephson Logic for Large Scale Integration (LSI) applications will be discussed.

2 Basic Gate Structure

The RCJLs are current switched logic gate, composed of multiple Josephson junctions and resistors. The RCJL family consists of OR, AND, and 2/3 MAJORITY gates as shown in Fig. 1. In the OR gate, the junction with a critical current of $0.8 \cdot I_0$ acts as an isolation device when all of three junctions are switched to the voltage state after the arrival of input signal. In the AND gate, all of the three junctions are switched if both inputs are applied, while two of the junctions remain in the superconducting state when either input is applied. The 2/3 MAJORITY gate is switched if more than two out of three inputs are activated. Using the RCJL gate family, a 4 bit adder [3], a 4 bit-16 bit decoder [4], and a latch circuit [5] have also been developed.

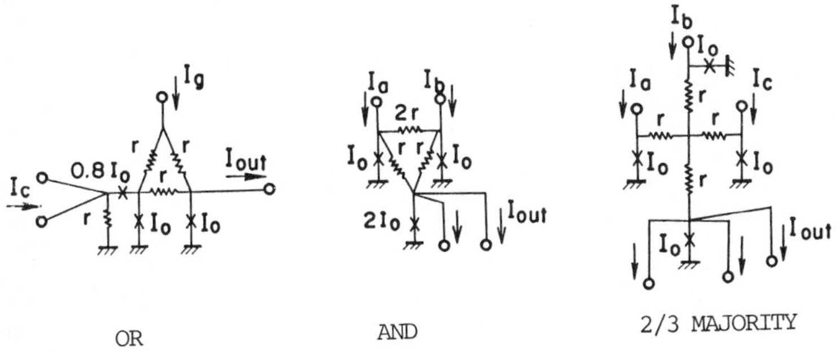

OR AND 2/3 MAJORITY

Fig.1 RCJL gate family

3 Medium Scale Integration

Figure 2 shows a 4 b x 4 b parallel multiplier chip's microphotograph. The multiplier circuit was fabricated on a silicon substrate, using the lead alloy Josephson IC processes with 5-μm minimum junction diameter and linewidth. The multiplier employs the dual rail logic aiming for the high-speed oparation. The circuit contains 249 gates consisting of 862 junctions. The RCJL gates are interconnected to form carry generators and sum generators, as shown in Fig. 3, which are basic elements of the multiplier.

The critical path delay was measured by performing an arithmetic operation (1, 1, 0, 0) x (1, 0, 1, 1). The time interval of 280 psec between the two output waveforms, observed on the sampling oscilloscope as in Fig. 4, corresponds to the critical path delay. The critical path contains 13 RCJL gates. Therefore, the

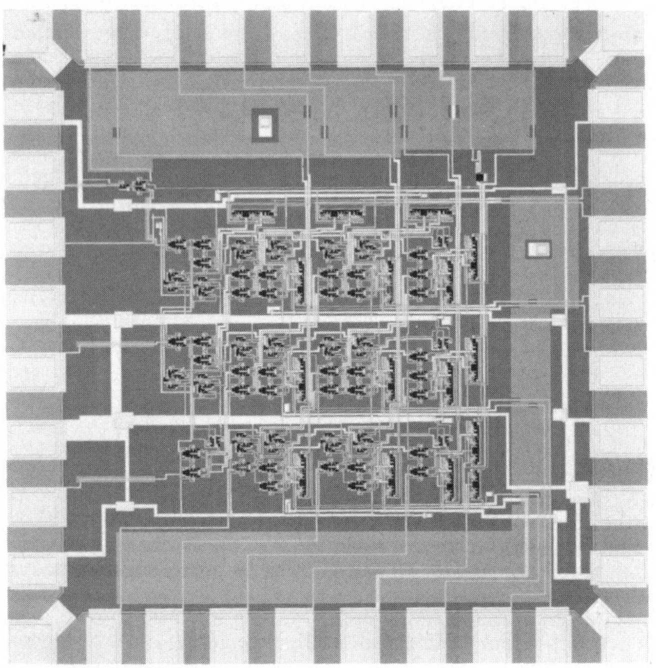

Fig.2 Microphotograph of the 4bX4b parallel multiplier chip. The device size is 2.7mmX2.7mm

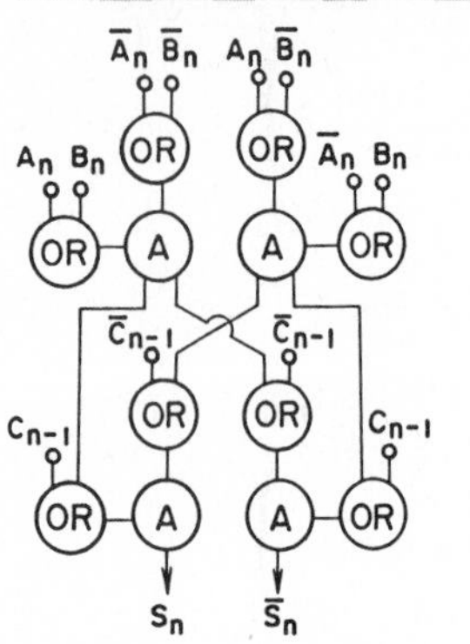

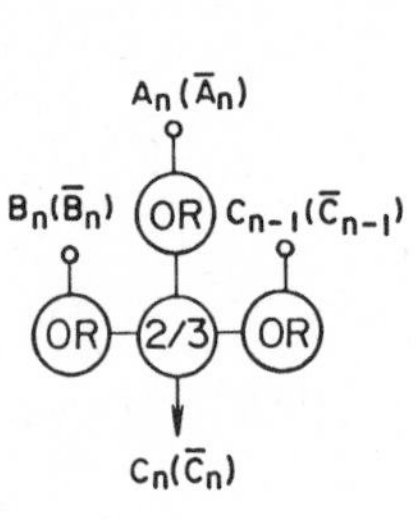

Fig.3 Schematic diagrams of a sum generator and a carry generator. A : AND gate OR : OR gate 2/3 : 2/3 MAJORITY gate

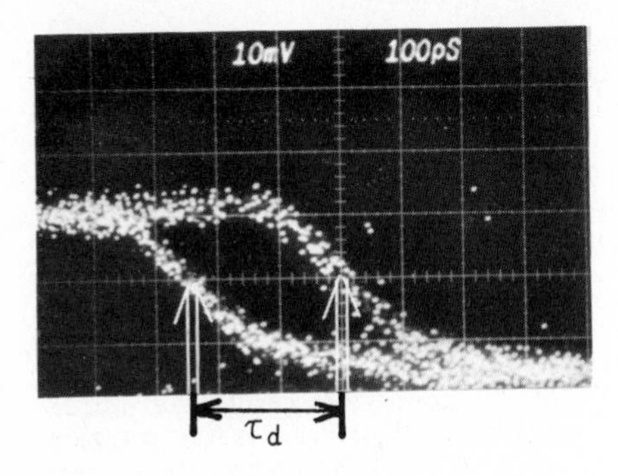

Fig.4 The critical path delay measurement of the multiplier. The time interval between the two waveforms give the critical path delay

averaged delay per gate is estimated to be 21 psec, which includes 3 psec signal propagation time along the transmission line.

4 Projected RCJL Performances

The essential factors for logic circuit performances are summarized in Tab. 1. The main interest in this paper is the projected logic delay in the future large scale integrated circuit chips, which depends on the fan-in and fan-out condition and on the averaged interconnection line length.

Table 1 Essential Factors for Josephson Logic Circuits

GATE	CHIP	SYSTEM
intrinsic delay	logic delay	system speed
power/gate	power/chip	cooling
drivability	gates/chip	packaging
size	I/O	reliability
	margin	

The performances of RCJL gates are discussed employing standard logic cell configuration. The standard cells are laid out on an LSI chip in an array, and are interconnected by superconducting transmission lines as shown in Fig. 5. The cell contains a gate unit, consisting of two OR gates and one AND gate, and an area to

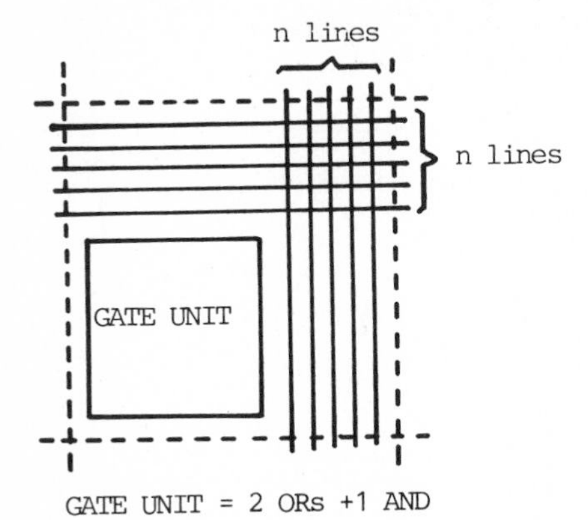

Fig.5 Standard cell layout

install n interconnection lines on X and Y directions. Each gate unit's output drives ℓ other gates. The averaged distance between the driving gate unit and driven gate unit is m times as long as the unit cell size.

Using the standard cell layout, the gate performance are projected along with the minimum feature size reduction. The RCJL logic delay is the sum of the turn-on delay τ_t, the voltage rise time τ_r and the signal propagation time τ_p. The junction size reduction results in τ_r reduction. The propagation time τ_p is in proportion to the unit cell size, which decreases proportionally with the minimum feature size. The logic delay for highly miniaturized RCJL gates is essentially determined by the quantum mechanical time-constant $\tau_q = h/2eV_g \sim 1$ psec and the velocity of light on the superconducting transmission line. Table 2 shows the results of RCJL performance projection. The fundamental condition in the projection is that the gate output drives 3 other gates, located 20 cells away from the driving gate. The projected logic delay per gate in a densely packed 1 μm-RCJL LSI is 10 psec per gate.

Table 2 1K gate chip performance
l=3 n=5 m=20

junction	5μm	1μm
line pitch	8μm	2μm
gate unit	110μm	22μm
cell	150μm	32μm
logic delay	35ps	10ps
device size	5mm	1mm
power	7.5mW	1.5mw

5 Conclusion

The advancement in Josephson IC fabrication technology has shown the possibility for fabricating Josephson LSIs with 1 μm minimum line width. Based on the realized performance in medium scale integrated circuits, the projected logic delay per gate in densely packed 1-μm-RCJL LSI is 10 psec or less.

References

1 J. Sone, T. Yoshida, H. Abe: Appl. Phys. Lett. 40, 741 (1982)
2 J. Sone, J.S. Tsai, S. Ema, H. Abe: ISSCC DIGEST OF TECHNICAL PAPERS, pp. 220-221(1985)
3 J. Sone, T. Yoshida, H. Abe: IEEE Electron Dev. Lett. EDL-4, 428 (1983)
4 Y. Wada, M. Hidaka, I. Ishida: IEEE Electron Dev. Lett. EDL-4, 455 (1983)
5 T. Yamada, J. Sone, S. Ema: Extended Abstracts of 16th Conference on Solid State Devices and Materials (Kobe, Japan 1984), pp. 623-626

Index of Contributors